The Physiology of Hemostasis

The Physiology of Hemostasis

Derek Ogston

HARVARD UNIVERSITY PRESS
Cambridge, Massachusetts 1983

Library of Congress Cataloging in Publication Data

Ogston, Derek.
    The physiology of hemostasis.

    Bibliography: p.
    Includes index.
    1. Hemostasis. I. Title. [DNLM: 1. Hemostasis.
WH 310 O35p]
QP93.5.O33   1983   612.′.115   83-125
ISBN 0-674-66660-7

Contents

Preface

The current widespread interest in the hemostatic mechanism stems largely from the probability that its inappropriate function may lead to thrombosis, but also for its relevance to the causation and management of bleeding disorders. Our understanding of the pathological events leading to thrombus formation or abnormal bleeding depends on knowledge of the basic physiology of the hemostatic system. A number of excellent multiauthor texts are available on the general area of hemostasis and thrombosis, and many symposia proceedings on specific topics in hemostasis have been published. The present volume aims to cover the normal function of hemostasis and explicitly excludes consideration of disease states and therapy. In addition, it is concerned with human hemostasis only although reference is made to studies on other mammalian species when the information supplements that available on man.

The book is divided broadly into two sections. The first covers the current knowledge of the principal components contributing to the hemostatic process; the second examines the changes in these components induced by physiological events, and details an accumulation of information not previously brought together in a single text. An introductory chapter, intended for the non-specialist, outlines the whole hemostatic process and provides an orientation for the later detailed information on individual components.

It is hoped that the volume will be of value as a general introduction to the study of hemostasis, to the specialist in a limited aspect of the field who wishes an account of related areas, and to those examining physiological influences on the hemostatic mechanism.

1 An Outline of the Hemostatic Mechanism

The term hemostasis refers to the processes involved in the arrest of bleeding following injury. These processes involve the participation of the vessel wall, blood platelets, the coagulation system and, in a less defined way, the fibrinolytic enzyme system. The complexity of the detail of the individual phases of the hemostatic mechanism has the hazard of obscuring its overall pattern and sequence of reactions. This brief introductory summary of the hemostatic processes is intended to provide a general framework for readers with limited previous knowledge in this area of physiology into which can be fitted the more detailed descriptions of its component stages in later chapters.

The rupture of a blood vessel conveying blood under pressure results in the escape of blood from the vessel until the breach is sealed or until the pressures inside and outside the vessels are equalised. Clearly the former mechanism is physiologically more important and to achieve it efficiently the hemostatic system requires to have the properties of speed, localisation of action and the provision of a seal with appropriate mechanical strength.

Vascular Responses to Injury

Injured arterioles and venules contribute to the initial phase of hemostasis by contraction of the smooth muscles of their wall. The vessel constriction reduces the pressure distally with resultant reduction of blood loss. The precise mechanism of blood vessel constriction in response to injury is not established since studies on vessels within the body pose considerable technical problems. The smooth muscle contraction could presumably be a direct result of trauma or could be mediated by the sympathetic nervous system. The vascular constriction after injury tends to be intermittent.

In contrast to arterioles and venules, capillaries have a wall composed of a single layer of endothelial cells without muscle fibres and, accordingly, are considered to be non-contractile, the blood flow through them being controlled by the smooth muscle of the pre-capillary sphincters. Obliteration of the capillary lumen through sticking of the walls together by endothelial adhesion may contribute to hemostasis in some instances (Sanders, 1970). There is also the possibility that capillaries themselves constrict by virtue of myofibril-like structures.

The endothelial cells lining blood vessels are likely to have an important function both in the prevention of inappropriate intra-vascular platelet and fibrin deposition and in the hemostatic process after injury. They contain, and presumably synthesise, a number of proteins relevant to these functions including the von Willebrand factor component of the coagulation factor VIII complex, the coagulation inhibitor antithrombin III and plasminogen activator. There is evidence that each of these can be released into the blood. Their individual properties and functions are considered in later chapters. Endothelial cells have also the capacity to synthesise prostacyclin, a vasodilator and inhibitor of platelet aggregation. Some of the agents involved in hemostasis and the prevention of intravascular thrombus formation that are associated with the endo-thelial cell are summarised in Figure 1.1.

Platelet Responses to Vessel Wall Injury

The next stage of the hemostatic process involves the formation of a plug of platelets to seal the breach in the vessel wall. Damage to a blood vessel exposes subendothelial structures, in particular, coll-agen. Platelets in the circulating blood make contact with, and adhere to, collagen fibres. The activation of the platelet induced by collagen and other agents such as thrombin leads through a sequence of cyto-plasmic events to a change in its shape from discoid to spherical with spiny processes, and to secretion of a number of substances from its granules. Adenosine diphosphate (ADP) is one such substance and has the property of inducing other platelets to adhere to each other, a process of cohesion termed aggregation. In addition to ADP, 5-hydroxytryptamine (serotonin) is secreted and through its vasocon-strictor properties may aid in the maintenance of the initial vessel constriction.

Figure 1.1: Components of the Hemostatic Mechanism Associated with the Endothelial Cell

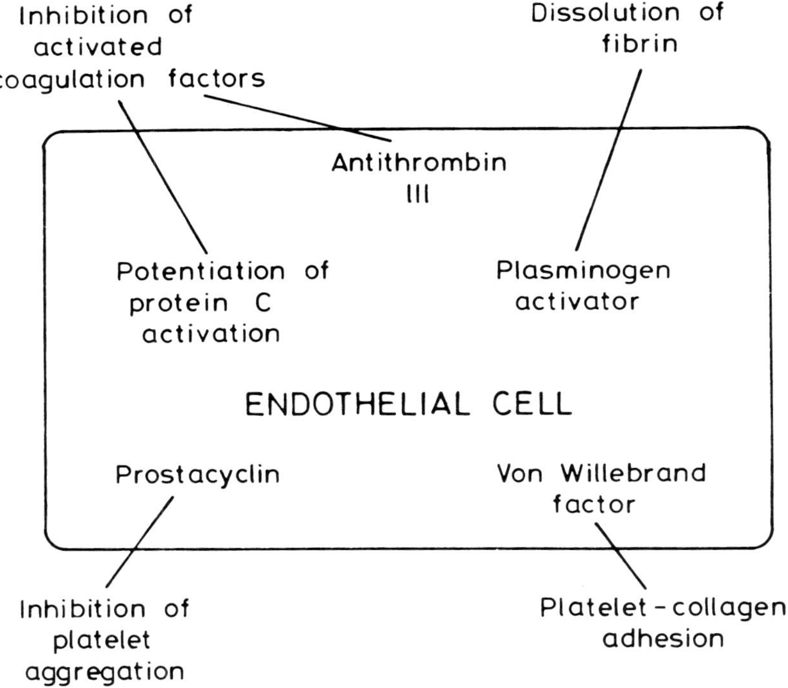

The process of platelet adhesion and activation also initiates the formation of arachidonic acid from the phospholipid of the platelet membrane. Cyclic endoperoxides are produced and in a further step thromboxane A_2, an unstable compound with potent platelet aggregating and vasoconstrictor properties, is formed. Through the two processes of ADP secretion and thromboxane A_2 formation, with resultant platelet aggregation, a mass of platelets is built up at the site of the vessel injury. This platelet mass seals the breach in the vessel wall and may extend into and occlude the vessel lumen itself.

Vascular constriction together with the formation of a platelet plug may be adequate to secure hemostasis in capillaries or venules where the hydrostatic pressure is low, but in vessels where the pressure is higher the platelet plug is inadequate to stem the escape of

3

blood, portions of the plug being easily dislodged and washed away. For adequate hemostasis a more secure and permanent plug is required.

The Coagulation System

The role of the coagulation system in hemostasis is to provide stability to the platelet plug through the formation of a network of insoluble fibrin at the site. It is evident that the processes of platelet adhesion and aggregation and that of fibrin formation must be closely integrated both in time and location; this is achieved by a number of mechanisms detailed later.

The coagulation system comprises a series of enzymatic reactions, the product of each acting as the enzyme in the succeeding reaction. In this way the system acts as a biological amplifier transforming a small initial stimulus into the explosive formation of large quantities of thrombin, the enzyme responsible for the conversion of soluble circulating fibrinogen into fibrin.

The coagulation system is traditionally divided into the intrinsic pathway in which all the components are present within the blood in an inactive precursor form, and the extrinsic pathway requiring a tissue factor which can gain access to the blood only when tissue is damaged. It is now recognised that these pathways are closely interdependent with a number of points of linkage. The general scheme of the coagulation system showing some of the points of interaction between the intrinsic and extrinsic pathways and feedback loops is shown in Figure 1.2.

The intrinsic pathway is initiated by the activation of factor XII (Hageman factor), a process in which prekallikrein (Fletcher factor) and high molecular weight kininogen (HMW kininogen) participate. Factor XII undergoes activation on contact with negatively charged surfaces: collagen provides such an activating surface and its exposure on tissue injury may be one starting point down the pathway towards fibrin formation. The absolute requirement for factor XII in the hemostatic process is open to question because of the absence of abnormal bleeding after injury in individuals with a severe congenital deficiency of the factor. Such individuals demonstrate that alternative pathways exist in the coagulation system which can by-pass activated factor XII.

Figure 1.2: The Coagulation Pathways (feedback loops shown by interrupted lines)

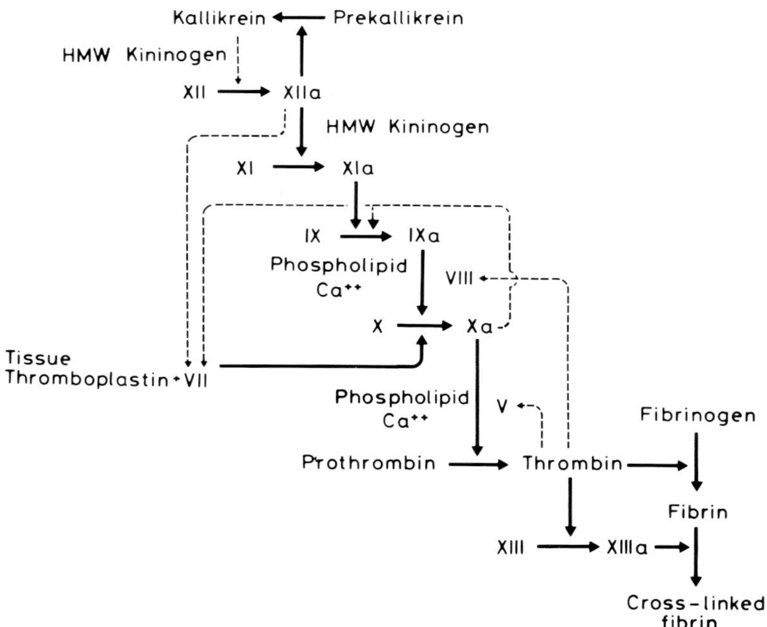

Factor XII in its activated form, designated factor XIIa, activates factor XI (plasma thromboplastin antecedent; PTA); this reaction probably requires HMW kininogen. Persons with a congenital deficiency of factor XI may have a hemorrhagic tendency, but this is relatively mild, inferring that this factor's contribution to coagulation is less important than that of such factors as VIII and IX.

In the next stage of the intrinsic pathway factor IX (Christmas factor) is transformed into its active coagulant state by factor XIa in the presence of calcium ions. Factor IX may also be activated via the extrinsic system since the activated form of factor VII is capable of achieving this in the presence of tissue factor. Factor Xa (activated Stuart factor), formed through either the intrinsic or extrinsic pathway, can also activate factor IX.

Factor IXa is the enzyme involved in the cleavage and activation of factor X through the intrinsic pathway, but it requires the presence of factor VIII (antihemophilic globulin: AHG) as a cofactor in addition to phospholipid and calcium. The phospholipid is normally provided by the platelet membrane. The factor VIII molecule requires to be altered by thrombin in order to acquire its cofactor properties. The presence of factor VIII greatly increases the rate of factor X activation, probably by aligning the enzymatic factor IXa and its substrate factor X in optimal position on the phospholipid surface.

The activation of factor X can also be effected through the extrinsic pathway. In the presence of tissue factor (tissue thromboplastin), found in most tissues, factor VII in its activated form is capable of splitting the specific bond of factor X to generate its enzymatically active form. The precise mode of initiation of the activation of factor VII is under debate: once activated, factor X itself is able to activate factor VII in a positive feedback mechanism while factor IXa, formed through the intrinsic mechanism, can also activate Factor VII.

Factor Xa now participates in the stage of thrombin formation, a reaction common to both the intrinsic and extrinsic pathways. The factor Xa molecule contains the enzymatic site required for the activation of prothrombin, but factor V (proaccelerin), phospholipid and calcium ions are also necessary for the reaction. In order to achieve its cofactor potential the factor V, present both in plasma and platelets, must be proteolytically altered, a further function of thrombin. Both factor Xa and prothrombin are bound to the phospholipid of the platelet membrane through calcium linkages while factor V appears to function in the activation process by facilitating the appropriate apposition of factor Xa and prothrombin. In the absence of factor V the activation of prothrombin is very slow, in its presence the reaction rate is enormously accelerated to allow thrombin formation to take place through the proteolytic scission of the prothrombin molecule at a rate which is physiologically useful.

Thrombin is able to cleave small fragments from the fibrinogen molecule termed fibrinopeptides; their removal allows the fibrin monomers formed to undergo polymerisation with the production of an insoluble fibrin network. Thrombin influences the hemostatic process in other ways. It is a potent inducer of platelet aggregation and secretion and thereby represents a further link between the coagulation system and platelet function. In addition, thrombin is required for the final stage of the coagulation process, the stabilisation of the polymerised fibrin through the formation of covalent bonds between the fibrin polymers. In order to allow this process to proceed factor XIII (fibrin-stabilising factor), present in plasma and platelets, must be activated by thrombin. Through the action of factor XIII the fibrin network is rendered stable and resistant to premature breakdown. A number of the effects of thrombin on the coagulation system are summarised in Figure 1.3.

Control of the coagulation system is complex and a variety of mechanisms probably operate to limit its initiation, localise its action and adjust its magnitude of effect to hemostatic requirements. The platelet makes an important contribution to the localisation of coagulation to the injury site by the provision of a number of clotting factors and a surface on which the coagulation process takes place after platelet activation. In addition, platelets have the capacity to protect activated clotting factors from inactivation by plasma in-hibitors.

A number of plasma proteins have the ability to inhibit one or more of the activated clotting factors and thereby help to limit fibrin formation which is inappropriate in site or extent. The principal plasma inhibitor of thrombin in antithrombin III which can also neutralise a number of the other activated clotting factors, particu-larly after its reaction with heparin. Cl inactivator has inhibitory activity on the initial phases of the intrinsic pathway with an action on factors XIIa and XIa and kallikrein. Alpha$_2$-macroglobulin and, possibly α-antitrypsin are also capable of inhibiting thrombin. The role of the more recently identified protein C, which inactivates the active forms of factors VIII and V, is presently uncertain. In addition to these proteins, it is possible that the by-products of some of the

Figure 1.3: Some Effects of Thrombin on the Coagulation Mechanism

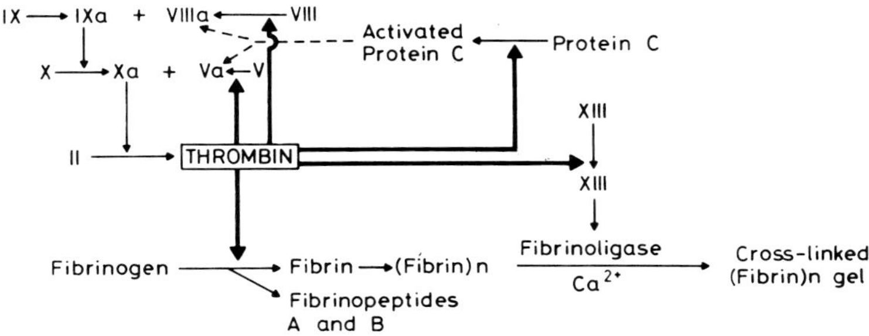

coagulation reactions may have a controlling influence on these reactions.

Areas of interaction between the coagulation system and platelets to produce a hemostatic plug are illustrated in Figure 1.4.

Morphology of the Hemostatic Plug

The hemostatic events following vessel injury have been demonstrated by light and electron microscopy. There is a pattern of platelet accumulation on the connective tissue exposed at the edge of the vessel with the later appearance of fibrin close to the connective tissue and around the periphery of the hemostatic plug. Early loss of organelles and pseudopod formation in those platelets making contact with collagen fibres is seen, and there is also marked lytic changes in the platelets at the periphery of the plug. Degranulation and lytic changes appear later in the more centrally located platelets. Fibrin increases in amount and there is eventual replacement of a large part of the plug by fibrin threads. A further late event is infiltra-

Figure 1.4: The Production of a Hemostatic Plug Through the Interactions Between the Coagulation System and Platelets

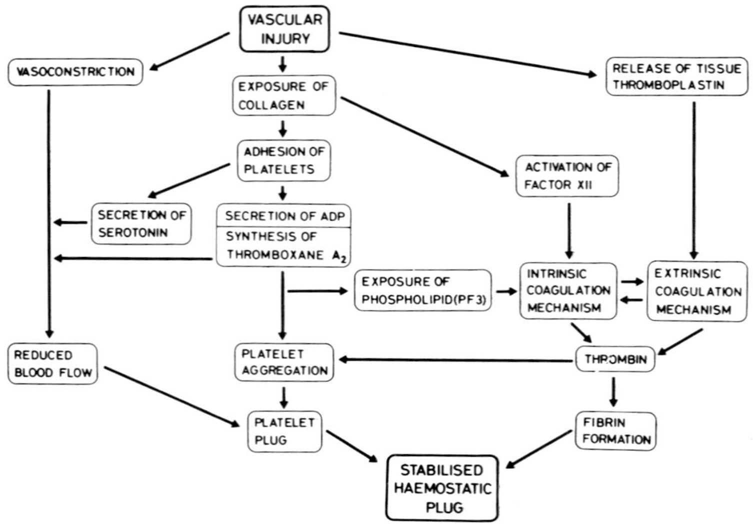

tion of the plug by leukocytes. Further detail of the morphological events in hemostasis may be found in the reviews by Sixma and Wester (1977) and by Sixma (1981).

The Fibrinolytic System

The final system, the fibrinolytic enzyme system, is usually considered along with the vessel wall, platelets and coagulation as part of the hemostatic mechanism. As its name implies it is capable of degrading fibrin and its function is presumed to be the removal of inappropriately formed fibrin or of fibrin excessive to the needs of hemostasis. The central reaction of the fibrinolytic system is the conversion of the zymogen plasminogen into the protease plasmin under the influence of agents collectively termed plasminogen activators (Figure 1.5). The precise number and relative importance of physiological activators is presently debated, but it is probable that the vascular endothelium is an important source of circulating activator. In addition, the activation of factor XII can lead to the generation of a plasminogen activator under *in vitro* conditions, but the physiological relevance of this pathway is presently uncertain. Urine contains an activator termed urokinase which is derived from the urinary tract: evidence has been provided recently that plasma and some tissues other than the kidney contain an activator, principally in precursor form, with similar properties and immunological identity

Figure 1.5: Outline of the Fibrinolytic System

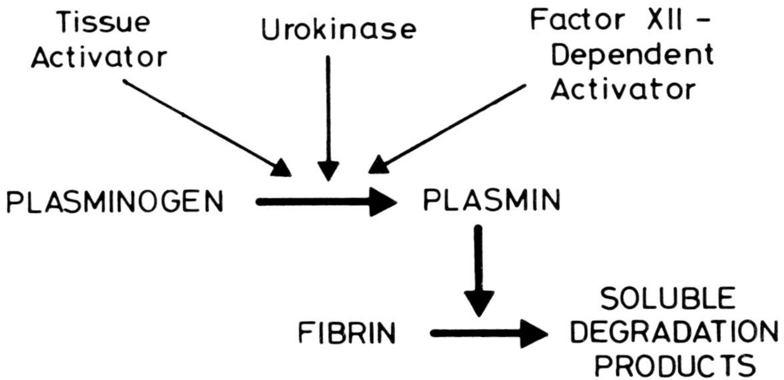

with urokinase. The vascular activator differs from urokinase in being strongly adsorbed to fibrin. Plasminogen is also adsorbed onto fibrin and it is likely that its activation to plasmin takes place at this site. Indeed, it is this adsorption of both plasminogen and activator to fibrin which dictates the localisation and specificity of physiological fibrinolysis. The plasmin formed progressively degrades the fibrin with the formation of well-characterised intermediate and final soluble polypeptides (fibrin degradation products).

A number of plasma protease inhibitors are capable of neutralising plasmin including α_2-macroglobulin, α_1-antitrypsin, Cl in-activator and antithrombin III, but of dominant importance is the rapidly-acting α_2-antiplasmin. The recently described histidine-rich glycoprotein has also been identified as an inhibitor of fibrinolysis through its action in reducing the binding of plasminogen to fibrin. The existence of separate inhibitors of plasminogen activators is not established.

References

Sanders, A.G. (1970) '*In vivo* observations on haemostasis in the hamster', in R.G. Macfarlane (ed.), *The Haemostatic Mechanism in Man and Other Animals*, Academic Press, London, pp. 109-20

Sixma, J.J. (1981) 'Role of blood vessel, platelet and coagulation interactions in haemostasis', in A.L. Bloom, D.P. Thomas (eds.), *Haemostasis and Thrombosis*, Churchill Livingstone, Edinburgh, pp. 253-57

Sixma, J.J. and Wester, J. (1977) 'The haemostatic plug', *Seminars in Hematology*, *14*, 265-99

  Platelet Structure and Function

Platelet Morphology and Structure

Normal human blood contains some 150,000 to 400,000 platelets per microliter. They circulate as anucleate discs two to five microns in diameter: detailed measurements of a number of parameters of normal platelets have been provided by Stahl and associates (1978). When examined by light microscopy platelets appear to consist of organelles randomly distributed or clumped together in a water-clear cytoplasm (hyaloplasm) enclosed by a membrane. Details of the ultrastructure of platelets have been revealed by electron microscopy studies and have provided information which helps to explain platelet behavior in the process of hemostasis. A diagrammatic representation of platelet structure is provided in Figure 2.1.

Circulating platelets have a smooth outline. The outer coat (glycocalyx) is 10 to 20 nm thick and is composed of acid mucopolysaccharides and mucoproteins (Behnke, 1968a). The platelet is bounded by a plasma membrane which has the trilaminar structure characteristic of mammalian cell membranes: it is rich in phospholipoprotein. Numerous invaginations of the membrane form channels to the interior of the cell which both expand the reactive surface of the platelet and allow the uptake of substances from the plasma and the extrusion of platelet contents. This surface-connected canalicular system has been demonstrated by the freeze-etching technique and the cannaliculi shown to make direct contact with organelles (Hoak, 1972).

Just below the plasma membrane of the platelet is a circumferential band of microtubules, each with a diameter of 200 to 250 Å. They act as a type of cytoskeleton which preserves the normal discoid shape of the platelet (Behnke, 1965). It is likely that the series of tubules seen on micrographs represent serial sections of the same microtubule encircling the cell many times. In addition to microtubules, platelets contain finer microfilaments of smooth muscle and the actomysin-like protein isolated from platelets. Such micro-filaments have a diameter range of 50 to 70 Å. It is presumed that they represent the morphological counterpart of the contractile protein of platelets. There is also a resemblance between the microfilaments and the subfilaments of microtubules. Submembranous filaments have been described which are also physically similar to micro-filaments (White, 1969). It was suggested that they may have a role in the maintenance of the platelet discoid shape, in extruding and stabilising pseudopods, and in platelet contraction.

There is a dense tubular system throughout the platelet cytoplasm which represents the sarcoplasmic reticulum of the platelet (White, 1972). It is a canalicular system with a capacity for calcium accumulation which forms a branched network, mainly underneath the plasma membrane, but also extending into the interior of the cell. The dense tubular system links up the surface-connected canalicular system and the storage organelles (Werner and Morgenstern, 1980).

Platelets contain a number of different organelles. The largest are the dense granules with a diameter of 0.2 to 0.3 microns. Each is bounded by a unit membrane and they contain a number of sub-stances which make important contributions to platelet function and hemostasis. Other organelles include the alpha-granules which contain an electron-dense nucleoid and store materials relevant to the hemostatic process, peroxisomes and lysosomes. Mitochondria are present in platelets; these are smaller than those in most other mammalian cells. Glycogen granules are also found in the cytoplasm.

Megakaryocytes

Platelets are formed by the fragmentation of the cytoplasm of the megakaryocyte. Megakaryocytes are large cells with a diameter up to 180 microns although the majority are much smaller (Levine, 1981): size variation appears to be related to cell maturation and ploidy state. They have three fairly distinct zones in their cytoplasm, a peripheral zone, an intermediate zone and a perinuclear zone. The peripheral zone consists of a fine fibrillar or granular matrix with scanty vesicles, ribosomes and granules. The intermediate zone is rich in granules, mitochondria and strands of rough-surfaced reticulum; in this zone there is a system of paired membranes termed demarcation membranes. It is believed that the demarcation membrane is derived

Figure 2.1: Diagrammatic Representation of the Structure of a Platelet

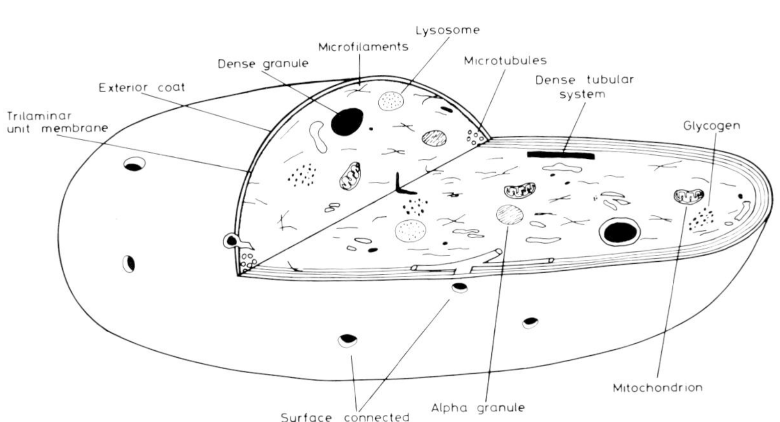

from the plasma membrane of the megakaryocyte by invagination, producing tubules within the cytoplasm of the megakaryocyte (Behnke, 1968b; Shaklai and Tavassoli, 1978), although others have proposed that it originates from endoplasmic reticulum of intracellular vesicles. The perinuclear zone contains a Golgi complex, mitochondria, free ribosomes and rough-surfaced endoplasmic reticulum.

Platelet Formation from Megakaryocytes

Megakaryocytes are found in the extravascular spaces of the bone marrow. It is probable that the platelets are partially delineated within the megakaryocytic cytoplasm by the demarcation membrane system (Behnke, 1968b; Shaklai and Tavassoli, 1978). Shaklai and Tavassoli (1978) suggested that the tubules formed by the invagination of the plasma membrane develop into membranes which subsequently enclose areas of the cytoplasm to form platelets.

The extravascular megakaryocytes extend cytoplasmic processes into the sinusoids of the marrow and it is their fragmentation which results in the formation of platelets. Megakaryocyte processes with constrictions between platelet-sized segments at the distal ends have been identified with the scanning electron microscope (Becker and De Bruyn, 1976). Radley and Scurfield (1980) also employed the

13

scanning electron microscope to study platelet formation: they observed that the cytoplasmic processes protruding from mega-karyocytes into sinusoids undergo attenuation and become constricted at irregular intervals along their lengths to form segments of platelet size. Microtubules were found orientated in the direction of the long axis of the processes and it was suggested that they have a role in the formation and maintenance of the shape of the cytoplasmic extensions and a role in the separation of the segments of cytoplasm.

The Regulation of Thrombopoiesis

The term 'thrombopoietin' was given by Kelemen and associates (1958) to a humoral factor which stimulates platelet production. Thrombopoiesis, as assessed by the circulating platelet count, has been shown to be stimulated by the removal or destruction of platelets and to be inhibited by the transfusion of viable platelets (de Gabriele and Penington, 1967a). Plasma from thrombocytopenic animals has been found repeatedly to induce thrombocytosis in normal recipients of the same species (Odell *et al.*, 1961; Spector, 1961; de Gabriele and Penington, 1967b). The incorporation of ^{75}Se-methionine or ^{35}S into developing platelets has been used to measure thrombopoiesis and experiments utilising these techniques have also provided evidence that a humoral agent acts on mega-karyocytes to regulate platelet production (Evatt and Levin, 1969; Harker, 1970). In man thrombocytopenia induced by platelet-phoresis is associated with thrombopoietically active plasma which was found to produce an increase in the percentage of immature megakaryocytes 24 hours after reinfusion and a peak rise in the peripheral platelet count at around six days (Adams *et al.*, 1978).

Plasma fractions from thrombocytopenic animals with thrombo-poiesis-stimulating activity have been isolated. Thrombopoietic activity was found in fractions of rabbit plasma precipitated at ammonium sulphate saturations of 60-70 per cent and 70-80 per cent (Evatt *et al.*, 1974). In a further study Evatt and associates (1979) isolated a plasma fraction with thrombopoietic activity from thrombocytopenic rabbits by ion-exchange chromatography and gel filtration: the active material was approximately 1000-fold purified and was found to be non-dialysable and stable over a pH range of 5.6 to 7.5. The precise chemical nature of thrombopoietin, however, has eluded definition to the present.

The site of production of thrombopoietin is uncertain. De Gabriele and Penington (1967a) observed that bilateral nephrectomy did not impair the response to platelet depletion. In contrast, Krizsa (1971) observed that thrombopoietin activity could not be induced in nephrectomised rats rendered thrombocytopenic by bleeding whereas it was present in the serum of rats with intact kidneys. Further evidence that the kidney is a site of production of thrombopoietin is provided by studies showing the presence of thrombopoietic activity in the medium of human embryonic kidney cell cultures (McDonald *et al.*, 1975). It is also possible that the liver is a source of thrombopoietin since partial hepatectomy in rats has been found to suppress the entry of platelets into the peripheral blood and to decrease the average diameter and ploidy of megakaryocytes (Siemensma *et al.*, 1975). Liver cell cultures, however, do not produce thrombopoietin (Ogle *et al.*, 1978).

The precise stimulus for thrombopoietin release is not established. It was concluded, on the basis of experiments suggesting that the humoral thrombopoietic factor is removed or inactivated by viable platelets, that a simple feedback mechanism controls platelet production (de Gabriele and Penington, 1967b). In this hypothesis the number of circulating platelets determines the quantity of thrombopoietin in the plasma and thereby platelet production. A different view was expressed by Podolsak and colleagues (1980) who found that the injection of platelet homogenates into rabbits produced a significant increase in the platelet count; they concluded that platelets contain a thrombopoiesis-stimulating factor which is released on platelet degradation.

In addition to thrombopoietin-stimulating activity, a humoral inhibitor of thrombopoiesis has been detected in some patients with low or normal platelet counts (Shreiner *et al.*, 1980). Thrombopoiesis may be controlled, therefore, by a balance between humoral stimulatory and inhibitory activities.

Platelet Heterogeneity

Circulating platelets show heterogeneity in size, density and function. One view is that this heterogeneity is due to platelet aging. McDonald and associates (1964) concluded on the basis of the change in platelet size distribution during recovery from thrombocytopenia that young platelets are larger. Evidence has been marshalled that heterogeneity in size and density results from age changes (Karpatkin, 1969a, 1972; Karpatkin and Garg, 1974): it was suggested that young platelets are of greater density and metabolic activity than older platelets and that they become lighter and smaller with age. The large platelets are termed megathrombocytes. An alternative explanation for platelet heterogeneity has been championed by Penington and associates (1976): they suggest that heterogeneity is unrelated to aging and results rather from the simultaneous release of platelets with differing properties from the three major types of megakaryocyte. This hypothesis, however, does not exclude the possibility that there is diminution in platelet size with aging.

The hemostatic properties of platelets also appear to vary with age. Young human platelets, identified by isotopic labelling, have been shown to have greater adherence to collagen, but no difference was seen between young and old platelets in their susceptibility to aggregation in response to ADP (Hirsch *et al.*, 1968). The latter finding, however, contrasts with that of Mannucci and Sharp (1967) who observed selective loss of large platelets on the addition of ADP, epinephrine or thrombin as well as collagen, while Karpatkin (1969b) reported that a megathrombocyte-rich population was more readily aggregated with the release of ADP and platelet factor 4. Additional evidence that platelet size is related to function has been provided by Thompson and colleagues (1982); they separated platelets on the basis of size by counterflow centrifugation and found that ADP-induced aggregation became progressively more rapid as platelet size increased together with increased uptake and thrombin-induced release of serotonin.

An extensive review on human platelet size and shape has been provided by Frojmovic and Milton (1982).

Platelet Plasma Membrane

The platelet membrane has a structure appropriate to its function of interacting with specific stimuli such as ADP, collagen and thrombin through the presence of externally orientated receptors in order to induce the hemostatic mechanisms of platelet adhesion and aggregation. The membrane contains functionally important lipids, proteins and carbohydrates.

Lipids

Most of the membrane lipid bilayer is composed of phospholipids; their fatty acid components are both saturated and unsaturated. The unsaturated fatty acids are derived from the dietary essential fatty acids. Arachidonic acid is abundant in the platelet membrane and in the non-activated platelet is esterified into the phospholipids. The arachidonic acid can be liberated from the platelet membrane phospholipids by the action of specific phospholipases, allowing it to take part in the prostaglandin synthetic pathways (see page 21). There is evidence that the phospholipids are distributed asymmetrically on the two sides of the lipid bilayer with the negatively charged phospholipids (phosphatidyethanolamine, phosphatidylserine and phosphatidylinositol) mainly on the cytoplasmic side of the membrane, and sphingomyelin mainly exposed at the platelet surface; the distribution of phosphatidylcholine on the two sides of the bilayer is approximately equal (Chap *et al.*, 1977; Zwaal, 1978; Perret *et al.*, 1979). On platelet activation there is a change in the phospholipid distribution in the membrane so that the negatively charged phosphatidylserine and phosphatidylinositol become available at the platelet surface. This transposition has been termed 'flip-flop'.

Cholesterol is also a significant platelet membrane constituent (Barber and Jamieson, 1970), and probably influences platelet membrane fluidity (Insel *et al.*, 1978).

Proteins

A proportion of the proteins of the platelet membrane comprise the contractile elements actin and myosin heavy chain. Tubulin, the subunit protein of microtubules, may also be a constituent of the membrane (Ikeda *et al.*, 1981).

Glycoproteins. A number of glycoproteins of functional importance have been identified on the external surface of the platelet membrane. A numbering system for these glycoproteins has been devised by Phillips and Agin (1977). Using a combination of radioactive labelling and two-dimensional electrophoresis Sixma and Schiphorst (1980) have identified some 25 surface glycoproteins, but the four major membrane glycoproteins appear to be Ib, IIb, IIIa and IIIb. A further glycoprotein, termed glycocalicin (Okumura and Jamieson, 1976) or Is (Nurden, 1977), shares structural and functional characteristics with glycoprotein Ib and there is evidence that it is derived from it by the action of a Ca^{2+}-dependent protease released from the platelet.

Some of the membrane glycoproteins have important roles in platelet function including mediation of platelet-platelet and platelet-surface interactions, stimuli reception and transfer and participation in membrane transport. It has been proposed that their loss may be a mechanism in platelet senescence *in vivo* (Greenberg *et al.*, 1979).

Glycoproteins IIb and IIIa, with molecular weights of 135,000 and 100,000 respectively, are grossly diminished or absent from the platelets of patients with Glanzmann's thrombasthenia (Nurden and Caen, 1974; Phillips *et al.*, 1975; Nurden and Caen, 1977), a condition in which platelet aggregation is defective. It has been estimated that there are normally some 40,000 glycoprotein IIb-IIIa complexes per platelet (McEver *et al.*, 1980); although structurally similar, glycoproteins IIb and IIIa have been shown to be distinct (Leung *et al.*, 1981). It has been reported that glycoproteins IIb and IIIa form a complex with purified plasma fibrinogen, suggesting that they act as the fibrinogen-binding site required for normal platelet aggregation (Nachman and Leung, 1981). In keeping with this finding, thrombasthenic platelets do not have the fibrinogen-binding property seen with normal platelets stimulated by ADP, epinephrine or collagen (Bennett and Vilaire, 1979). There is, in addition, evidence that these glycoproteins interact with the contractile elements of the platelet following thrombin-induced aggregation (Phillips *et al.*, 1980). It is possible, therefore, that on platelet stimulation the glycoprotein IIb-IIIa complex undergoes a change allowing the binding of fibrinogen to the surface and their interaction with the contractile proteins. The bound fibrinogen molecules may interact with those on adjacent platelets to bring about aggregation while fibrin clot retraction may be induced through the interaction with the contractile system. Glycoprotein IIIa was tentatively identified as platelet α-actinin (Gerrard *et al.*, 1979), and it was proposed that it had a role in anchoring actin to the platelet membrane, thereby linking the platelet contractile apparatus to the membrane. More recent studies, however, have demonstrated that α-actinin and glycoprotein IIIa are not identical (Sixma *et al.*, 1982).

The Bernard-Soulier syndrome is characterised by giant platelets, a prolonged bleeding time, reduced platelet adhesion to the subendothelium and decreased agglutination of platelets by ristocetin, but without an abnormality in the plasma factor VIII:vWF. The membranes of platelets from patients with this syndrome have been shown to have a deficiency of glycoprotein Ib (Nurden and Caen, 1975; Hagen *et al.*, 1980). It is probable that this glycoprotein serves as a receptor for factor VIII: vWF (Cooper *et al.*, 1979), and is therefore required for platelet adhesion to the subendothelium (see page 38).

It has been suggested that glycoprotein V acts as a substrate for α-thrombin (Phillips and Agin, 1977). Subsequent experiments showed that the purified glycoprotein, of molecular weight 82,000, is a thrombin substrate yielding a major 69,000 daltons new fragment (Berndt and Phillips, 1981), but its role as a thrombin receptor is not established. Glycocalicin has also been suggested as a possible thrombin receptor on the basis of a relationship between the amount of glycoprotein I complex (glycalicin/glycoprotein) on platelets and the quantity of thrombin which will bind to them (Okumura *et al.*, 1978), while reduced thrombin binding to platelets has been noted in Bernard-Soulier syndrome patients (Jamieson and Okumura, 1978). Specific binding of thrombin to glycoprotein I was reported by Ganguly and Gould (1979) and they proposed that it acts as the high affinity receptor for thrombin on platelets.

Carbohydrate

The external plasma membrane coat (platelet glycocalyx) is rich in carbohydrate. The carbohydrates comprise glycolipid, glycoprotein and glycosaminoglycan. While the functions of the cell surface glycosaminoglycans are not established, it is believed that they are important in cell-cell and cell-surface interactions (Lindahl and Hook, 1978).

Platelet Intracellular Membrane

Substantial differences exist between the composition of the platelet plasma membrane and the intracellular membranes. Phosphatidylcholine and phosphatidylinositol are more enriched in the intracellular membranes while phosphatidylethanolamine is more enriched in the plasma membrane (Lagarde *et al.*, 1982). Phospholipase A_2 and diglyceride lipase are associated predominantly with the intracellular membrane vesicles (Lagarde *et al.*, 1981). The major proportion of the platelet sphingomyelin is located in the outer leaf of the surface membrane; the intracellular membranes are substantially depleted (Chap *et al.*, 1977).

Platelet Prostaglandins

Arachidonic acid and its oxygenation products have an important role in platelet function. On stimulation of the platelet, arachidonic acid is released from the membrane phospholipids and the biosynthesis of biologically active products ensues. It was believed that the principal source of arachidonic acid was membrane phosphatidylcholine and phosphatidylinositol from which it was released after the activation of endogenous phospholipase A_2 (Bills *et al.*, 1976, 1977). The activation of human platelet phospholipase A_2 requires calcium and is mediated through the Ca^{2+}-binding protein calmodulin (Wong and Cheung, 1979). Other pathways for the release of arachidonic acid have been proposed subsequently (Figure 2.2). Evidence has been assembled for a phosphatidylinositol-specific phospholipase C which provides 1,2-diacylglycerol as a transient intermediate (Rittenhouse-Simmons, 1979; Broekman *et al.*, 1980); this diglyceride may be phosphorylated to phosphatidic acid by 1,2-diacylglycerol kinase (Billah *et al.*, 1979) or degraded by 1,2-diacylglycerol lipase with the release of arachidonic acid (Bell *et al.*, 1979; Bry *et al.*, 1979). A further pathway has been suggested by Billah and associates (1981) involving the action of a calcium-sensitive, phosphatidic acid-specific phospholipase A_2 in the formation of arachidonic acid. It has been proposed that the platelet phospholipases can hydrolyse phospholipids at limited domains only within the intact platelet membrane where the structure of the lipid bilayer has been disarranged by receptor activation (Kannagi *et al.*, 1981), even when sufficient calcium is available.

The mechanism of phospholipase activation following stimulation of the platelet by, for example, thrombin is uncertain. Mobilisation of intracellular calcium is probably one important factor (Pickett *et al.*, 1977). Lapetina and associates (1981) provide evidence that thrombin induces the degradation of a thrombin-receptor fraction of phosphatidylinositol to phosphatidic acid which mediates the subsequent production of arachidonic acid from various phospholipids, possibly through the activation of phospholipase A_2.

It has been recognised recently that human platelets contain an arachidonyl coenzyme A synthetase which is highly specific for

Figure 2.2: Pathways Leading to the Release of Arachidonic Acid from Platelet Membrane Phospholipids

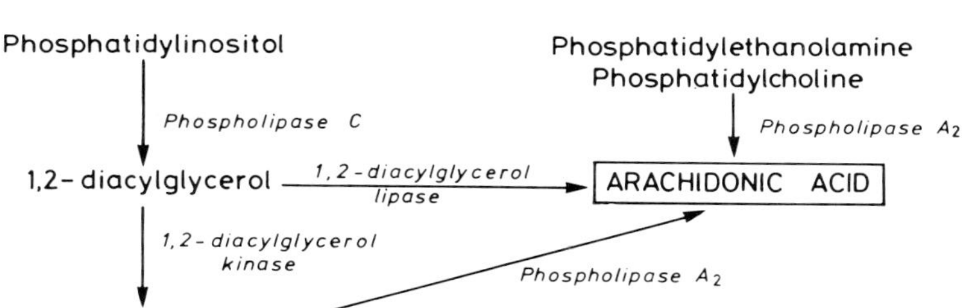

arachidonic acid and 8,11,14-eicosatrienoic acid: its activity may maintain a low level of free arachidonate in unstimulated platelets with efficient incorporation of arachidonic acid into the platelet phospholipid (Wilson *et al.*, 1982).

Following liberation of arachidonic acid by the phospholipases its further transformation may proceed down one of two pathways (Figure 2.3). One leads to the thromboxanes and is catalysed by thromboxane synthetase while the other, catalysed by lipoxygenase, gives rise to a 12-hydroxy-derivative and other compounds. In the latter pathway arachidonic acid is converted into 12L-hydroperoxyl-5,8,10,14-eicosatetraenoic acid (HPETE) and subsequently reduced to 12L-hydroxy-5,8,10,14-eicosatetraenoic acid (HETE). The function of HPETE and HETE is uncertain, but there is evidence that HETE is chemotactic for polymorphonuclear leukocytes (Turner *et al.*, 1975). In addition, it appears that thromboxane synthetase is inhibited by HPETE (Hammarström and Falardeau, 1977) and it has been proposed that HPETE may modulate platelet aggregation by reducing thromboxane production (Aharony *et al.*, 1981).

The other pathway of arachidonic acid metabolism leads to the formation of the cyclic endoperoxides PGG_2 and PGH_2 (Hamberg *et al.*, 1974), a reaction catalysed by cyclo-oxygenase. These cyclic endoperoxides are converted into thromboxane A_2 by thromboxane synthetase (Hamberg *et al.*, 1975; Smith *et al.*, 1976). Thromboxane A_2 is a potent platelet aggregating agent (Willis and Kuhn, 1973; Vargaftig and Zirmis, 1973) and induces granule secretion. In addition, it has vasoconstrictor properties (Piper and Vane, 1969). It is an unstable material with a half-life of some 30 seconds in aqueous solution at 37°C; in plasma the half-life is approximately five minutes

Figure 2.3: Derivatives of Arachidonic Acid

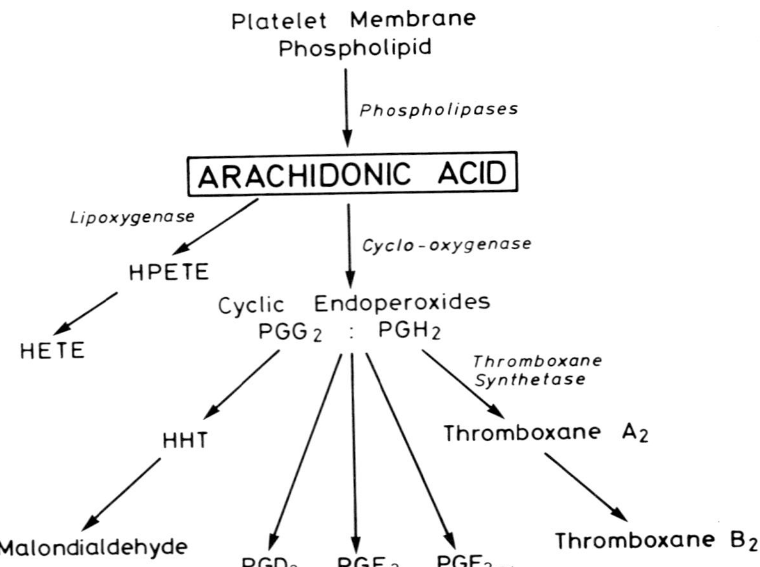

(Smith *et al.*, 1976). Thromboxane A_2 hydrolyses to the stable thromboxane B_2 (Hamberg *et al.*, 1975).

The mechanism by which thromboxane A_2 causes platelet aggregation or vessel constriction is not wholly understood. Earlier studies indicated that ADP release is required for thromboxane A_2-induced platelet aggregation (Malmsten *et al.*, 1975), but other investigations have not supported this conclusion (Charo *et al.*, 1977; Kinlough-Rathbone *et al.*, 1976).

Part of the cyclic endoperoxides formed is converted into the degradation products 12L-hydroxy-5,8,10-heptadecatrienoic acid (HHT) and malondialdehyde (MDA). In addition, small quantities of the stable prostaglandins PGE_2, $PGF_2\alpha$ and PGD_2 are formed. Their physiological significance is again uncertain, but PGD_2 is a potent inhibitor of platelet aggregation (Smith *et al.*, 1974), probably through its ability to stimulate adenylate cyclase and raise intracellular cyclic AMP levels.

Figure 2.4: Structure of Arachidonic Acid, Endoperoxides and Thromboxanes

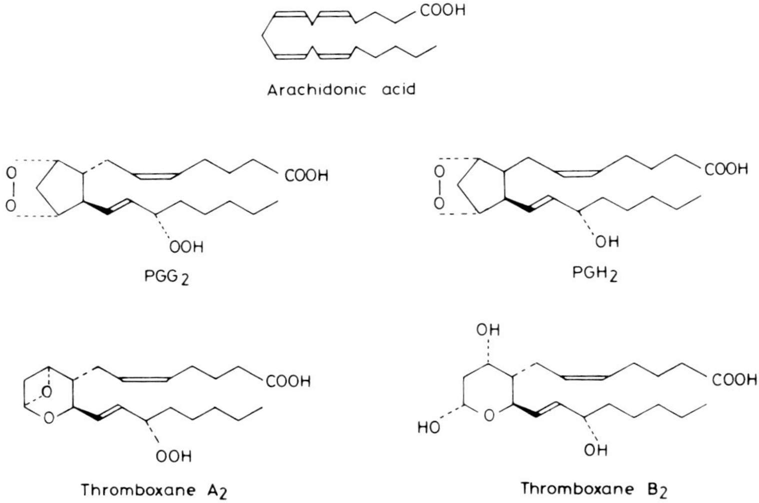

COOH

Arachidonic acid

O / O — COOH — OOH — PGG$_2$

O / O — COOH — OH — PGH$_2$

O — COOH — OOH — Thromboxane A$_2$

OH — HO — O — COOH — OH — Thromboxane B$_2$

Prostacyclin

The description of an enzyme in the blood vessel wall that converted the cyclic endoperoxides into an unstable agent with vasodilator and platelet aggregation inhibitory effects came from Moncada and colleagues (1976a, 1976b), while Kulkarni and associates (1976) demonstrated that a dilating substance is produced when coronary arteries are perfused with arachidonic acid. It was shown subsequently that arachidonic acid is converted into an unstable substance which is degraded to 6-keto prostaglandin $F_{1\alpha}$ (Raz *et al.*, 1977): this material was designated as PGI_2 (Johnson *et al.*, 1976) and is also known as prostacyclin. The structure of prostacyclin and its degradation product 6-keto $PGF_{1\alpha}$ is shown in Figure 2.5.

Vascular rings were shown to produce PGI_2 when incubated with PGH_2 or platelet-rich plasma (Bunting *et al.*, 1976), and it was demonstrated that endothelial cells in tissue culture also produce prostacyclin (Weksler *et al.*, 1977). It was postulated that the substrate for PGI_2 production by the vascular wall is the endoperoxides synthesised in platelets and secreted during aggregation (Moncada and Vane, 1979), and that significant vascular synthesis of PGI_2

23

Figure 2.5: Structure of Prostacyclin and its Degradation Product 6-Keto PGF$_{1\alpha}$

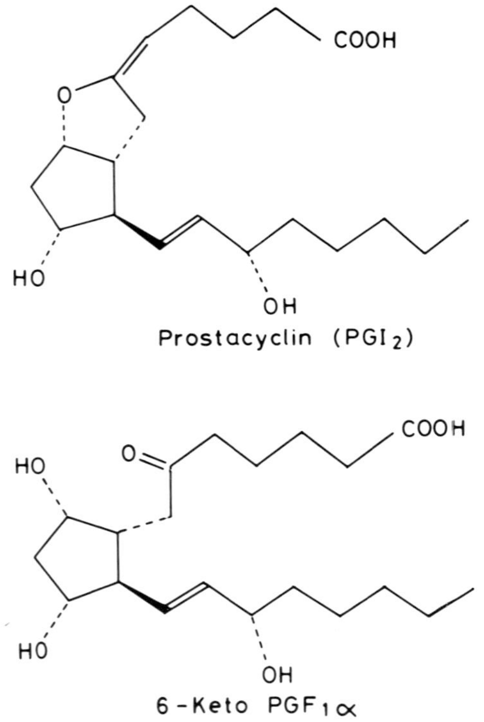

occurs only in the intact endothelium. The hypothesis proposed that platelet adhesion to the subendothelial structures during hemostasis takes place after disruption of the endothelium and that the plug of aggregated platelets grows through the influence of thromboxane A$_2$. Areas of the vessel wall where the endothelium is intact convert the endoperoxides secreted by platelets into prostacyclin which inhibits platelet aggregation. In this way the platelet plug formation is limited to the site of vessel damage. The hypothesis, however, has not gone unchallenged. Prostacyclin production is not limited to the endo-thelium and has been shown to be achieved by vascular smooth muscle cells and skin fibroblasts (Baenziger *et al.*, 1977). It has also been shown that cultured endothelial cells are capable of forming prostacyclin from arachidonic acid as well as from PGH_2 (Weksler *et al.*, 1977), and there is experimental evidence to suggest that the endothelium is unable to synthesise prostacyclin from extracellular cyclic endoperoxides. Nevertheless, whatever the major substrate for prostacyclin production, it is likely that its synthesis by intact endothelium confers on it, wholly or partially, its established ability to prevent platelet activation and thereby serve as a non-thrombogenic surface. In addition, prostacyclin appears to be able to inhibit the process by which thrombin and collagen make platelet procoagulant activity available, although it does not interfere directly with factor Xa or thrombin formation (Bevers *et al.*, 1981; Brox and Østerud, 1981).

Platelet Granules

The contents of the platelet dense and α-granules are detailed in Table 2.1. Early studies suggested that the α-granules were storage sites for acid hydrolases, but later observations suggested that these are stored in smaller vascular structures distinct from typical α-granules (Broekman *et al.*, 1974; Bentfield and Bainton, 1975).

Dense Granules

The dense granules, less numerous and of greater electron density than the α-granules, contain nucleotides, 5-hydroxytryptamine (serotonin), calcium and pyrophosphates; they may also contain an antiplasmin.

Adenine Nucleotides. Human platelets contain substantial quantities of adenine nucleotides with a particularly high proportion in the form of ADP, resulting in an ATP/ADP ratio of around 1:7 (Mills and Thomas, 1969). There are two major pools of platelet adenine nucleotides — the granular (storage) pool and the metabolic pool which is distributed between the mitochondria and cytosol. Part of the metabolic pool ADP exists as a complex with actin and behaves as though it is unavailable for energy metabolism (Daniel *et al.*, 1979). In human platelets there is about six times more granule-stored than extragranular ADP (Holmsen *et al.*, 1972).

The addition of ADP to stirred platelet-rich plasma induces platelet aggregation, while platelet stimulation by ADP, thrombin or collagen results in the release of endogenous ADP from the dense

Table 2.1: Contents of Platelet Dense and Alpha-Granules

Dense granules	Alpha-granules
ADP	Fibrinogen
ATP	Factor V
5-Hydroxytryptamine	Factor VIII:vWF
Calcium	Platelet factor 4
Pyrophosphate	β-thromboglobulin
	Albumin
	Fibronectin
	Thrombospondin
	Platelet-derived growth factor

granules of the storage pool. This sequence of ADP → release reaction → aggregation by released ADP is responsible for the biphasic aggregation response seen after the addition of certain concentrations of ADP to platelet rich plasma (see p. 42), and is probably an integral part of the hemostatic process.

5-Hydroxytryptamine. 5-Hydroxytryptamine (serotonin) is present in higher concentration in platelets than in any other mammalian tissue. Using a bioassay Hardisty and Stacey (1955) found a mean value of 57 ± 18 ng/10^8 human platelets. A gradual decrease with aging has been reported (Kita *et al.*, 1981). Platelets do not synthesise serotonin, but it is readily transported across the platelet membrane to be stored in the dense granules.

5-Hydroxytryptamine is capable of aggregating platelets *in vitro*: its ability to do this *in vivo* is suggested by experiments showing that mast cell degranulation in rats, a species with a significant content of serotonin in their mast cells, results in the formation of platelet aggregates (De Clerck and Van Gorp, 1981). There is evidence that circulating platelets can lose their 5-hydroxytryptamine, presumably as a result of secretion (Osim and Wyllie, 1982).

Serotonin has an action on smooth muscle and the vessel constriction induced by this effect has been thought to contribute to hemostasis. However, it is a long-standing observation that gross depletion of platelet 5-hydroxytryptamine by reserpine does not prolong the bleeding time (Shore *et al.*, 1956), while platelet serotonin concentration was reported to be unrelated to the incidence of purpuric

26

symptoms in patients with a variety of blood disorders (Hardisty and Stacey, 1957).

Measurement of the amount of radioactivity set free from platelets preincubated with ^{14}C-labelled serotonin has been a frequently used technique for the assessment of platelet secretion.

Alpha-Granules

Platelet Factor 4. Platelet factor 4 is a heat-stable protein with anti-heparin activity which is released from the α-granules on platelet stimulation (Deutsch *et al.*, 1955; Niewiarowski and Thomas, 1969; Broekman *et al.*, 1975; Fukami *et al.*, 1979; Ryo *et al.*, 1980a). It has also been found within tissue mast cells (McLaren *et al.*, 1980), but it is uncertain whether it is produced by such cells or is formed elsewhere and bound to the heparin of the mast cell granules. Platelet factor 4 antigen has been detected in bone marrow megakaryocytes, suggesting that it may be synthesised in these cells (Ryo *et al.*, 1980b).

Platelet factor 4 has been purified to homogeneity and its primary structure established (Deuel *et al.*, 1977; Hermodson *et al.*, 1977; Walz *et al.*, 1977). The molecule is composed of a single chain with an amino-terminal glutamic acid and a total of 70 amino acids. The molecular weight is 7780 daltons. It has a lysine-rich carboxyl-terminal region which is probably of importance in its interaction with heparin. On release from platelets it is in the form of a proteoglycan-platelet factor 4 complex, the proteoglycan carrier consisting of chondroitin 4-sulphate and having a molecular weight of 59,000 (Barber *et al.*, 1972). The characterisation of this proteoglycan has been described recently (Huang *et al.*, 1982). Heparin displaces platelet factor 4 from the complex.

Intravenous injections of heparin induce the release of platelet factor 4 (Dawes *et al.*, 1978). It is rapidly cleared from the circulation and small quantities appear in the urine. The plasma concentration has been assayed at 13.9 ± 6.1 ng/ml (Dawes *et al.*, 1978), while a mean level of 15.2 ng/ml was found by Zahavi *et al.* (1980). Rucinski and colleagues (1979) reported a level of 12.4 ± 3.2 µg/10^9 intact platelets.

The physiological function of platelet factor 4 is uncertain. It binds tightly to heparin and thereby prevents it from complexing with antithrombin III. Heparin has been shown to possess multiple binding regions for both platelet factor 4 and antithrombin III (Jordan *et al.*, 1982). However, the possible role of platelet factor 4 as an anti-heparin agent of physiological significance hinges on the existence of functional quantities of endogenous heparin in the circulation: this remains to be established. Radiolabelled platelet factor 4 has been shown to bind to cultured human endothelial cells (Busch *et al.*, 1980), and it was suggested that it might favor hemostasis by interfering with a heparin-like function of the endothelial cell surface in the reaction between antithrombin III and thrombin.

Some preliminary data have raised the possibility that platelet factor 4 can inhibit the activation of factor XII by dextran sulphate. Platelet factor 4 did not inhibit XIIa or kallikrein; its effect on dextran sulphate-induced contact activation is probably due to its ability to bind to negatively charged polysaccharides (Weerasinghe *et al.*, 1981). A further possible function for platelet factor 4 stems from the finding that it is chemotactic for human polymorphonuclear leukocytes and monocytes at concentrations found in human serum and attained locally in injured tissue (Deuel *et al.*, 1981).

Beta-Thromboglobulin. Moore and colleagues (1975) described the isolation of a platelet-specific β-globulin from the supernatant of thrombin-aggregated platelets. It has been purified and found to have a molecular weight of 8800 (Begg *et al.*, 1978). The amino acid sequence of this protein shows marked homology with that of platelet factor 4, including four cysteine residues and two pairs of lysine residues near the carboxyl-terminus.

A further protein with anti-heparin activity was isolated by Rucinski and colleagues (1979) and termed low-affinity platelet factor 4. It was shown to be immunologically identical to β-thromboglobulin, but distinct from platelet factor 4. It had an apparent molecular weight of 7800 daltons on sodium dodecyl sulphate polyacrylamide gels and 9070 daltons as calculated by amino acid analysis. Beta-thromboglobulin appears to be derived from low-affinity platelet factor 4 by the removal of four amino acids from the amino-terminal end of the molecule. Niewiarowski and associates (1980) have provided evidence that platelets secrete low-affinity platelet factor 4 which is then converted into β-thromboglobulin by plasmin or a protease from platelets.

The mean plasma level of β-thromboglobulin has been found to be 30.7 ng/ml by Dawes *et al.* (1978) and 33.9 ng/ml by Zahavi *et al.* (1980). Rucinski *et al.* (1979) obtained a value of 24.2 µg/10^9 platelets for low-affinity platelet factor 4.

Beta-thromboglobulin has relatively low anti-heparin activity and its biological function is unknown. Hope and colleagues (1979) reported that human β-thromboglobulin reduces the production of prostaglandin-like activity in cultured bovine aortic endothelial cells and proposed that it might act locally to favor platelet aggregation by diminishing prostacyclin production. Others, however, have been unable to demonstrate an effect of β-thromboglobulin or low-affinity platelet factor 4 on prostacyclin production by aortic endothelial cells (Poggi *et al.*, 1981).

Although the function of the heparin-binding proteins is obscure, the levels of platelet factor 4 and β-thromboglobulin in platelet-rich plasma, assessed by radioimmunoassays, have been used as a test of platelet activation *in vivo*.

Fibronectin. The fibronectins are a group of high molecular weight glycoproteins which are present on the surface of many cells, in connective tissues and most basement membranes, and in the extra-cellular fluid. It is present in plasma as cold-insoluble globulin. Fibronectin has a molecular weight of 440,000 and consists of two equal polypeptide chains linked by a disulphide bond. Its structure and biological properties have been reviewed by Mosesson and Amrani (1980).

Fibronectin is found in the α-granules of platelets (Zucker *et al.*, 1979; Plow *et al.*, 1979). Part of the platelet fibronectin is released on activation with collagen (Zucker *et al.*, 1979) or thrombin (Zucker *et al.*, 1979; Ginsberg *et al.*, 1980). It is present on the platelet surface, but only after thrombin stimulation (Ginsberg *et al.*, 1980).

Fibronectin is a substrate for factor XIIIa; the outcome of the reaction is the formation of covalent linkages with other fibronectin molecules, fibrin or collagen (Mosher *et al.*, 1979): the site on fibronectin which mediates strong binding to collagen is distinct from the site of cross-linking (Mosher *et al.*, 1980).

It has been proposed that fibronectin on the platelet surface has a role in platelet aggregation and adhesion to connective tissue and fibrin thrombi. One suggestion is that surface-bound platelet fibronectin is the platelet membrane receptor for collagen (Bensusan *et al.*, 1978), but this is unlikely since treatment of platelets with an antibody to fibronectin has little effect on the adhesion of the platelets to collagen (Santoro and Cunningham, 1979).

Platelet-Derived Growth Factors. One or more growth factors are secreted from the α-granules which have the property of stimulating human cells to initiate DNA synthesis and cell growth. These cells include arterial smooth muscle cells (Ross *et al.*, 1974), fibroblasts (Kaplan *et al.*, 1979) and glial cells (Heldin *et al.*, 1977). Platelet-derived growth factor has been shown to stimulate tyrosine-specific phosphorylation of two membrane components of human fibroblasts and glial cells (Ek *et al.*, 1982).

A number of platelet-derived growth factors of differing molecular weights have been found in platelet extracts. The most highly purified component of the preparation of Antoniades and associates (1975) had a molecular weight of 13,000, close to that reported by Paul *et al.* (1980). This latter protein appeared to be immunologically identical to low-affinity platelet factor 4 and β-thromboglobulin. A cationic platelet growth factor of molecular weight 26,000 to 33,000 daltons was isolated by Heldin and colleagues (1979). Deuel *et al.* (1981) purified growth factor and obtained two equally active fractions of molecular weights 31,000 and 28,000; both were glycoproteins of similar amino acid composition, and it was suggested that one may be a proteolytic cleavage product of the other. A two-polypeptide chain molecule was proposed for the platelet-derived growth factor purified from human plasma by Raines and Ross (1982), the chains being of 14,400 daltons and either 17,500 or 16,000 daltons.

The function of platelet-derived growth factor is uncertain, but it may be involved in tissue repair and wound healing after injury. It has been proposed that growth factor may be implicated in the pathogenesis of atherosclerosis by stimulating arterial smooth muscle proliferation in response to endothelial injury (Ross and Glomset, 1976).

Thrombospondin. Thrombospondin is a high molecular weight heparin-binding glucoprotein released from platelets in response to thrombin (Lawler *et al.*, 1978). It is composed of three disulphide-linked polypeptides each of 140,000 daltons (Margossian *et al.*, 1981). Its biological function is unknown.

Platelet Calcium

The concentration of calcium in the cytoplasm of the resting platelet is about 10^{-7} M: raising the concentration has a number of functionally important sequelae. In particular, the contractile process of the platelet is calcium-mediated, phospholipases A_2 and C require calcium for their activity, the disassembly of microtubules is provoked by calcium and the activation of factor XIII requires calcium. Accordingly, calcium ion is intimately involved in the specific platelet processes of shape change, aggregation and secretion.

The mobilisation of calcium probably takes place in a sequential reaction starting with the liberation of membrane-bound calcium followed by release from specific intracellular structures and, finally, by influx from the extracellular fluid. The dense tubular system is the source of the greater part of the intracellular calcium; mitochondria may make a minor contribution. The dense granules constitute an important calcium reservoir, but exchange of calcium through the organelle membrane is not believed to take place. Calcium pumps, stimulated by cyclic AMP, exist in the platelet membrane and the dense tubular system so that calcium ion can be pumped out of the cytoplasm into the dense tubular system or into the extracellular fluid. Inhibitors of platelet activity stimulate adenylate cyclase and thereby increase the level of cyclic AMP with the result that cytoplasmic calcium levels are reduced.

Calmodulin, a calcium-binding protein, is present in the cytosol in considerable quantity and regulates a number of calcium-dependent platelet processes. Its involvement has been demonstrated in the activation of phospholipase A_2 (Wong and Cheung, 1979), myosin light chain kinase (Hathaway and Adelstein, 1979), factor XIII (Kahn *et al.*, 1981) and in the release reaction (Nishikawa *et al.*, 1980).

Platelet Cytoskeleton

There is now considerable evidence that the circumferential band of microtubules acts as a flexible cytoskeleton which maintains the discoid shape of the platelet (Behnke, 1965; White and Gerrard, 1979). It is likely that it also has a role in platelet secretion, centripetal movement of the microtubules and granules taking place before the release reaction (White, 1968a; Menche *et al.*, 1980).

The microtubules are composed mainly of heterodimers of α and β tubulin subunits non-covalently polymerised to form hollow tubular cylinders (Castle and Crawford, 1977). There is dynamic equilibrium between polymerised and depolymerised forms of tubulin. In the resting platelet most of the tubulin is polymerised as microtubules; on stimulation transient depolymerisation takes place

(Steiner and Ikeda, 1979). When repolymerisation ensues the peripheral microtubule band becomes constricted and microtubules appear in pseudopods, lying parallel to their long axes.

Tubulin is also a constituent of the plasma membrane of platelets. Platelets pre-incubated with a monospecific antibody against tubulin have been reported to show reduced aggregation and shape change in response to collagen, suggesting that membrane-associated tubulin may have a role in collagen-platelet interactions (Ikeda *et al.*, 1981).

Contractile Apparatus of Platelets

Platelets contain a contractile apparatus analogous to that found in muscle, in particular, smooth muscle. The major component was originally termed thrombosthenin and represents the actomyosin of muscle (Bettex-Galland and Lüscher, 1959, 1961). It is probable that this apparatus functions in the production of the characteristic shape change with accompanying granule secretion which takes place on platelet stimulation, and it is also involved in clot retraction.

Actin is the most abundant of the platelet proteins comprising some 10 per cent of the total protein mass. It has a molecular weight of 43,000 daltons. Platelet actin differs slightly in structure from the α-actin form of skeletal muscle: isoelectric focusing gel analysis has shown that it exists in β and γ forms in a ratio of five to one (Landon *et al.*, 1977). In the resting platelet actin is present in a stable monomeric non-filamentous form termed profilactin. The maintenance of actin in its non-filamentous form appears to be due to its association with profilin, a small protein which binds to monomeric actin with resultant stabilisation (Carlsson *et al.*, 1977). Profilin has been shown to be present in platelets (Markey *et al.*, 1978; Harris and Weeds, 1978). A small proportion of the platelet actin is firmly adherent to the surface membrane (Bennett *et al.*, 1981), but most is located in the cytoplasm.

Myosin is present in platelets at a much lower concentration than in skeletal muscle. The platelet myosin has structural differences from muscle myosin, being composed of two non-covalently-linked heavy chains with a molecular weight of 200,000 daltons, each of which binds a pair of myosin light chains of molecular weights 20,000 and 15,000 at its globular head region (Adelstein *et al.*, 1973; Pollard *et al.*, 1974, 1977). It is the globular regions of the myosin molecules which are responsible for both the binding to actin filaments and for the ATPase activity required for the generation of contraction.

Platelets contain a tropomyosin-like protein, smaller in size than muscle tropomyosin (Cohen and Cohen, 1972), which appears to bind to platelet actin. Troponin has not been isolated from platelets, although some evidence for its presence has been presented (Cohen *et al.*, 1973).

Following thrombin-stimulated platelet activation there is polymerisation of actin with a marked increase in the amount of the filamentous form (Jennings *et al.*, 1981). Platelets contain an actin-binding protein with a molecular weight of around 250,000 daltons which appears to act with α-actinin to facilitate the organisation of actin filaments (Schollmeyer *et al.*, 1978). The association of the actin-binding protein with actin may be initiated by its phosphorylation (Carroll and Gerrard, 1981). The interaction of α-actinin with actin is inhibited by calcium, perhaps allowing the rapid organisation of the actin filaments in the presence of calcium. Evidence has been provided that thrombin-induced platelet secretion is associated with the appearance of actin on the platelet surface (George *et al.*, 1980).

During platelet activation polymerisation of the soluble myosin takes place; this results in the formation of bipolar myosin filaments with exposed globular heads near each end of the filaments. Phosphorylation of the larger of the myosin light chains controls the amount of tension generated by the actomyosin threads (Lebowitz and Cooke, 1978; Daniel *et al.*, 1981), and it has been shown that phosphorylation of this chain increases the actin-activated ATPase activity of the myosin (Adelstein and Conti, 1975).

The contractile mechanism of platelets is, as in skeletal muscle, dependent on calcium ions. Both tropomyosin and troponin are required for the control of actin-myosin interaction in skeletal muscle: calcium binds to the troponin moiety of the tropomyosin-troponin complex, this binding inducing a conformational change in the complex and allowing myosin to bind to actin with initiation of myosin ATPase contractile activity. In the absence of identifiable troponin in platelets precise analogy with the control mechanism of the contractile apparatus of skeletal muscle is not possible. Other calcium-sensitive control mechanisms may be more important in platelets, in particular, and analogous to the situation in smooth muscle, is the action of a myosin light chain kinase (Adelstein *et al.*, 1973; Dabrowska and Hartshorne, 1978). The role of the myosin light chain kinase is to activate the contractile mechanism via the phosphorylation of myosin. The molecular weight of the myosin light chain kinase is around 105,000 (Hathaway and Adelstein, 1979) and calcium ions are required for its activity (Waisman *et al.*, 1978; Dabrowska *et al.*, 1978). Calmodulin is also implicated in the activity of the myosin light chain kinase (Dabrowska and Hartshorne, 1978; Hathaway and Adelstein, 1979). The possibility of a platelet calcium-independent myosin light chain kinase of lower molecular weight has been raised (Daniel and Adelstein, 1976).

The principal source of the calcium required for the platelet contractile mechanism is the dense tubular system while small quantities may be released from the platelet membrane. The mechanism of production of the calcium flux is considered elsewhere.

Alpha-actinin functions as a membrane attachment site for the actin microfilaments. It has been proposed that platelet membrane glycoprotein IIIa is α-actinin (Gerrard *et al.*, 1979), but purified preparations of the two proteins have been shown to migrate differently on SDS-electrophoresis (Godstad *et al.*, 1981). It has been suggested that α-actinin undergoes a conformational change after platelet activation resulting in exposure of the bound actin to the external environment; the actin so exposed on the platelet surface could provide a receptor site for polymerising fibrin and the contractile power for clot retraction (George *et al.*, 1980).

A schematic representation of the platelet contractile mechanism is shown in Figure 2.6.

Platelet Activation

Platelet activation, integral to the hemostatic process, involves a number of distinguishable steps. These follow the adhesion of platelets to collagen, but can also be induced by the addition of a variety of agents to platelet-containing plasma. The sequential steps of shape change, aggregation and secretion (release reaction) can be recognised.

The initiation of the platelet activation sequence involves the binding of an agonist to a receptor on the platelet surface. It is likely that such binding causes a local conformational change in the membrane with release of small quantities of calcium. Activation of phospholipases allows labilisation of the membrane to proceed while the membrane phospholipids undergo the 'flip-flop' movement with the negatively-charged phospholipids becoming available at the platelet surface with subsequent participation in the coagulation

Figure 2.6: Schematic Representation of Platelet Contractile Mechanism

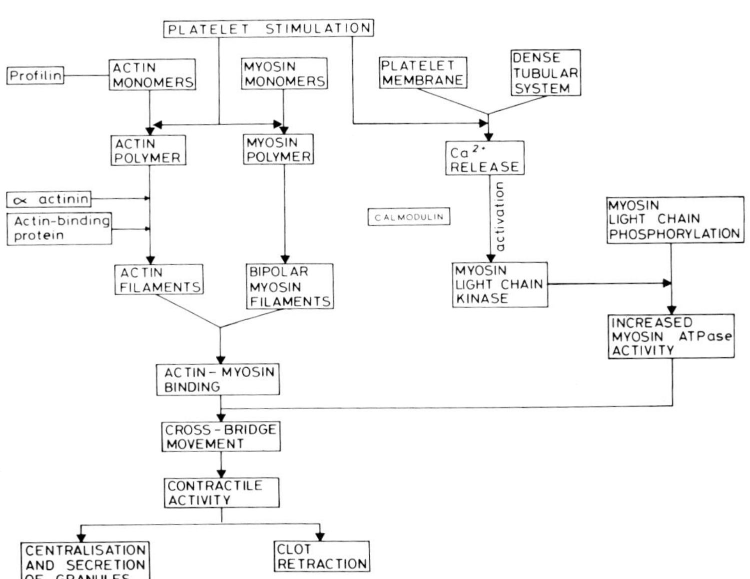

mechanism; it is likely that the coagulant activity of platelets termed platelet factor 3 is an expression of this process. Calcium channels open allowing the influx of calcium into the cytoplasm of the platelet with consequent induction of the various calcium-dependent events (see page 30). A diagrammatic representation of processes taking place on platelet activation is provided in Figure 2.7.

Platelet activation inducers may be classified as strong or weak. Strong inducers, for example, thrombin cause a general membrane disturbance and activation can proceed without feed-back mechanisms to make more calcium available. Weak inducers such as serotonin cause a local membrane disturbance with local contraction producing shape change. If the calcium is removed the membrane structure may be restored and aggregation will not take place; alternatively, with the feed-back mechanisms, the process will proceed to aggregation and secretion.

Figure 2.7: Schematic Representation of Events Following Platelet Activation

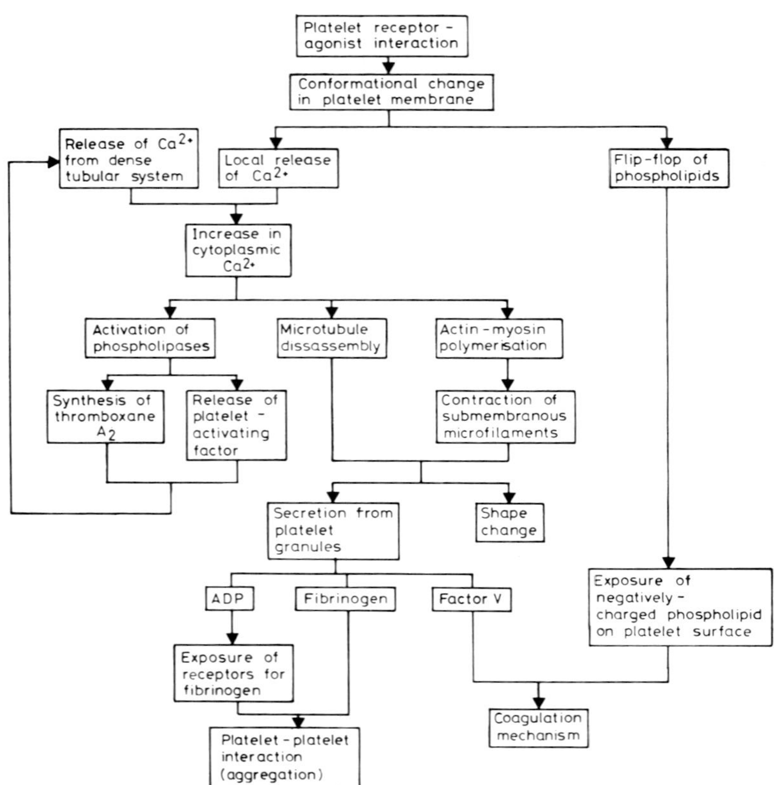

Platelet Adhesion

The term platelet adhesion refers to the attachment of platelets to the subendothelial layer of the blood vessel wall or to a foreign surface. In the process of hemostasis platelets adhere principally to collagen which is exposed in the wall of the damaged vessel. Of particular functional relevance is the lack of adherence to endothelial cells: the endothelium not only forms a non-reactive lining, but also synthesises the platelet function inhibitor prostacyclin.

Subendothelium. The subendothelium of blood vessels is composed of a reticular network of collagen fibrils with some elastin microfibrils resting on the internal elastic lamina which separates the intima from the media; the fibrils are embedded in amorphous basement membrane-like material. The basement membrane contains at least one collagenous protein and non-collagenous glycoproteins such as fibronectin (see page 29) and laminin, a large molecule of 880,000 daltons (Timpl *et al.*, 1979). Heparan sulphate appears to be the predominant glycosaminoglycan of the subendothelium. The various components of the subendothelium are synthesised largely by the endothelial cells.

There are five distinct types of collagen with differing amino acid composition and sequence and carbohydrate content. Types I, III, IV and A-B are found in the subendothelium, type II being the most abundant. Agreement on the relative activity of the collagen types in inducing platelet adhesion and aggregation has not been reached. Collagens I, II and III appear to be equally active (Santoro and Cunningham, 1977; Balleisen *et al.*, 1979). Trelstad and Carvalho (1979) found that types IV and A-B did not cause aggregation.

The physical state of collagen is important in the induction of platelet activation. The findings of Meyer and Weisman suggest that adhesion of platelets to a surface of triple helical collagen *per se* does not lead to their activation; the surface must, in addition, be of sufficiently high curvature with a fibre diameter of less than that of the platelet (Meyer and Weisman, 1981). The chemical determinants recognised by human platelets on the triple helical surface involve proline/hydroxyproline and arginine residues (Meyer and Frojmovic, 1979).

Elastic tissue microfibrils and elastin do not interact with platelets; the few platelets which are seen to adhere to such elements of the subendothelium do so by simple cellular contact.

Mechanism of Adhesion. Studies using preparations in which the endothelium was removed by a balloon catheter or perfusion chambers to examine the interaction of platelets with the subendothelium of animal or human vessels have demonstrated that platelets adhere through contact with part of the membrane of a discoid platelet or through pseudopods after shape change. The platelets then spread along the subendothelium, usually accompanied by the release reaction and adherence of other platelets to the first ones. A number of variables have been found to influence the interaction of platelets with the subendothelium, including shear rate, red cell concentration and von Willebrand factor (Baumgartner, 1973; Weiss *et al.*, 1977; Turitto and Baumgartner, 1979; Baumgartner *et al.*, 1980). The rate of platelet adherence to human artery sub-endothelium has been shown to be greater in steady flow than in pulsatile flow (Sakariassen *et al.*, 1980).

Von Willebrand factor, synthesised by megakaryocytes and endothelial cells, is required for the adhesion of platelets to connective tissue and appears to be the only protein in plasma which facilitates platelet-vessel wall interaction (Sakariassen *et al.*, 1979). Under physiological rates of shear von Willebrand factor acts as a link between a glycoprotein receptor on the platelet surface and the subendothelial collagen. The binding of von Willebrand factor to platelet receptors can be induced *in vitro* by the antibiotic ristocetin; binding can also be induced by thrombin (Fujimoto *et al.*, 1982) and by ADP (Fujimoto and Hawiger, 1982). It appears that the ADP secreted from the dense granules by thrombin is responsible for part, at least, of the subsequent binding of von Willebrand factor, and it has been proposed that the activation of platelets by ADP results in exposure of the receptors for von Willebrand factor and fibrinogen.

Assessment of Platelet Adhesiveness. While glass provides a foreign surface which can be standardised to some extent, its differences from collagen may diminish the physiological relevance of measurements of platelet adhesiveness to glass. Additionally, many of the techniques used to assess platelet adhesiveness measure both aggregation and adhesion since, on adherence to glass, platelet activation takes place with release of ADP so that further platelets adhere to those sticking to the glass surface. This is true of the widely employed glass bead column technique where blood is pumped through a narrow plastic tube containing fine glass beads. The difference in the platelet count in the blood or plasma before and after passage through the column is taken as a measure of the adhesion to the glass beads. The original technique was described by Hellem (1960). Many variations in the basic technique have been used including drawing native blood directly from the venepuncture needle through the glass beads (Salzman, 1963). Such factors as the concentration of the anticoagulant used, rate of passage through the column, hematocrit, filter age and number of beads affect the percentage of platelets retained. Platelet-rich plasma may be used in place of whole blood, but ADP must be added to the plasma.

Techniques have been described for the measurement of the adhesion of platelets to surfaces of glass or glass previously exposed to plasma or fibrinogen (Zucker and Vroman, 1969; George, 1972): these are not influenced by secondary platelet aggregation, but have not been adopted to any extent for studies on the physiology of platelet function.

Measurement of the ability of platelets to adhere to the walls of damaged vessels would have obvious physiological relevance. Borchgrevink (1960) described a technique in which a standard skin incision is made and the platelet count of venous blood compared with that of the blood oozing from the incision.

The assay of platelet adhesion to monolayers of cultured vascular endothelial cells using [^{3}H]adenine-labelled platelets has been described recently (Curwen *et al.*, 1982).

Platelet Shape Change

On activation the disc-shaped resting platelet is transformed into a spherical structure with long, filiform pseudopodia. The disc-shaped platelet is characterised by a marginal bundle of microtubules; shape change involves the decay of the microtubules and a process responsible for pseudopod formation. Shape change may be due to the contraction of the submembranous network of actomyosin filaments and requires the availability of Ca^{2+} ions which lead both to the depolymerisation of the microtubules and to actomyosin contraction. On receiving the activation stimulus the ring of microtubules reduces in diameter, contracting towards the center of the platelet. The microtubules then rapidly disassemble (depolymerise) and exist in the fully activated platelet in a newly polymerised state and in a different conformation, lying in parallel array within the pseudopodia.

Platelet Secretion

Platelet secretion, also known as the platelet release reaction, involves the specific discharge of the contents of the platelet granules to the outside, and can be detected within seconds of stimulation. The quantity and type of substance secreted depends on the nature and concentration of the stimulating agent. ADP, epinephrine and low concentrations of thrombin and collagen induce secretion from the dense granules only, while higher concentrations of thrombin and collagen evoke, in addition, secretion from the α-granules.

Secretion is achieved through the process of exocytosis. The granule membranes fuse with the surface membrane of the platelet or its invaginations and the contents of the granules are discharged into the extracellular space (Fukami and Salganikoff, 1977). Secretion involves the utilisation of energy: there is conversion of ATP into hypoxanthine although this is not necessary for secretion, and stimulation of glycolysis is shown by increased glucose uptake and lactate production. A burst in oxygen uptake is also detectable. Details of these metabolic events may be found in the review by Holmsen and colleagues (1977).

Platelet secretion follows the binding of agonists to their plasma membrane receptors. This event probably induces an increase in the cytoplasmic calcium concentration and the resultant activation of the contractile mechanism causes the membranes of the platelet and the granules to move towards each other and allow fusion and discharge of the granule contents.

Platelet Aggregation

The process of aggregation, defined as the cohesion of platelets to each other, is integral to the build-up of a hemostatic platelet plug. It may be induced by a number of agents including ADP, epinephrine, 5-hydroxytryptamine, thrombin and collagen: the platelet possesses receptors for each of these. It has been proposed that platelet aggregation may be mediated by three pathways — the release of ADP from granules, the synthesis of thromboxane A_2 and the release of platelet-activating factor (Vargaftig *et al.*, 1981a), but their relative importance is unknown and may vary with the stimulating agent.

Platelet-Activating Factor. Platelet-activating factor (PAF) is a phospholipid which was shown to be released from IgE-stimulated rabbit basophils and found to have a platelet-aggregating action (Benveniste *et al.*, 1972). It was proposed that PAF is a 2-acyl-glycerophospholipid (Benveniste *et al.*, 1977), and found subsequently that 1-*O*-alkyl-2-acetyl-sn-glyceryl-3-phosphorylcholine has the biological properties of PAF (Demopoulos *et al.*, 1979). Hanahan and associates (1980) have identified acetyl glyceryl ether phosphorylcholine as the naturally occurring PAF from rabbit basophils.

Platelet-activating factor is released from platelets stimulated by thrombin, collagen or a calcium ionophore: its formation is calcium-dependent and phospholipase A_2 may be involved (Chignard *et al.*, 1981). It acts as a potent aggregating agent on both rabbit and human

40

platelets (Benveniste, 1974; Benveniste *et al.*, 1975), and rabbit basophil-derived PAF has been shown to induce the release of serotonin from human platelets (O'Donnell *et al.*, 1978). The action of PAF on platelets is independent of the ADP and thromboxane A_2 mechanisms of platelet aggregation and release (Chignard *et al.*, 1979). In addition to its platelet aggregating action it has been claimed that PAF can also stimulate prostacyclin release from cultured human endothelial cells (Test and Bang, 1981).

Further information on the origin, chemistry and pharmacology of PAF may be found in the review by Vargaftig and colleagues (1981b).

Mechanism of Aggregation. Fibrinogen is essential for platelet aggregation (Mustard *et al.*, 1978). In addition to the plasma source, fibrinogen may be made available from the platelet α-granules for aggregation. It has been shown that platelet stimulation by ADP or epinephrine exposes fibrinogen receptors on the surface of the platelet (Bennett and Vilaire, 1979). There is evidence, as discussed on page 18, that platelet membrane glycoprotein IIb/IIIa acts as a receptor for fibrinogen. The precise mechanism by which platelet-bound fibrinogen is involved in aggregation is not established; one suggestion is that the dimeric molecule binds to receptors on adjacent platelets with the formation of cross-links. The possibility of a further intermediary is raised by the finding that thrombin-activated platelets secrete an endogenous hemagglutinin which interacts with receptors on other platelets to cause aggregation: the receptor for this hemagglutinin may be fibrinogen (Gartner *et al.*, 1981). It has been suggested that thrombospondin is the endogenous lectin of human platelets (Jaffe *et al.*, 1982), and it has been shown that fibrinogen can form a complex with adsorbed platelet thrombospondin (Leung and Nachman, 1982).

The simple explanation of the build-up of a platelet plug through adherence of platelets to collagen, release of ADP from the dense granules and synthesis of thromboxane A_2, and the consequent accumulation of aggregates of platelets has been questioned. The doubt stems partly from the rapidity with which hemostatic aggregates of platelets grow, whereas aggregation following adhesion of platelets to collagen only occurs after a lag period of 15 to 30 seconds (Wilner *et al.*, 1968). The other evidence is that platelets tend to aggregate as mural thrombi when anti-coagulated blood flows through plastic vessels, oxygenators or dialysis apparatus which do not contain collagen or other material capable of activating platelets, suggesting that platelets may be activated in the blood by agents other than constituents of vessel walls. It has been proposed that ADP derived from red cells by the action of fluid-mechanical effects is one such agent; platelets could also be activated by shear stresses. The evidence for a dominant influence of the ADP released from red cells on platelet activation in the hemostatic process has been marshalled by Born (1980).

Assessment of Platelet Aggregability. Platelet aggregation can be measured *in vitro* by recording the changes in light transmission through stirred platelet-rich plasma in a cuvette at 37°C; as platelet aggregation proceeds the optical density decreases (Born, 1962; O'Brien, 1962; Born and Cross, 1963). The addition of an aggregating agent to platelet-rich plasma causes a transient increase in optical density due to shape change which is followed by an increase in light transmission as the platelets form aggregates. The type of response obtained depends on the aggregating agent used and its concentration. With a low concentration of an aggregating agent such as ADP the platelets may undergo deaggregation so that the light transmission decreases again. There may be two phases, the initial aggregation being followed by a phase of reversal before further secondary aggregation takes place as a result of ADP release and the formation of thromboxane A_2 in the stimulated platelets. Aggregation is monophasic and irreversible when higher concentrations of the aggregating agent are used (Figure 2.8).

Figure 2.8: Responses of Platelet-Rich Plasma to the Addition of an Aggregating Agent

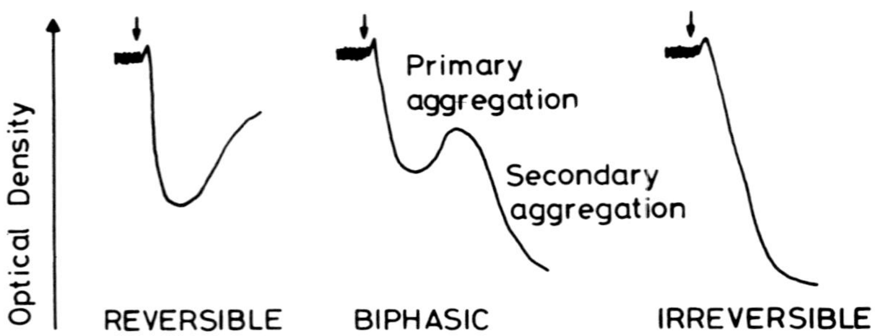

A number of factors lead to variability in the rate and extent of aggregation and require careful control if a meaningful assessment of platelet aggregability is to be obtained. These include the platelet number in the platelet-rich plasma (Born and Cross, 1963), the plasma fibrinogen concentration (Bloom and Evans, 1969), the calcium ion concentration and pH (Han and Ardlie, 1974), the rate of stirring, the temperature (Frojmovic, 1973; Praga *et al.*, 1973) and the time interval between venepuncture and measurement of aggregation (Warlow *et al.*, 1974).

A different approach to the assessment of platelet aggregability is the screen filter pressure technique. Blood is passed through microfilters with a pore size which allows erythrocytes and leukocytes to pass, but excludes platelet aggregates (Swank *et al.*, 1964). The pressure gradient across the filter is measured and provides an index of platelet aggregation. The blood may be passed through the filter directly from arterial or venous cannulae (Hornstra, 1970; Hornstra *et al.*, 1973) or anticoagulated before being forced back through the filter.

Chandler (1958) demonstrated that a thrombus with a histological structure resembling that formed *in vitro* is produced when whole blood is made to flow round a closed loop of plastic tubing. This technique has been modified to use platelet-rich plasma which allows visual assessment of platelet aggregation as the 'snow-storm' effect before formation of the white head of the thrombus by the coalescing of the platelet clumps (Cunningham *et al.*, 1965).

References

Adams, W.H., Liu, Y.K. and Sullivan, L.W. (1978) 'Hormonal regulation of thrombopoiesis in man', *Journal of Laboratory and Clinical Medicine, 91*, 141-7

Adelstein, R.S. and Conti, M.A. (1975) 'Phosphorylation of platelet myosin increases actin-activated myosin ATPase activity', *Nature, 256*, 597-8

Adelstein, R.S., Conti, M.A. and Anderson, W. (1973) 'Phosphorylation of human platelet myosin', *Proceedings of the National Academy of Sciences, 70*, 3115-9

Aharony, D., Smith, J.B. and Silver, M.J. (1981) 'Inhibition of platelet aggregation, malonyldialdehyde and thromboxane formation by hydroperoxides of arachidonic acid', *Thrombosis and Haemostasis, 46*, 265

Antoniades, H.N., Stathakos, D. and Scher, C.D. (1975) 'Isolation of a cationic polypeptide from human serum that stimulates proliferation of 3T3 cells', *Proceedings of the National Academy of Sciences, 72*, 2635-9

Baenziger, N.L., Dillender, M.J. and Majerus, P.W. (1977) 'Cultured skin fibroblasts and arterial cells produce a labile platelet-inhibitory prostaglandin', *Biochemical and Biophysical Research Communications, 78*, 294-301

Balleisen, L., Nowack, H., Gay, S. and Timpl, R. (1979) 'Inhibition of collagen-induced platelet aggregation by antibodies to distinct types of collagens', *Biochemical Journal, 184*, 683-7

Barber, A.J. and Jamieson, G.A. (1970) 'Isolation and characterization of plasma membranes from human blood platelets', *Journal of Biological Chemistry, 45*, 6357-65

Barber, A.J., Käset-Glanzmann, R., Jakabova, M. and Lüscher, E.F. (1972) 'Characterization of a chondroitin 4-sulphate proteoglycan carrier for heparin neutralizing activity (platelet factor 4) released from human blood platelets', *Biochimica et Biophysica Acta, 286*, 312-29

Baumgartner, H.R. (1973) 'The role of blood flow in platelet adhesion, fibrin deposition, and formation of mural thrombi', *Microvascular Research, 5*, 167-79

Baumgartner, H.R., Tschopp, T.B. and Meyer, D. (1980) 'Shear rate dependent inhibition of platelet adhesion and aggregation on collagenous surfaces by antibodies to human factor VIII/von Willebrand factor', *British Journal of Haematology, 44*, 127-39

Becker, R.P. and De Bruyn, P.P.H. (1976) 'The transmural passage of blood cells into myeloid sinusoids and the entry of platelets into the sinusoidal circulation; a scanning electron microscopic investigation', *American Journal of Anatomy, 145*, 183-206

Begg, S., Pepper, D.S., Chesterman, C.N. and Morgan, F.T. (1978) 'Complete covalent structure of human β-thromboglobulin', *Biochemistry, 17*, 1739-44

Behnke, O. (1965) 'Further studies on microtubules. A marginal bundle in human and rat thrombocytes', *Journal of Ultrastructure Research, 13*, 469-77

Behnke, O. (1968a) 'Electron microscopical observations on the surface coating of human blood platelets', *Journal of Ultrastructure Research, 24*, 51-69

Behnke, O. (1968b) 'An electron microscope study of the megacaryocyte of the rat bone marrow 1. The development of the demarcation membrane system and the platelet surface coat', *Journal of Ultrastructure Research, 24* 412-33

Bell, R.L., Kennerly, D.A., Stanford, N. and Majerus, P.W. (1979) 'Diglyceride lipase: A pathway for arachidonate release from human platelets', *Proceedings of the National Academy of Sciences, 76*, 3238-41

Bennett, J.S. and Vilaire, G. (1979) 'Exposure of platelet fibrinogen receptors by ADP and epinephrine', *Journal of Clinical Investigation, 64*, 1393-401

Bennett, J.S., Vilaire, G., Colman, R.F. and Colman, J.W. (1981) 'Localisation of human platelet membrane-associated actomyosin using the affinity label 5′-*p*-fluorosulfonylbenzoyl adenosine', *Journal of Biological Chemistry, 256*, 1185-90

Bensusan, H.B., Koh, T.L., Henry, K.G., Murray, B.A. and Culp, L.A. (1978) 'Evidence that fibronectin is the collagen receptor on platelet membranes', *Proceedings of the National Academy of Sciences, 75*, 5864-8

Bentfield, M.E. and Bainton, D.F. (1975) 'Cytochemical localization of lysosomal enzymes in rat megakaryocytes and platelets', *Journal of Clinical Investigation, 56*, 1635-9

Benveniste, J. (1974) 'Platelet-activating factor, a new mediator of anaphylaxis and immune complex deposition from rabbit and human basophils', *Nature, 249*, 581-2

Benveniste, J., Henson, P.M. and Cochrane, C.G. (1972) 'Leucocyte-dependent histamine release from rabbit platelets. The role of IgE, basophils and a platelet-activating factor', *Journal of Experimental Medicine, 136*, 1356-77

Benveniste, J., Le Couedic, J.P. and Kamoun, P. (1975) 'Aggregation of human platelets by platelet-activating factor', *Lancet, i*, 344-5

Benveniste, J., Le Couedic, J.P., Polonsky, J. and Tence, M. (1977) 'Structural analysis of purified platelet-activating factor by lipases', *Nature, 269,* 170-1

Berndt, M.C. and Phillips, D.R. (1981) 'Purification and preliminary physicochemical characterization of human platelet membrane glycoprotein V', *Journal of Biological Chemistry, 256,* 59-65

Bettex-Gallard, M. and Lüscher, E.F. (1959) 'Extraction of an actomyosin-like protein from human thrombocytes', *Nature, 184,* 276-7

Bettex-Gallard. M. and Lüscher, E.F. (1961) 'Thrombosthenin — a contractile protein from thrombocytes. Its extraction from human blood platelets and some of its properties', *Biochimica et Biophysica Acta, 49,* 536-47

Bevers, E., v. Dieijen, G., Rosing, J., Hornstra, G. and Zwaal, R.F.A. (1981) 'The effect of prostacyclin on the participation of platelets in X-activation and thrombin formation', *Thrombosis and Haemostasis, 46,* 271

Billah, M.M., Lapetina, E.G. and Cuatrecasas, P. (1979) 'Phosphatidylinositol-specific phospholipase-C of platelets: association with 1,2-diacylglycerol-kinase and inhibition by cyclic-AMP', *Biochemical and Biophysical Research Communications, 90,* 92-8

Billah, M.M., Lapetina, E.G. and Cuatrecasas, P. (1981) 'Phospholipase A_2 activity specific for phosphatidic acid', *Journal of Biological Chemistry, 256,* 5399-403

Bills, T.K., Smith, J.B. and Silver, M.J. (1976) 'Metabolism of ^{14}C arachidonic acid by human platelets', *Biochimica et Biophysica Acta, 424,* 303-14

Bills, T.K., Smith, J.B. and Silver, M.J. (1977) 'Selective release of arachidonic acid from the phospholipids of human platelets in response to thrombin', *Journal of Clinical Investigation, 60,* 1-6

Bloom, A.L. and Evans, E.P. (1969) 'Plasma-fibrinogen and the aggregation of platelets by adenosine diphosphate', *Lancet, i,* 349-50

Borchgrevink, C.F. (1960) 'A method for measuring platelet adhesiveness *in vivo*', *Acta Medica Scandinavia, 168,* 157-64

Born, G.V.R. (1962) 'Aggregation of blood platelets by adenosine diphosphate and its reversal', *Nature, 194,* 927-9

Born, G.V.R. (1980) 'Platelets in haemostasis and thrombosis', in A. Rotman, F.A. Meyer, C. Gitler and A. Silberberg (eds.) *Platelets: Cellular Response Mechanisms and their Biological Significance,* John Wiley, Chichester, pp. 3-15

Born, G.V.R. and Cross, M.J. (1963) 'The aggregation of blood platelets', *Journal of Physiology, 168,* 178-95

Broekman, M.J., Westmoreland, N.P. and Cohen, P. (1974) 'An improved method for isolating alpha granules and mitochondria from human platelets', *Journal of Cell Biology, 60,* 507-19

Broekman, M.J., Handin, R.I. and Cohen, P. (1975) 'Distribution of fibrinogen and platelet factors 4 and XIII in subcellular fractions of human platelets', *British Journal of Haematology, 31,* 51-5

Broekman, M.J., Ward, J.W. and Marcus, A.J. (1980) 'Phospholipid metabolism in stimulated human platelets', *Journal of Clinical Investigation, 66,* 275-83

Brox, J. and Østerud, B. (1981) 'The effect of prostacyclin (PGI_2) on the procoagulant activity of human platelets', *Thrombosis and Haemostasis, 46,* 271

Bry, K., Andersson, L.C., Kuvsi, T. and Kinnunen, P.K.J. (1979) 'Monoacylglycerol hydrolase in human platelets', *Biochimica et Biophysica Acta, 575,* 121-7

Bunting, S., Gryglewski, R., Moncada, S. and Vane, J.R. (1976) 'Arterial walls generate from prostaglandin endoperoxides a substance (prostaglandin X) which relaxes strips of mesenteric and coeliac arteries and inhibits platelet aggregation', *Prostaglandins, 12,* 897-913

Busch, C., Dawes, J., Pepper, D.S. and Wasteson, A. (1980) 'Binding of platelet factor 4 to cultured human umbilical vein endothelial cells', *Thrombosis Research, 19,* 129-37

Carlsson, L., Nyström, L.-E., Sundkvist, I., Markey, F. and Lindberg, U. (1977) 'Actin polymerizability is influenced by profilin, a low molecular weight protein in non-muscle cells', *Journal of Molecular Biology, 115,* 465-83

Carroll, R.C. and Gerrard, J.M. (1981) 'Phosphorylation of platelet actin-binding protein during platelet activation', *Thrombosis and Haemostasis, 46,* 267

Carroll, R.C., Gerrard, J.M. and Gilliam, J.M. (1981) 'Clot retraction facilitates clot lysis', *Blood, 57,* 44-8

Castle, A.G. and Crawford, N. (1977) 'The isolation and characterisation of platelet microtubule proteins', *Biochimica et Biophysica Acta, 494,* 76-91

Cazenave, J.-P., Sutter, A., Hemmendinger, S., Wiesel, M.-L., Lanza, F. and Daver, J. (1981) 'Adrenaline activates human platelets but does not cause primary aggregation if thrombin generation is inhibited by hirudin', *Thrombosis and Haemostasis, 46,* 95

Chandler, A.B. (1958) '*In vitro* thrombotic coagulation of the blood', *Laboratory Investigation, 7,* 110-4

Chap, H.J., Zwaal, R.P.A. and van Deenen, L.L.M. (1977) 'Action of highly purified phospholipases on blood proteins. Evidence for an asymmetric distribution of phospholipids in the surface membrane', *Biochimica et Biophysica Acta, 467,* 146-64

Charo, I.F., Feinman, R.D., Detwiler, T.C., Smith, J.B., Ingerman, C.M. and Silver, M.J. (1977) 'Prostaglandin endoperoxides and thromboxane A_2 can induce platelet aggregation in the absence of secretion', *Nature, 269,* 66-9

Chignard, M., Le Couedic, J.-P., Tencé, M., Vargaftig, B.B. and Benveniste, J. (1979) 'The role of platelet activating factor in platelet aggregation', *Nature, 279,* 799-800

Chignard, M., Vargaftig, B.B., Le Couedic, J.P. and Benveniste, J. (1981) 'Is phospholipase A_2 (PLA_2) involved in PAF-acether formation by platelets?' *Thrombosis and Haemostasis, 46,* 270

Cohen, I. and Cohen, C. (1972) 'A tropomyosin-like protein from human platelets', *Journal of Molecular Biology, 68,* 383-7

Cohen, I., Kaminski, E. and De Vries, A. (1973) 'Actin-linked regulation of the human platelet contractile system', *FEBS Letters, 34,* 315-7

Cooper, H.A., Clemetson, K.J. and Lüscher, E.F. (1979) 'Human platelet membrane receptor for bovine von Willebrand factor (platelet aggregating factor): An integral membrane glycoprotein', *Proceedings of the National Academy of Sciences, 76,* 1069-73

Cunningham, G.M., McNicol, G.P. and Douglas, A.S. (1965) 'Effect of anticoagulant drugs on platelet aggregation in the Chandler's tube', *Lancet, i,* 729-30

Curwen, K.D., Hee-Young, K., Vazquez, M., Handin, R.I. and Gimbone, M.A. (1982) 'Platelet adhesion to cultured vascular endothelial cells', *Journal of Laboratory and Clinical Medicine, 100,* 425-36

Dabrowska, R. and Hartshorne, D.J. (1978) 'A Ca^{2+}- and modulator-dependent myosin light chain kinase from non-muscle cells', *Biochemical and Biophysical Research Communications, 85,* 1352-9

Dabrowska, R., Sherry, J.M.F., Aromatorio, D.K. and Hartshorne, D.J. (1978) 'Modulator protein as a component of the myosin light chain kinase from chicken gizzard', *Biochemistry, 17,* 253-8

Daniel, J.L. and Adelstein, R.S. (1976) 'Isolation and properties of platelet myosin light chain kinase', *Biochemistry, 15,* 2370-7

Daniel, J.L., Robkin, L., Molish, I.R. and Holmsen, H. (1979) 'Determination of the ADP concentration available to participate in energy metabolism in an actin-rich cell, the platelet', *Journal of Biological Chemistry, 254*, 7870-3

Daniel, J., Sevy, R. and Selganicoff, L. (1981) 'Correlation of tension generation and myosin light chain phosphorylation in a thrombin activated platelet contractile model', , *Thrombosis and Haemostasis, 46*, 95

Dawes, J., Smith, R.C. and Pepper, D.S. (1978) 'The release, distribution, and clearance of human β-thromboglobulin and platelet factor 4', *Thrombosis Research, 12*, 851-61

De Clerck, F. and Van Gorp, L. (1981) 'Induction of circulating platelet aggregates by release of endogenous 5-hydroxytryptamine in the rat', *Thrombosis and Haemostasis, 46*, 29

Demopoulos, C.A., Pinckard, R.N. and Hanahan, D.J. (1979) 'Platelet-activating factor. Evidence for 1.*O*-alkyl-2-acetyl-*sn*-glyceryl-3-phosphorylcholine as the active component (a new class of lipid chemical mediators)', *Journal of Biological Chemistry, 254*, 9355-8

Deuel, T.F., Keim, P.S., Farmer, M. and Heinrikson, R.L. (1977) 'Amino acid sequence of human platelet factor 4', *Proceedings of the National Academy of Sciences, 74*, 2256-8

Deuel, T.F., Huang, J.S., Proffitt, R.T., Baenziger, J.U., Chang, D. and Kennedy, B.B. (1981) 'Human platelet-derived growth factor. Purification and resolution into two active protein fractions', *Journal of Biological Chemistry, 256*, 8896-9

Deutsch, E., Johnson, S.A. and Seegers, W.H. (1955) 'Differentiation of certain platelet factors related to blood coagulation', *Circulation Research, 3*, 110-5

Ek, B., Westermark, B., Wasteson, Å. and Heldin, C.-H. (1982) 'Stimulation of tyrosine-specific phosphorylation by platelet-derived growth factor', *Nature, 295*, 419-20

Evatt, B.L. and Levin, J. (1969) 'Measurement of thrombopoiesis in rabbits using [75]Selenomethionine', *Journal of Clinical Investigation, 48*, 1615-26

Evatt, B.L., Shreiner, D.P. and Levin, J. (1974) 'Thrombopoietic activity of fractions of rabbit plasma: studies in rabbits and mice', *Journal of Laboratory and Clinical Medicine, 83*, 364-71

Evatt, B.L., Levin, J. and Algazy, K.M. (1979) 'Partial purification of thrombopoietin from the plasma of thrombocytopenic rabbits', *Blood, 54*, 377-88

Feinstein, M.B. and Fraser, C. (1975) 'Human platelet secretion and aggregation induced by calcium ionophores', *Journal of General Physiology, 66*, 561-81

Frojmovic, M.M. (1973) 'Quantitative parameterization of the light transmission properties of citrated, platelet-rich plasma as a function of platelet and adenosine diphosphate concentration and temperature', *Journal of Laboratory and Clinical Medicine, 82*, 137-53

Frojmovic, M.M. and Milton, J.G. (1982) 'Human platelet size, shape, and related functions in health and disease', *Physiological Reviews, 62*, 185-261

Fujimoto, T. and Hawiger, J. (1982) 'Adenosine diphosphate induces binding of von Willebrand factor to human platelets', *Nature, 297*, 154-6

Fujimoto, T., Ohara, S. and Hawiger, J. (1982) 'Thrombin-induced exposure and prostacyclin inhibition of the receptor for factor VIII/von Willebrand factor in human platelets', *Journal of Clinical Investigation, 69*, 1212-22

Fukami, M.H. and Salganikoff, L. (1977) 'Human platelet storage organelles: a review', *Thrombosis and Haemostasis, 38*, 963-70

Fukami, M.H., Niewiarowski, S., Rucinski, B. and Salganicoff, L. (1979) 'Subcellular localization of human platelet antiheparin proteins', *Thrombosis Research, 14*, 433-43

de Gabriele, G. and Penington, D.G. (1967a) 'Physiology of the regulation of platelet production', *British Journal of Haematology, 13*, 202-9

de Gabriele, G. and Penington, D.G. (1967b) 'Regulation of platelet production:

"thrombopoietin" ', *British Journal of Haematology, 13,* 210-5

Ganguly, P. and Gould, N.L. (1979) 'Thrombin receptors of human platelets: thrombin binding and antithrombin properties of glycoprotein I', *British Journal of Haematology, 42,* 137-45

Gartner, T.K., Gerrard, J.M., White, J.G. and Williams, D.C. (1981) 'Fibrinogen is the receptor for the endogenous lectin of human platelets', *Nature, 289,* 688-90

George, J.N. (1972) 'Direct assessment of platelet adhesion to glass: a study of the forces of interaction and the effects of plasma and serum factors, platelet functions, and modification of the glass surface', *Blood, 40,* 862-74

George, J.N., Lyons, R.M. and Morgan, R.K. (1980) 'Membrane changes associated with platelet activation. Exposure of actin on the platelet surface after thrombin-induced secretion', *Journal of Clinical Investigation, 66,* 1-9

Gerrard, J.M., Schollmeyer, J.V., Phillips, D.R. and White, J.G. (1979) 'α-actinin deficiency in thrombasthenia. Possible identity of α-actinin and glycoprotein III', *American Journal of Pathology, 94,* 509-28

Ginsberg, M.H., Pointer, R.G., Forsyth, J., Birdwell, C. and Plow, E.F. (1980) 'Thrombin increases expression of fibronectin antigen on the platelet surface', *Proceedings of the National Academy of Sciences, 77,* 1049-53

Godstad, G.O., Solum, N.O. and Hagen, I. (1981) 'Platelet glycoprotein III and α-actinin are different proteins', *Thrombosis Research, 24,* 157-62

Greenberg, J.P., Packham, M.A., Guccione, M.A., Rand, M.L., Reimers, H.J. and Mustard, J.F. (1979) 'Survival of blood platelets treated *in vitro* with chymotrypsin, plasmin, trypsin, or neuraminidase', *Blood, 53,* 916-27

Hagen, I., Nurden, A., Bjerrum, O.J., Solum, N.O. and Caen, J. (1980) 'Immunochemical evidence for protein abnormalities in platelets from patients with Glanzmann's thrombasthenia and Bernard-Soulier syndrome', *Journal of Clinical Investigation, 65,* 722-31

Hamberg, M., Svensson, J., Wakabayashi, T. and Samuelsson, B. (1974) 'Isolation and structure of two prostaglandin endoperoxides that cause platelet aggregation', *Proceedings of the National Academy of Sciences, 71,* 345-9

Hamberg, M., Svensson, J. and Samuelsson, B. (1975) 'Thromboxanes: A new group of biologically active compounds derived from prostaglandin endoperoxides', *Proceedings of the National Academy of Sciences, 72,* 2994-8

Hammarström, S. and Falardeau, P. (1977) 'Resolution of prostaglandin endoperoxide synthase and thromboxane synthase of human platelets', *Proceedings of the National Academy of Sciences, 74,* 3691-5

Han, P. and Ardlie, N.G. (1974) 'The influence of pH, temperature and calcium on platelet aggregation: maintenance of environmental pH and platelet function for *in vitro* studies in plasma', *British Journal of Haematology, 26,* 373-89

Hanahan, D.J., Demopoulos, C.A., Lichr, J. and Pinckard, R.N. (1980) 'Identification of platelet activating factor isolated from rabbit basophils as acetyl glyceryl ether phosphorylcholine', *Journal of Biological Chemistry, 255,* 5514-6

Hardisty, R.M. and Stacey, R.S. (1955) '5-hydroxytryptamine in normal human platelets', *Journal of Physiology, 130,* 711-20

Hardisty, R.M. and Stacey, R.S. (1957) 'Platelet 5-hydroxytryptamine (HT) in disorders of the blood', *British Journal of Haematology, 3,* 292-8

Harker, L.A. (1970) 'Regulation of thrombopoiesis', *American Journal of Physiology, 218,* 1376-80

Harris, H.E. and Weeds, A.G. (1978) 'Platelet actin: sub-cellular distribution and association with profilin', *FEBS Letters, 90,* 84-8

Hathaway, D.E, and Adelstein, R.S,. (1979) 'Human platelet myosin light chain kinase requires the calcium-binding protein calmodulin for activity', *Proceedings of the National Academy of Sciences, 76,* 1653-7

Heldin, C.-H., Wasteson, A. and Westermark, B. (1977) 'Partial purification and characterization of platelet factors stimulating the multiplication of normal human glial cells', *Experimental Cell Research, 109,* 429-37

Heldin, C.H., Westermark, B. and Wasteson, A. (1979) 'Platelet-derived growth factor: purification and partial characterization', *Proceedings of the National Academy of Sciences, 76,* 3722-6

Hellem, A.J. (1960) 'The adhesiveness of human blood platelets *in vitro*', *Scandinavian Journal of Clinical and Laboratory Investigation,* Suppl. 51, 1-117

Hermodson, M., Schmer, G. and Kurachi, K. (1977) 'Isolation, crystallization, and primary amino acid sequence of human platelet factor 4', *Journal of Biological Chemistry, 252,* 6276-9

Hirsch, J., Glynn, M.F. and Mustard, J.F. (1968) 'The effect of platelet age on platelet adherence to collagen', *Journal of Clinical Investigation, 47,* 466-73

Hoak, J.C. (1972) 'Freeze-etching studies of human platelets', *Blood, 40,* 514-22

Holmsen, H., Day, H.J. and Setkowsky, C.A. (1972) 'Secretory mechanisms. Behaviour of adenine nucleotides during the platelet release reaction induced by adenosine diphosphate and adrenaline', *Biochemical Journal, 129,* 67-82

Holmsen, H., Salganicoff, L. and Fukami, M.H. (1977) 'Platelet behaviour and biochemistry', in D. Ogston and B. Bennett (eds.) *Haemostasis: Biochemistry, Physiology and Pathology,* Wiley, London, pp. 239-319

Hope, W., Martin, T.J., Chesterman, C.N. and Morgan, F.J. (1979) 'Human β-thromboglobulin inhibits PGI_2 production and binds to a specific site in bovine aortic endothelial cells', *Nature, 282,* 210-2

Hornstra, G. (1970) 'Method to determine the degree of ADP-induced platelet aggregation in circulating rat blood ('filter-loop technique')', *British Journal of Haematology, 19,* 321-9

Hornstra, G., Lewis, B., Chait, A., Turpeinen, O., Karvonen, M.J. and Vergraesen, A.J. (1973) 'Influence of dietary fat on platelet function in men', *Lancet, i,* 1155-7

Huang, S.S., Huang, J.S. and Deuel, T.F. (1982) 'Proteoglycan carrier of human platelet factor 4. Isolation and characterization', *Journal of Biological Chemistry, 257,* 11546-50

Ikeda, Y., Handa, M., Yoshii, Y., Imai, M., Sugiura, K. and Toyama, K. (1981) 'The role of membrane-bound tubulin in platelet functions', *Thrombosis and Haemostasis, 46,* 267

Insel, P.A., Nirenberg, P., Turnbull, J. and Shattil, S.J. (1978) 'Relationships between membrane cholesterol, α-adrenergic receptors, and platelet function', *Biochemistry, 17,* 5269-74

Jaffe, E.A., Leung, L.L.K., Nachman, R.L., Levin, R.I. and Mosher, D.F. (1982) 'Thrombospondin is the endogenous lectin of human platelets', *Nature, 295,* 246-8

Jamieson, G.A. and Okumura, T. (1978) 'Reduced thrombin binding and aggregation in Bernard-Soulier platelets', *Journal of Clinical Investigation, 61,* 861-4

Jennings, L.K., Fox, J.E.B., Edwards, H.H. and Phillips, D.R. (1981) 'Changes in the cytoskeletal structure of human platelets following thrombin activation', *Journal of Biological Chemistry, 256,* 6927-32

Johnson, R.A., Morton, D.R., Kinner, J.H., Gorman, R.R., McGuire, J.C. and Sun, F.F. (1976) 'The chemical structure of prostaglandin X (prostacyclin)', *Prostaglandins, 12,* 915-28

Jordan, R.E., Favreau, L.V., Braswell, E.H. and Rosenberg, R.D, (1982) 'Heparin with two binding sites for antithrombin or platelet factor 4', *Journal of Biological Chemistry, 257,* 400-6

Kahn, D., Crawford, N. and Cohen, I. (1981) 'Platelet calmodulin mediates the calcium dependent activation of factor XIII', *Thrombosis and Haemostasis, 46,* 172

Kannagi, R., Koizumi, K. and Masuda, T. (1981) 'Limited hydrolysis of platelet membrane phospholipids', *Journal of Biological Chemistry, 256,* 1177-84

Kaplan, D.R., Chao, F.C., Stiles, C.D., Antoniades, H.N. and Scher, C.D. (1979) 'Platelet α-granules contain a growth factor for fibroblasts', *Blood, 53,* 1043-52

Karpatkin, S. (1969a) 'Heterogeneity of human platelets. I. Metabolic and kinetic evidence suggestive of young and old platelets', *Journal of Clinical Investigation, 48,* 1073-82

Karpatkin, S. (1969b) 'Heterogeneity of human platelets. II. Functional evidence suggestive of young and old platelets', *Journal of Clinical Investigation, 48,* 1083-7

Karpatkin, S. (1972) 'Human platelet senescence', *Annual Reviews of Medicine, 23,* 101-28

Karpatkin, S. and Garg, S.K. (1974) 'The megathrombocyte as an index of platelet production', *British Journal of Haematology, 26,* 307-11

Kelemen, E., Cserháti, I. and Tanos, B. (1958) 'Demonstration of some properties of human thrombopoietin in thrombocythaemic serum', *Acta Haematologica, 20,* 350-5

Kinlough-Rathbone, R.L., Reimers, H.J. and Mustard, J.F. (1976) 'Sodium arachidonate can induce platelet shape change and aggregation which are independent of the release reaction', *Science, 192,* 1011-2

Kita, E., Horiike, N., Kyoi, K., Utsumi, S. and Umegaki, K. (1981) 'Intraplatelet and plasma serotonin level in normal subjects and in patients with cerebrovascular diseases', *Thrombosis and Haemostasis, 46,* 144

Krizsa, F. (1971) 'Study on the development of posthaemorrhagic thrombosis in rats', *Acta Haematologica, 46,* 228-31

Kulkarni, P.S., Roberts, R. and Needleman, P. (1976) 'Paradoxical endogenous synthesis of a coronary dilating substance from arachidonate', *Prostaglandins, 12,* 337-53

Lagarde, M., Menashi, S. and Crawford, N. (1981) 'Localisation of phospholipase A_2 and diglyceride lipase activities in human platelet intracellular membranes', *FEBS Letters, 124,* 23-6

Lagarde, M., Guichardant, M., Menashi, S. and Crawford, N. (1982) 'The phospholipid and fatty acid composition of human platelet surface and intracellular membranes isolated by high voltage free flow electrophoresis', *Journal of Biological Chemistry, 257,* 3100-4

Landon, F., Huc., C., Thome, F., Oriol, C. and Olomucki, A. (1977) 'Human platelet actin. Evidence of β and γ forms and similarity of properties with sarcomeric actin', *European Journal of Biochemistry, 81,* 571-7

Lapetina, E.G., Billah, M.M. and Cuatrecasas, P. (1981) 'The initial action of thrombin on platelets. Conversion of phosphatidylinositol to phosphatidic acid preceding the production of arachidonic acid', *Journal of Biological Chemistry, 256,* 5037-40

Lawler, J.W., Slayter, H.S. and Coligan, J.E. (1978) 'Isolation and characterization of a high molecular weight glycoprotein from human blood platelets', *Journal of Biological Chemistry, 253,* 8609-16

Lebowitz, E.A, and Cooke, R. (1978) 'Contractile properties of actomyosin from human blood platelets', *Journal of Biological Chemistry, 253,* 5443-7

Leung, L.L.K. and Nachman, R.L. (1982) 'Complex formation of platelet thrombospondin with fibrinogen', *Journal of Clinical Investigation, 70,* 542-9

Leung, L.L.K., Kinoshita, T. and Nachman, R.L. (1981) 'Isolation, purification, and partial characterization of platelet membrane glycoproteins IIb and IIIa',

Journal of Biological Chemistry, 256, 1994-7

Levine, R.F. (1981) 'Megakaryocyte sizes', *Thrombosis and Haemostasis, 46,* 227

Lindahl, U. and Höök, M. (1978) 'Glycosaminoglycans and their binding to biological macromolecules', *Annual Reviews of Biochemistry, 47,* 385-417

Lynch, C.J. and Steer, M.L. (1981) 'Evidence for high and low affinity α_2-receptors. Comparison of [^{3}H]norepinephrine and [^{3}H]phentolamine binding to human platelet membranes', *Journal of Biological Chemistry, 256,* 3298-303

Malmsten, C., Hamberg, M., Svensson, J. and Samuelsson, B. (1975) 'Physiological role of an endoperoxide in human platelets: Haemostatic defect due to platelet cyclo-oxygenase deficiency', *Proceedings of the National Academy of Sciences, 72,* 1446-50

Mannucci, P.M. and Sharp, A.A. (1967) 'Platelet volume and shape in relation to aggregation and adhesion', *British Journal of Haematology, 13,* 604-17

Margossian, S.S., Lawler, J.W. and Slayter, H.S. (1981) 'Physical characterization of platelet thrombospondin', *Journal of Biological Chemistry, 256,* 7495-500

Markey, F., Lindberg, U. and Eriksson, L. (1978) 'Human platelets contain profilin, a potential regulator of actin polymerisability', *FEBS Letters, 88,* 75-9

McDonagh, J. and McDonagh, R.P. (1972) 'Factor XIII from human platelets: effect on fibrin cross-linking', *Thrombosis Research, 1,* 147-60

McDonald, T.P., Clift, R., Lange, R.D., Nolan, C., Tribby, I.I.E. and Barlow, G.H. (1975) 'Thrombopoietin production by human embryonic kidney cells in culture', *Journal of Laboratory and Clinical Medicine, 85,* 59-66

McDonald, T.P., Odell, T.T. and Gosslee, D.G. (1964) 'Platelet size in relation to platelet age', *Proceedings of the Society for Experimental Biology and Medicine, 115,* 684-9

McEver, R.P., Baenziger, L. and Majerus, P.W. (1980) 'Isolation and quantitation of the platelet membrane glycoprotein deficient in thrombasthenia using a monoclonal hybridoma antibody', *Journal of Clinical Investigation, 66,* 1311-8

McLaren, K.M., Holloway, L. and Pepper, D.S. (1980) 'Human platelet factor 4 and tissue mast cells', *Thrombosis Research, 19,* 293-7

Menche, D., Israel, A. and Karpatkin, S. (1980) 'Platelets and microtubules. Effect of colchicine and D_2O on platelet aggregation and release induced by calcium ionophore A 23187', *Journal of Clinical Investigation, 66,* 284-91

Meyer, F.A. (1980) 'Determinants for platelet adhesion and aggregation on collagen', in A. Rotman, F.A. Meyer, C. Gitler and A. Silberberg (eds.) *Platelets: Cellular Response Mechanisms and their Biological Significance,* Wiley, Chichester, pp. 51-66

Meyer, F.A. and Frojmovic, M.M. (1979) 'Characteristics of the major platelet membrane site used in binding to collagen', *Thrombosis Research, 15,* 755-67

Meyer, F.A. and Weisman, Z. (1981) 'Collagen-induced platelet aggregation. Dependence on triple helical structure and fiber diameter', *Thrombosis Research, 22,* 1-4

Mills, D.C.B. and Thomas, D.P. (1969) 'Blood platelet nucleotides in man and other species', *Nature, 222,* 991-2

Moncada, S. and Vane, J.R. (1979) 'Arachidonic acid metabolites and the interactions between platelets and blood-vessel walls', *New England Journal of Medicine, 300,* 1142-7

Moncada, S., Gryglewski, R., Bunting, S. and Vane, J.R. (1976a) 'An enzyme isolated from arteries transforms prostaglandin endoperoxides to an unstable substance that inhibits platelet aggregation', *Nature, 263,* 663-5

Moncada, S., Gryglewski, R.J., Bunting, S. and Vane, J.R. (1976b) 'A lipid peroxide inhibits the enzyme in blood vessel microsomes that generates from prostaglandin endoperoxides the substance (prostaglandin X) which prevents

platelet aggregation', *Prostaglandins, 12*, 715-37
Moore, S., Pepper, D.S. and Cash, J.D. (1975) 'The isolation and characterisation of a platelet-specific β-globulin (β-thromboglobulin) and the detection of antiurokinase and antiplasmin released from thrombin-aggregated washed human platelets', *Biochimica et Biophysica Acta, 379*, 360-9
Mosesson, M.W. and Amrani, D.L. (1980) 'The structure and biologic activities of plasma fibronectin', *Blood, 56*, 145-58
Mosher, D.F., Schad, P.E. and Kleinman, H.K. (1979) 'Cross-linking of fibronectin to collagen by blood coagulation factor XIIIa', *Journal of Clinical Investigation, 64*, 781-7
Mosher, D.F., Schad, P.E. and Vann, J.M. (1980) 'Cross-linking of collagen and fibronectin by factor XIIIa', *Journal of Biological Chemistry, 255*, 1181-8
Mustard, J.F., Packham, M.A., Kinlough-Rathbone, R.L., Perry, D.W. and Regoeczi, E. (1978) 'Fibrinogen and ADP-induced platelet aggregation', *Blood, 52*, 453-66
Nachman, R. and Leung, L. (1981) 'Complex formation of platelet glycoproteins IIb and IIIa with fibrinogen', *Thrombosis and Haemostasis, 46*, 253
Niewiarowski, S. and Thomas, D.P. (1969) 'Platelet factor 4 and adenosine diphosphate release during human platelet aggregation', *Nature, 222*, 1269-70
Niewiarowski, S., Waltz, D.A., James, P., Rucinski, B. and Kueppers, F. (1980) 'Identification and separation of secreted platelet proteins by isoelectric focusing. Evidence that low affinity platelet factor 4 is converted to β-thromboglobulin by limited proteolysis', *Blood, 55*, 453-6
Nishikawa, M., Tanaka, T. and Hidaka, H. (1980) 'Ca^{2+}-calmodulin-dependent phosphorylation and platelet secretion', *Nature, 287*, 863-5
Nurden, A.T. (1977) 'A peripheral high molecular weight glycoprotein located at the surface of human platelets', *Experientia, 33*, 331-2
Nurden, A.T. and Caen, J.P. (1974) 'An abnormal platelet glycoprotein pattern in three cases of Glanzmann's thrombasthenia', *British Journal of Haematology, 28*, 253-60
Nurden, A.T. and Caen, J.P. (1975) 'Specific roles for platelet surface glycoproteins in platelet function', *Nature, 255*, 720-2
Nurden, A.T. and Caen, J.P. (1977) 'Further studies on the glycoprotein composition of normal human, Bernard-Soulier and thrombasthenic platelets', *Thrombosis and Haemostasis, 38*, 200
O'Brien, J.R. (1962) 'Platelet aggregation. II. Some results from a new method of study', *Journal of Clinical Pathology, 15*, 452-5
Odell, T.T., McDonald, T.P. and Detwiler, T.C. (1961) 'Stimulation of platelet production by serum of platelet-depleted rats', *Proceedings of the Society for Experimental Biology and Medicine, 107*, 428-31
O'Donnell, M.C., Henson, P.M. and Fiedel, B.A. (1978) 'Activation of human platelets by platelet activating factor (PAF) derived from sensitized rabbit basophils', *Immunology, 35*, 953-8
Ogle, J.W., Dunn, C.D., McDonald, T.P. and Lange, R.D. (1978) 'The *in vitro* production of erythropoietin and thrombopoietin', *Scandinavian Journal of Haematology, 21*, 188-96
Okumura, T. and Jamieson, G.A. (1976) 'Platelet glycocalicin. I. Orientation of glycoproteins of platelet surface', *Journal of Biological Chemistry, 251*, 5944-9
Okumura, T., Hasitz, M. and Jamieson, G.A. (1978) 'Platelet glycocalicin. Interaction with thrombin and role as thrombin receptor of the platelet surface', *Journal of Biological Chemistry, 253*, 3435-43
Osim, E.E. and Wyllie, J.H. (1982) 'Evidence for loss of 5-hydroxytryptamine from circulating platelets', *Journal of Physiology, 326*, 25-6P
Østerud, B. and Rapaport, S.I. (1977) 'Activation of factor IX by the reaction

product of tissue factor and factor VII: additional pathway for initiating blood coagulation', *Proceedings of the National Academy of Sciences, 74,* 5260-4

Paul, D., Niewiarowski, S., Varma, K.G., Rucinski, B., Rucker, S. and Lange, E. (1980) 'Human platelet basic protein associated with antiheparin and mitogenic activities: purification and partial characterization', *Proceedings of the National Academy of Sciences, 77,* 5914-8

Penington, D.G., Streatfield, K. and Roxburgh, A.E. (1976) 'Megakaryocytes and the heterogeneity of circulating platelets', *British Journal of Haematology, 34,* 639-53

Perret, B., Chap, H.J. and Douste-Blazy, L. (1979) 'Asymmetric distribution of arachidonic acid in the plasma membrane of human platelets. A determination using purified phospholipases and a rapid method for membrane isolation', *Biochimica et Biophysica Acta, 556,* 434-6

Phillips, D.R. and Agin, P.P. (1977) 'Platelet plasma membrane glycoproteins. Identification of a proteolytic substrate for thrombin', *Biochemical and Biophysical Research Communications, 75,* 940-7

Phillips, D.R., Jenkins, C.S.P., Lüscher, E.P. and Larrieu, J.M. (1975) 'Molecular differences of exposed surface proteins on thromboasthenic platelet plasma membranes', *Nature, 257,* 559-60

Phillips, D.R., Jennings, L.K. and Edwards, H.H. (1980) 'Identification of membrane proteins mediating the interaction of human platelets', *Journal of Cell Biology, 86,* 77-86

Pickett, W.C., Jesse, R.L. and Cohen, P. (1977) 'Initiation of phospholipase A_2 activity in human platelets by the calcium ion ionophore A 23187', *Biochimica et Biophysica Acta, 486,* 209-13

Piper, P.J. and Vane, J.R. (1969) 'Release of additional factors in anaphylaxis and its antagonism by anti-inflammatory drugs', *Nature, 223,* 29-35

Plow, E.F., Birdwell, C. and Ginsberg, M.H. (1979) 'Identification and quantitation of platelet-associated fibronectin antigen', *Journal of Clinical Investigation, 63,* 540-3

Podolsak, B., Mingers, A.-M. and Engelke, A. (1980) 'The influence of platelet homogenates on thrombopoiesis in rabbits', *Thrombosis Research, 20,* 191-7

Poggi, A., Niewiarowski, S., Holt, J.C., Dall'Olio, A., Stewart, G.T. and Smith, J.B. (1981) 'Human platelet-secreted proteins (LA-PF4, βTG) do not modify prostacyclin (PGI_2) production by bovine aortic endothelial cells', *Thrombosis and Haemostasis, 46,* 37

Pollard, T.D., Thomas, S.M. and Niederman, R. (1974) 'Human platelet myosin. I. Purification by a rapid method applicable to other nonmuscle cells', *Analytical Biochemistry, 60,* 258-66

Pollard, T.D., Fujiwara, K., Handin, R. and Weiss, G. (1977) 'Contractile proteins in platelet activation and contraction', *Annals of the New York Academy of Sciences, 283,* 218-36

Praga, C.A., Pogliani, E.M. and Ferrari, R. (1973) 'Effect of temperature on ADP-induced platelet aggregation. Its significance in studying anti-aggregating drugs', *Thrombosis et Diathesis Haemorrhagica, 29,* 183-9

Radley, J.M. and Scurfield, G. (1980) 'The mechanism of platelet release', *Blood, 56,* 996-9

Raines, E.W. and Ross, R. (1982) 'Platelet-derived growth factor I. High yield purification and evidence for multiple forms', *Journal of Biological Chemistry, 257,* 5154-60

Raz, A., Isakson, P.C., Minkes, M.S. and Needleman, P. (1977) 'Characterization of a novel metabolic pathway of arachidonate in coronary arteries which generates a potent endogenous coronary vasodilator', *Journal of Biological Chemistry, 252,* 1123-6

Rittenhouse-Simmons, S. (1979) 'Production of diglyceride from phosphatidylinositol in activated human platelets', *Journal of Clinical Investigation, 63*, 580-7

Ross, R. and Glomset, J.A. (1976) 'The pathogenesis of atherosclerosis', *New England Journal of Medicine, 295*, 369-77

Ross, R., Glomset, J., Kariya, B. and Harker, L. (1974) 'A platelet-dependent serum factor that stimulates the proliferation of arterial smooth muscle cells *in vitro*', *Proceedings of the National Academy of Sciences, 71*, 1207-10

Rucinski, B., Niewiarowski, S., James, P., Walz, D.A. and Budzynski, A.Z. (1979) 'Antiheparin proteins secreted by human platelets. Purification, characterization and radioimmunoassay', *Blood, 53*, 47-62

Ryo, R., Proffitt, R.T. and Deuel, T.F. (1980a) 'Human platelet factor 4: subcellular localization and characteristics of release from intact platelets', *Thrombosis Research, 17*, 629-44

Ryo, R., Proffitt, R.T., Poger, M.E., O'Bear, R. and Deuel, T.F. (1980b) 'Platelet factor 4 in megakaryocytes', *Thrombosis Research, 17*, 645-52

Sakariassen, K.S., Bolhuis, P.A. and Sixma, J.J. (1979) 'Human blood platelet adhesion to artery subendothelium is mediated by factor VIII-von Willebrand factor bound to subendothelium', *Nature, 279*, 636-8

Sakariassen, K.S., Bolhuis, P.A. and Sixma, J.J. (1980) 'Platelet adherence to subendothelium of human arteries in pulsatile and steady flow', *Thrombosis Research, 19*, 547-9

Salzman, E.W. (1963) 'Measurement of platelet adhesiveness. A simple *in vitro* technique demonstrating an abnormality in von Willebrand's disease', *Journal of Laboratory and Clinical Medicine, 62*, 724-35

Santoro, S.A. and Cunningham, L.W. (1977) 'Collagen-mediated platelet aggregation', *Journal of Clinical Investigation, 60*, 1054-60

Santoro, S.A. and Cunningham, L.W. (1979) 'Fibronectin and the multiple interaction modes for platelet-collagen adhesion', *Proceedings of the National Academy of Sciences, 76*, 2644-8

Schneider-Trip, M.D., Jenkins, C.S.P., Kahlé, L.H., Sturk, A. and ten Cate, J.W. (1979) 'Studies on the mechanism of ristocetin-induced platelet aggregation: binding of factor VIII to platelets', *British Journal of Haematology, 43*, 99-112

Schollmeyer, J.V., Rao, G.H.R. and White, J.G. (1978) 'An actin-binding protein in human platelets', *American Journal of Pathology, 93*, 433-46

Shaklai, M. and Tavassoli, M. (1978) 'Demarcation membrane system in rat megakaryocyte and the mechanism of platelet formation: A membrane reorganization process', *Journal of Ultrastructure Research, 62*, 270-85

Shore, P.A., Pletscher, A. and Brodie, B.B. (1956) 'Release of blood platelet serotonin by reserpine and lack of effect on bleeding time', *Journal of Pharmacology and Experimental Therapeutics, 117*, 232-6

Shreiner, D.P., Weinberg, J. and Enoch, D. (1980) 'Plasma thrombopoietic activity in humans with normal and abnormal platelet counts', *Blood, 56*, 183-8

Siemensma, N.P., Bathal, P.S. and Penington, D.G. (1975) 'The effect of massive liver resection on platelet kinetics in the rat', *Journal of Laboratory and Cllinical Medicine, 86*, 817-33

Sixma, J.J. and Schiphorst, M.E. (1980) 'Identification of ectoproteins of human platelets. Combination of radioactive labelling and two-dimensional electrophoresis', *Biochimica et Biophysica Acta, 603*, 70-83

Sixma, J.J., Schiphorst, M.E., Verhoeckx, C. and Jockusch, B.M. (1982) 'Peripheral and integral proteins of human blood platelet membranes. α-actinin is not identical to glycoprotein III', *Biochimica et Biophysica Acta, 704*, 333-44

Smith, J.B., Silver, M.J., Ingerman, C.M. and Kocsis, J.J. (1974) 'Prostaglandin D_2 inhibits the aggregation of human platelets', *Thrombosis Research, 5*, 291-9

Smith, J.B., Ingerman, C. and Silver, M.J. (1976) 'Persistence of thromboxane A$_2$-like material and platelet release-inducing activity in plasma', *Journal of Clinical Investigation, 58*, 1119-22

Spector, B. (1961) '*In vivo* transfer of a thrombopoietic factor', *Proceedings of the Society for Experimental Biology and Medicine, 108*, 146-9

Stahl, K., Themann, H. and Dame, W.R. (1978) 'Ultrastructural morphometric investigations on normal human platelets', *Haemostasis, 7*, 242-51

Steiner, M. and Ikeda, Y. (1979) 'Quantitative assessment of polymerized and depolymerized platelet microtubules. Changes caused by aggregating agents', *Journal of Clinical Investigation, 63*, 443-8

Swank, R.L., Roth, J.C. and Jansen, J. (1964) 'Screen filtration pressure method and adhesiveness and aggregation of blood cells', *Journal of Applied Physiology, 19*, 340-6

Test, S.T. and Bang, N.U. (1981) 'Platelet-activating factor (l-*O*-alkyl-2-acetyl-*sn*-glyceryl-3-phosphorylcholine) stimulates prostacyclin synthesis by cultured human and endothelial cells', *Thrombosis and Haemostasis, 46*, 269

Thompson, C.B., Eaton, K.A., Princiotta, S.M., Rushin, C.A. and Valeri, C.R. (1982) 'Size dependent platelet subpopulations: relationship of platelet volume to ultrastructure, enzymatic activity, and function', *British Journal of Haematology, 50*, 509-19

Thunberg, L., Backström, G., Grundberg, H., Riesenfeld, J. and Lindahl, U. (1980) 'The molecular size of the antithrombin-binding sequence of heparin', *FEBS Letters, 117*, 203-6

Timpl, R., Rohde, H., Robey, P.G., Rennard, S.I., Foidart, J.-M. and Martin, G.R. (1979) 'Laminin — a glycoprotein from basement membranes', *Journal of Biological Chemistry, 254*, 9933-7

Trelstad, R.L. and Carvalho, A.C.A. (1979) 'Type IV and type "A-B" collagens do not elicit platelet aggregation or the serotonin release reaction', *Journal of Laboratory and Clinical Medicine, 93*, 499-505

Turitto, V.T. and Baumgartner, H.R. (1979) 'Platelet interaction with subendothelium in flowing rabbit blood: effect of blood shear rate', *Microvascular Research, 17*, 38-54

Turner, S.R., Tainer, J.A. and Lynn, W.S. (1975) 'Biogenesis of chemotactic molecules by the arachidonate lipoxygenase system of platelets', *Nature, 257*, 680-1

Vargaftig, B.B. and Zirmis, P. (1973) 'Platelet aggregation induced by arachidonic acid is accompanied by release of potential inflammatory mediators distinct from PGE$_2$ and PGF$_2$', *Nature New Biology, 244*, 114-6

Vargaftig, B.B., Chignard, M. and Benveniste, J. (1981a) 'Present concepts on the mechanisms of platelet aggregation', *Biochemical Pharmacology, 30*, 263-71

Vargaftig, B.B., Chignard, M. and Benveniste, J. (1981b) 'Background and present status of research on platelet-activating factor (PAF-acether)', *Annals of the New York Academy of Sciences, 370*, 119-37

Waisman, D.M., Singh, T.J. and Wang, J.H. (1978) 'The modulator-dependent protein kinase. A multifunctional protein kinase activatable by the Ca^{2+}-dependent modulator protein of the cyclic nucleotide system', *Journal of Biological Chemistry, 253*, 3387-90

Walsh, P.N. and Griffin, J.H. (1981) 'Contributions of human platelets to the proteolytic activation of blood coagulation factors XII and XI', *Blood, 57*, 106-18

Walz, D.A., Wu, V.Y., Delamo, R. and McCoy, L.E. (1977) 'Primary structure of human platelet factor 4', *Thrombosis Research, 11*, 893-8

Warlow, C., Corina, A., Ogston, D. and Douglas, A.S. (1974) 'The relationship between platelet aggregation and time interval after venepuncture', *Thrombosis et Diathesis Haemorrhagica, 31*, 133-41

Weerasinghe, K., Scully, M.F. and Kakkar, V.V. (1981) 'Inhibition of contact activation by platelet factor 4', *Thrombosis and Haemostasis, 46*, 258

Weiss, H.J., Tschoff, T.B., Baumgartner, H.R., Sussman, I.I., Johnson, M.M. and Egan, J.J. (1974) 'Decreased adhesion of giant (Bernard-Soulier) platelets to subendothelium. Further implications on the role of the von Willebrand factor in hemostasis', *American Journal of Medicine, 57*, 920-5

Weiss, H.J., Baumgartner, H.R., Tschopp, T.B. and Turitto, V.T. (1977) 'Interaction of platelets with subendothelium. A new method for identifying and classifying abnormalities of platelet function', *Annals of the New York Academy of Sciences, 283*, 293-309

Weksler, B.B., Marcus, A.J. and Jaffe, E.A. (1977) 'Synthesis of prostaglandin I_2 by cultured human and bovine endothelial cells', *Proceedings of the National Academy of Sciences, 74*, 3922-6

Werner, G. and Morgenstern, E. (1980) 'Three-dimensional reconstruction of human blood platelets using serial sections', *European Journal of Cell Biology, 20*, 276-82

White, J.G. (1968a) 'Effects of colchicine and vinca alkaloids on human platelets I. Influence on platelet microtubules and contractile function', *American Journal of Pathology, 53*, 281-92

White, J.G. (1968b) 'Fine structural alterations induced in platelets by adenosine diphosphate', *Blood, 31*, 604-22

White, J.G. (1969) 'The submembrane filaments of blood platelets', *American Journal of Pathology, 56*, 267-71

White, J.G. (1972) 'Interaction of membrane systems in blood platelets', *American Journal of Pathology, 66*, 295-305

White, W.G. and Gerrard, J.M. (1979) 'Interaction of microtubules and microfilaments in platelet contractile physiology', *Methods and Achievements in Experimental Pathology, 9*, 1-39

Willis, A.L. and Kuhn, D.C. (1973) 'A new potential mediator of arterial thrombosis whose biosynthesis is inhibited by aspirin', *Prostaglandins, 4*, 127-30

Wilner, G.D., Nossel, H.L. and Le Roy, E.C. (1968) 'Aggregation of platelets by collagen', *Journal of Clinical Investigation, 47*, 2616-21

Wilson, D.B., Prescott, S.M. and Majerus, P.W. (1982) 'Discovery of an arachidonyl coenzyme A synthetase in human platelets', *Journal of Biological Chemistry, 257*, 3510-5

Wong, P.Y.-K. and Cheung, W.Y. (1979) 'Calmodulin stimulates human platelet phospholipase A_2', *Biochemical and Biophysical Research Communications, 90*, 473-80

Zahavi, J., Jones, N.A.G., Leyton, J., Dubiel, M. and Kakkar, V.V. (1980) 'Enhanced *in vivo* platelet "release reaction" in old healthy individuals', *Thrombosis Research, 17*, 329-36

Zucker, M.B. and Vroman, L. (1969) 'Platelet adhesion induced by fibrinogen adsorbed onto glass', *Proceedings of the Society for Experimental Biology and Medicine, 131*, 318-20

Zucker, M.B., Mosesson, M.W., Broekman, M.J. and Kaplan, K.L. (1979) 'Release of platelet fibronectin (cold-insoluble globulin) from alpha granules induced by thrombin or collagen; lack of requirement for plasma fibronectin in ADP-induced platelet aggregation', *Blood, 54*, 8-12

Zwaal, R.F.A. (1978) 'Membrane and lipid involvement in blood coagulation', *Biochimica et Biophysica Acta, 515*, 163-205

Zwaal, R.F.A., Comfurius, P. and van Deenen, L.L.M. (1977) 'Membrane asymmetry and blood coagulation', *Nature, 268*, 358-60

3 The Coagulation System

Contact Activation System

Factor XII

The classical intrinsic coagulation pathway is initiated by the activation of factor XII. This unique protein is also known as Hageman factor from the name of the patient in whom its deficiency was first recognised (Ratnoff and Colopy, 1955; Ratnoff and Rosenblum, 1958). Factor XII is also involved in one pathway leading to the generation of an activator of plasminogen (Niewiarowski and Prou-Wartelle, 1959), discussed in Chapter 4, and in the formation of kinins (Margolis, 1958).

Factor XII, both human and bovine, is a single-chain glycoprotein. The human protein has been found to have a molecular weight of 80,000 (Revak *et al.*, 1974; Chan and Movat, 1976), while sizes of 74,000 and 78,000 daltons have been obtained for bovine factor XII (Fujikawa *et al.*, 1977a; Claeys and Collen, 1978). The partial amino acid sequence of bovine factor XII was determined by Fujikawa and associates (1977a).

The mean concentration of factor XII in normal human plasma has been determined by radial immunodiffusion and a value of 29 µg/ml reported (Revak *et al.*, 1974), while a radioimmunoassay provided a higher mean value of 40 µg/ml (Saito *et al.*, 1976).

The site of synthesis of factor XII is not yet secure, but the decreased plasma concentration in patients with hepatic cirrhosis suggests the liver (Saito *et al.*, 1976).

Activation of Factor XII. Factor XII in plasma is activated on exposure to a wide variety of negatively charged surfaces such as glass or kaolin (Ratnoff and Rosenblum, 1958), ellagic acid (Ratnoff and Crum, 1964), L-homocystine (Ratnoff, 1968), chondroitin sulphate (Moskowitz *et al.*, 1970) and sodium urate (Ginsberg *et al.*, 1980).

Materials of possible physiological relevance which can activate factor XII in plasma include collagen (Wilner *et al.*, 1968; Harpel, 1972), platelet.membranes (Walsh, 1972b) and sebum (Nossel, 1966; Ogston *et al.*, 1969). Homogenates of cultured rabbit endothelial cells have also been found to be capable of activating rabbit factor XII (Wiggins *et al.*, 1980), but whether the factor XII-activating agent can be released from stimulated or damaged endothelial cells *in situ* is unknown. There is evidence from dichroism spectroscopy and fluorescence spectroscopy that a conformational change in the factor XII molecule with exposure of active sites takes place on its binding to a negatively charged surface (McMillin *et al.*, 1974; Fair *et al.*, 1977).

The binding of factor XII to a surface has been shown to be associated with its cleavage into two fragments with molecular weights of 52,000 and 28,000 (Revak *et al.*, 1977). Cleavage takes place at two closely spaced sites; one is within a disulphide loop with the formation of a two-chain molecule. Cleavage external to the disulphide bond releases the 28,000 dalton fragment which contains the enzymatic site (Figure 3.1). The surface binding site is located in the heavy chain (Revak and Cochrane, 1976). Bovine factor XIIa is also composed of a heavy and light chain linked by one or more disulphide bonds (Fujikawa *et al.*, 1977b): conversion into the two-chain form results from cleavage of a specific arginyl-valine bond.

Figure 3.1: The Activation of Factor XII

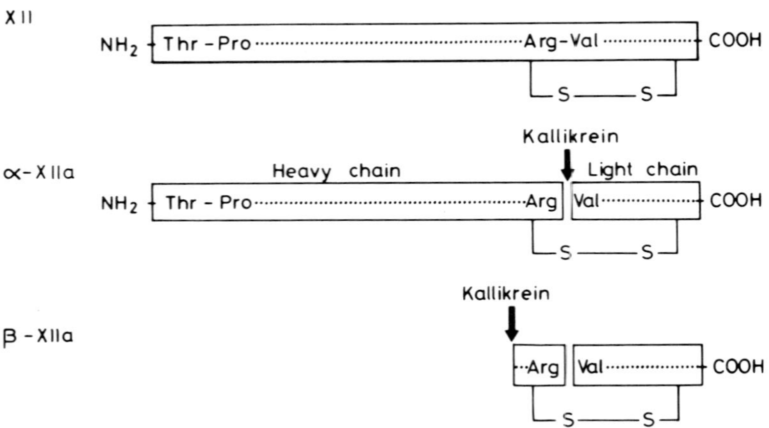

Activation of factor XII can also be achieved in the fluid phase by incubation with kallikrein, plasmin and trypsin. The specific arginyl-valine bond cleavage requires a lipid component: sulphatide, a sulphonated glycolipid from brain, and stearic acid were found to have this property, but phospholipids were inactive (Fujikawa *et al.*, 1980). It was suggested that some lipids may be important in the initiation of the coagulation process, analogous to the requirement for phospholipid in subsequent stages of the pathway. Further cleavage results in the formation of an active 28,000 molecular weight fragment (Revak *et al.*, 1974). Autoactivation of human factor XII by the two-chain form of factor XIIa, but not by the 28,000 dalton fragment, has been demonstrated (Silverberg *et al.*, 1980), while rabbit single-chain factor XII has been shown to cleave other factor XII molecules in the presence of kaolin (Wiggins and Cochrane, 1979). The rate of autoactivation of factor XII is, however, much slower than its activation by kallikrein (Dunn *et al.*, 1982).

Both human and bovine activated factor XII can hydrolyse *N*-α-acetylglycyl-lysine methyl ester, and have proteolytic activity as manifested by the ability to cleave prekallikrein and factor XI (Ulevitch *et al.*, 1974; Kurachi *et al.*, 1980). Amidolytic activity has also been demonstrated (Fujikawa *et al.*, 1977b; Ratnoff and Saito, 1979b). Both esterase and protease activities are inhibited by di-isopropylfluorophosphate (DFP), indicating that factor XIIa is a serine protease (Fujikawa *et al.*, 1977b).

The ability of the surface bound two-chain factor XIIa to cleave and activate prekallikrein and factor XI has been compared with that of the free 28,000 molecular weight fragment (Revak *et al.*, 1978). Both forms of factor XII could activate prekallikrein, but only the surface bound two-chain enzyme cleaved factor XI, leading to the conclusion that the initiation of the intrinsic coagulation pathway is a localised event taking place at the site of contact activation.

The necessity for cleavage of factor XII into its two-chain form before acquisition of clot-promoting properties has been vigorously debated. On one side is the view that the conformational changes induced by adsorption onto negatively charged surfaces expose enzymatically active sites on the factor XII molecule. Supporting such a hypothesis is the finding that purified factor XII, activated by exposure to ellagic acid adsorbed to Sephadex gels, has amidolytic properties in the absence of cleavage as determined by SDS-poly-acrylamide gel electrophoresis (Ratnoff and Saito, 1979b). Evidence has been presented that the single-chain factor XII can activate prekallikrein and factor XI in the presence of kaolin and HMW kininogen; in the case of prekallikrein the initial rate of this activation was almost identical to that achieved by the two-chain molecule (Heimark *et al.*, 1980; Fujikawa *et al.*, 1980). The alternative view points to the need for proteolytic cleavage of the factor XII molecule to achieve significant activation: consistent with this proposal are the results of experiments showing that surface bound single-chain factor XII does not react with DFP or exhibit esterolytic activity (Griffin, 1977; Meier *et al.*, 1977; Fujikawa *et al.*, 1977a).

Factor XI

Factor XI (plasma thromboplastin antecedent) is a glycoprotein with a molecular weight of about 124,000: human factor XI has some five per cent carbohydrate, the bovine molecule has eleven per cent (Kurachi and Davie, 1977; Koide *et al.*, 1977). Both bovine and human factor XI are composed of two identical polypeptide chains linked by one or more disulphide bonds (Wuepper, 1972; Koide *et al.*, 1977; Kurachi and Davie, 1977). The partial amino acid sequence of human factor XI has been determined by Kurachi and Davie (1977).

The plasma concentration of factor XI has been reported as 4 µg/ml (Bouma and Griffin, 1977) while Saito and Goldsmith (1977) obtained a value of 6 µg/ml in normal plasma by means of a radioimmunoassay: the reduced plasma level in liver disease is compatible with a hepatic site of synthesis.

The activation of factor XI by factor XIIa takes place by cleavage of a specific peptide bond in each of its two polypeptide chains, probably an arginyl-isoleucine bond (Kurachi and Davie, 1977). Activated factor XI is therefore composed of four polypeptide chains linked by disulphide bonds (Figure 3.2). The estimated molecular weights of the individual heavy and light chains of factor XIa are 35,000 and 25,000 respectively; the active site serine residues are located in each of the light chains. Factor XIa has esterase activity on *p*-toluene-sulphonyl-L-arginine methyl ester and weak amidase activity towards *N*-benzoyl-L-phenylalanyl-L-valyl-L-arginyl-*p*-nitroanilide (Kurachi and Davie, 1977).

Prekallikrein

Prekallikrein is the zymogen form of kallikrein, a protease which has the ability to generate kinin from its precursor (kininogen; prokinin). Appreciation of the involvement of prekallikrein in the coagulation

Figure 3.2: The Activation of Factor XI

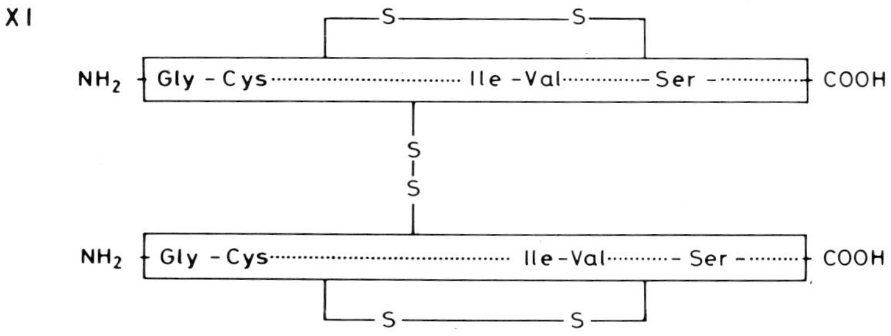

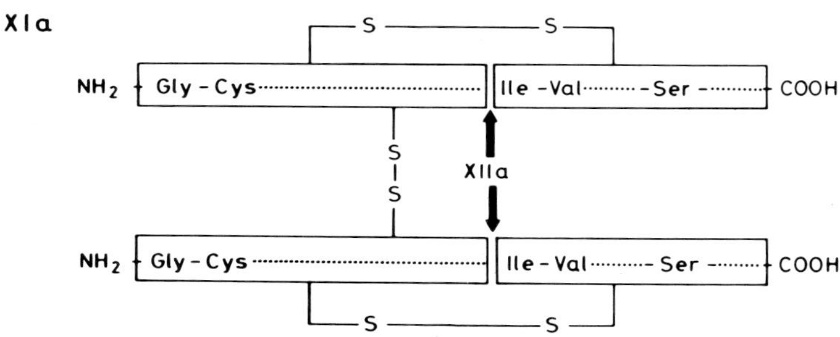

mechanism stemmed from the discovery of a coagulation defect characterised by an abnormality of the intrinsic system and ascribed to deficiency of an agent termed Fletcher factor (Hathaway *et al.*, 1965): this agent was subsequently identified as plasma prekallikrein (Wuepper, 1973).

There are a number of naturally occurring kallikreins: they are of two basic types, plasma and glandular. These differ in the type of kininogen cleaved preferentially, in their behavior towards inhibitors, in physicochemical properties, and in their immunological characteristics. The plasma form of kallikrein circulates as its inactive precursor prekallikrein.

The molecular weight of human plasma prekallikrein has been estimated to be around 100,000 (Wuepper and Cochrane, 1972; Gallimore *et al.*, 1978), or about 85,000 (Mandle and Kaplan, 1977; Bouma *et al.*, 1980). Bovine plasma prekallikrein was found to consist of a single polypeptide chain of molecular weight 82,000, containing some 13 per cent carbohydrate (Heimark and Davie, 1979): its amino-terminal sequence shows considerable homology with that of human factor XI.

The mean prekallikrein level in human plasma has been estimated to lie close to 100 µg/ml using a radial immunodiffusion assay (Bagdasarian *et al.*, 1974; Gallimore *et al.*, 1978; Heber *et al.*, 1978), but this value is too high due to complex formation between pre-kallikrein and HMW kininogen: using a radioimmunoassay a mean level of 50 µg/ml for plasma prekallikrein has been reported (Saito *et al.*, 1978).

The activation of prekallikrein involves hydrolysis of an internal peptide bond and the formation of a two-chain molecule linked by one of more disulphide bonds: the active site serine of kallikrein resides in the light chain of molecular weight around 35,000 (Mandle and Kaplan, 1977).

High Molecular Weight Kininogen

Plasma contains two forms of kininogen, a high molecular weight type (HMW kininogen) and a lower molecular weight form (LMW kininogen). Glandular kallikrein preferentially cleaves LMW kinin-ogen with the release of lysyl-bradykinin (kallidin), while plasma kallikrein has HMW kininogen as its preferred substrate and the kinin released is bradykinin.

Following the recognition that prekallikrein is involved in the early stages of the intrinsic coagulation pathway, a number of families were discovered with a different defect in contact-mediated coagulation and identified as the result of a deficiency of HMW kininogen. The disorders due to deficiency of HMW kininogen have been variously referred to as Fitzgerald trait (Saito *et al.*, 1975), Williams trait (Colman *et al.*, 1975), Flaujeac trait (Wuepper *et al.*, 1975) and Reid trait (Lutcher, 1976).

Human HMW kininogen is a single-chain glycoprotein with a molecular weight in the region of 110,000 to 120,000 (Saito, 1977; Thompson *et al.*, 1978; Kerbiriou and Griffin, 1979; Nakayasu and Nagasawa, 1979). Bovine HMW kininogen is also a single-chain glycoprotein, but has a molecular weight of 76,000 (Komiya *et al.*, 1974). The amino acid composition of the kininogens from the two species shows considerable similarity (Nakayasu and Nagasawa, 1979).

The presence of complexes of prekallikrein and HMW kininogen in plasma has been demonstrated (Mandle *et al.*, 1976; Donaldson *et al.*, 1977), and evidence has been provided to support the concept that free prekallikrein exists in equilibrium with such complexes (Scott and Colman, 1980).

The mean level of HMW kininogen in human plasma has been found to be close to 90 µg/ml as assessed by both a bioassay based on kinin release and a radioimmunoassay (Proud *et al.*, 1980).

HMW kininogen undergoes cleavage by kallikrein with progressive release of kinin and the formation of a two-chain disulphide-linked kinin-free molecule (Thompson *et al.*, 1978). The heavy chain has an apparent molecular weight of 65,000 and the light chain 37,000 to 44,000 (Thompson *et al.*, 1978; Kerbiriou and Griffin, 1979). The light chain retains the procoagulant activity of HMW kininogen, while the heavy chain was found to share antigenic determinants with LMW kininogen (Kerbiriou *et al.*, 1980). In the light chain of human HMW kininogen resides a histidine-rich region which is not readily liberated by human plasma kallikrein and remains bound to the kinin-free protein (Nakayasu and Nagasawa, 1979; Kerbiriou and Griffin, 1979): it appears that HMW kininogen binds to negatively charged surfaces through this histidine-rich region. It has been found that the single binding site for prekallikrein or kallikrein in HMW kininogen also resides in the light chain (Kerbiriou *et al.*, 1980).

Significant differences exist between human and bovine HMW kininogen in respect of their cleavage to plasma kallikrein. On incubation of bovine HMW kininogen with bovine plasma kallikrein equimolar amounts of bradykinin, kinin-free kininogen and a peptide (fragment 1.2) are released (Han *et al.*, 1975b). On further incubation fragment 1.2 is cleaved into fragment 1, a glycoprotein of molecular weight 8000 (Han *et al.*, 1976) and fragment 2, a histidine-rich peptide with a molecular weight close to 4600 (Han *et al.*, 1975a) which is analogous to the histidine-rich region in the light chain of human HWM kininogen. The function-structure relationships in bovine HMW kininogen appear to be very similar to those of the human protein: adsorption to a surface is dependent on the fragment 1.2 region while complex formation with kallikrein is through the light chain of the kinin-free kininogen (Ikari *et al.*, 1981). As a result both the light chain and fragment 1.2 are necessary for the coagulant cofactor activity of HMW kininogen (Sugo *et al.*, 1980). Fragment 1.2 alone has been noted to have inhibitory activity on the contact-mediated activation of factor XII (Oh-ishi *et al.*, 1977; Sugo *et al.*,

1980): this is attributed to competition with HMW kininogen for binding sites on the negatively charged surface.

Interactions Between Components of the Contact Activation System

The original concept of the initiation of the intrinsic coagulation pathway was that factor XII, activated by adsorption onto a surface, converted factor XI into its enzymatic form. It was found, however, that a mixture of purified factor XII and factor XI incubated with kaolin did not induce the generation of clot-promoting activity (Schiffman and Lee, 1974). With the identification of families with a hereditary deficiency of prekallikrein and HMW kininogen came the recognition that both these components of the kallikrein-kinin system are required for the initiation of surface-mediated reactions. Their precise role in the early phases of the intrinsic pathway has been increasingly clarified over the last few years.

There is now general acceptance for the belief that reciprocal activation of factor XII and prekallikrein is an important mechanism in contact activation of the intrinsic coagulation system. It has been proposed that factor XII binds to a negatively charged surface where it undergoes a conformational change rendering it susceptible to proteolytic cleavage by kallikrein. The factor XIIa activates further prekallikrein and the kallikrein generated reciprocally augments the quantity of factor XIIa formed. It has been shown that the kallikrein produced can rapidly dissociate from the surface and is then available to activate further surface-bound factor XII (Wiggins *et al.*, 1977). What remains unclear is the mode of initiation of the reciprocal interaction between factor XII and prekallikrein. One explanation, in accord with the findings of Ratnoff and Saito (1979a, b), is that the binding of factor XII to a surface alone can induce some activation. Wiggins and Cochrane (1979) have demonstrated that single-chain factor XII, incubated with kaolin, can cleave other factor XII molecules in the absence of prekallikrein. Other data from Heimark and colleagues (1980) suggests that surface-bound single-chain factor XII can activate prekallikrein as readily as does factor XIIa, and thereby initiate the reciprocal activation process in the absence of even trace quantities of factor XIIa, kallikrein or other protease.

HMW kininogen is required to attain maximal rates of production of the coagulant activity of factor XII: it was found to accelerate the activation of factor XI by kallikrein, the activation of prekallikrein by factor XIIa, and the activation of factor XI by factor XIIa (Griffin and

Cochrane, 1976). HMW kininogen accelerates these reactions by forming complexes with prekallikrein or factor XI through a binding site on the light chain of the molecule and by adsorbing to a negatively charged surface through the histidine-rich region. It acts, therefore, as a non-enzymatic cofactor localising the prekallikrein and factor XI to the surface so that reciprocal interaction with the bound factor XII can take place. The accelerating effect of HMW kininogen on kaolin-mediated activation of bovine factor XII has been found to be enhanced by brief incubation with kallikrein (Kato *et al.*, 1979; Sugo *et al.*, 1981). It is probable that this is the result of a limited proteolytic modification of the kininogen molecule. This increase in cofactor activity by proteolytic action is analogous to the effect of thrombin on factors VIII and V.

Contact Activation and Platelets

There is evidence that factor XI activity is firmly bound to the platelet membrane, being retained after successive washes, whereas factor XII is not significantly adsorbed onto the platelet surface (Horowitz and Fujimoto, 1965; Walsh, 1972a). The factor XI activity resides in the plasma membrane fraction of platelets (Lipscomb and Walsh, 1979), and has been shown to differ from plasma factor XI, existing as a disulphide-linked 52,000 dalton tetramer (Tuszynski *et al.*, 1982). HMW kininogen has also been identified in platelets and its secretion induced by collagen (Schmaier *et al.*, 1981): it was suggested that it may be secreted locally and modulate reactions of the contact phase of plasma proteolysis.

Walsh (1972b) described experiments which suggested that the addition of ADP to platelets results in the formation of an agent which is capable of activating plasma factor XII and thereby promoting the clotting of platelet-poor plasma. In a further report Walsh (1972c) raised the possibility of an alternative pathway for the activation of factor XI, bypassing factor XII, by showing that collagen can induce coagulant activity in washed platelets in the absence of factor XII. In later studies it was shown that ADP or collagen-treated platelets can promote the proteolytic activation of factor XII in the presence of kallikrein and HMW kininogen (Walsh and Griffin, 1980). It was concluded that platelets promote the activation of factor XII by kallikrein and the activation of factor XI by both a factor XII-dependent and a factor XII-independent mechanism. Others, however, have failed to detect any role for platelets in the activation of factor XII (Vecchione and Zucker, 1975; Vicic *et al.*, 1979), and

Østerud and associates (1979) were unable to confirm the platelet-dependent activation of factor XI by collagen. The role of platelets in the contact activation system remains to be clarified.

Vitamin K

Dam reported that chicks fed a synthetic diet developed a bleeding tendency which could be prevented by agents in cereals or seeds: the name vitamin K was given to this agent (Dam, 1935). Vitamin K exists naturally in two forms, vitamin K_1 (2-methyl-3-phytyl-1,4-naphthoquinone) and vitamin K_2 (2-methyl-3-difarnesyl-1,4-naphthoquinone). Their structures are shown in Figure 3.3.

Investigations into a severe hemorrhagic disorder of cattle fed on spoiled sweet clover revealed that it was associated with a profound deficiency of prothrombin (Link, 1959). The toxic component of clover was identified by Link as bishydroxycoumarin. The coumarin anticoagulant drugs have structural similarities to compounds with vitamin K-like activity and were believed to act by competitive inhibition at binding sites on hepatocytes, but it is probable that coumarin drugs antagonise a reaction that regenerates vitamin K

Figure 3.3: The Structure of Vitamin K

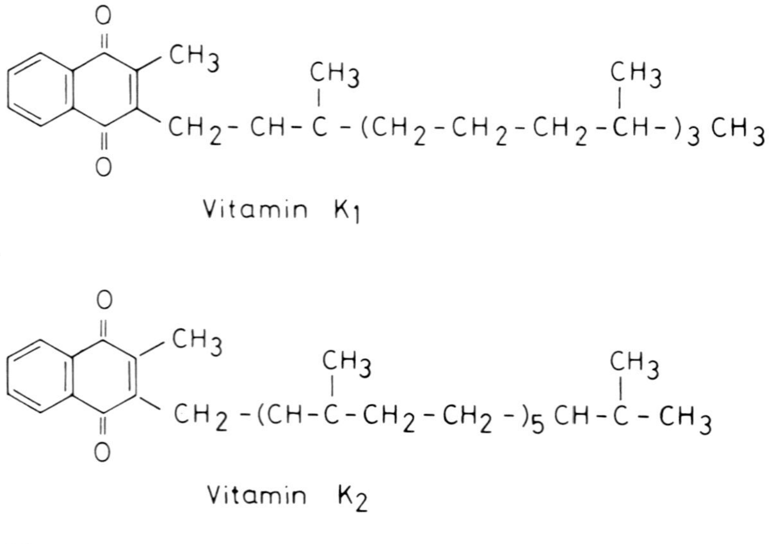

CH3
CH2 - CH - C - (CH2 - CH2 - CH2 - CH -)3 CH3
CH3
CH3
Vitamin K1

CH3
CH2 - (CH - C - CH2 - CH2 -)5 CH - C - CH3
CH3
CH3
Vitamin K2

from its 2,3-epoxide, the formation of which is coupled to carboxy lation.

It was assumed originally that a lack of vitamin K or the administration of a coumarin drug results in the shut-down of the biosynthesis of prothrombin in the hepatic cells. Hemker and colleagues (1963) demonstrated, however, the presence of an abnormal protein in the plasma of patients receiving oral anticoagulant drugs by kinetic studies: it was proposed that the abnormal protein was an inactive precursor of prothrombin, and later termed PIVKA, an abbreviation for Protein Induced by Vitamin K Antagonists (Hemker and Muller, 1968). This protein cross-reacted with a prothrombin antiserum, but could not be activated to thrombin and also differed from normal prothrombin in its lack of adsorption to barium citrate and in altered binding of Ca^{2+} ions (Ganrot and Niléhn, 1968). In addition, there was a difference in electrophoretic mobility between prothrombin from warfarin-anticoagulated and untreated cattle (Ganrot and Niléhn, 1968; Stenflo, 1970).

The structural difference between normal and abnormal prothrombin was found to lie in the content of γ-carboxyglutamic acid residues in normal prothrombin, located in the amino-terminal end of the molecule (Stenflo *et al.*, 1974; Magnusson *et al.*, 1974), whereas coumarin prothrombin contained normal glutamic acid in the same positions (Figure 3.4). It has been concluded, therefore, that the function of vitamin K in the biosynthesis of prothrombin and the other vitamin K-dependent coagulation factors is to facilitate the γ-carboxylation of the glutamic acid residues. It is the γ-carboxyglutamic acid residues which are responsible for the stronger calcium-binding of normal prothrombin in its interaction with phospholipids; it is believed that the calcium ions form bridges between the γ-carboxyl groups and the phosphate groups on the phospholipids.

Six vitamin K-dependent factors have been identified; these are prothrombin, factor VII, factor X, factor IX, protein C and protein S. Recently a seventh, protein Z, has been described. Prothrombin, factor VII, factor IX and protein S are single-chain molecules whereas factor X and protein C have two polypeptide chains. The calcium-binding vitamin K-dependent structures are sited in the amino-terminal portions of the molecules while those of factor X and protein C are in the amino-terminal parts of their light chains. In these parts of the molecules there are strong amino acid sequence homologies between the proteins.

Figure 3.4: The Structure of Glutamic Acid and γ-Carboxy-glutamic Acid

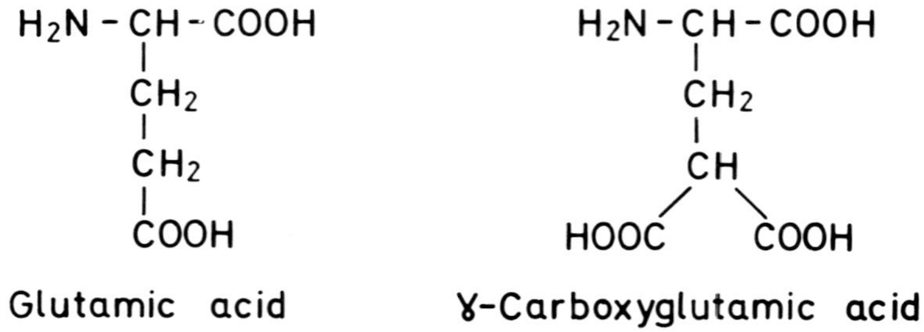

$$H_2N - CH - COOH$$
$$|$$
$$CH_2$$
$$|$$
$$CH_2$$
$$|$$
$$COOH$$

Glutamic acid

$$H_2N - CH - COOH$$
$$|$$
$$CH_2$$
$$|$$
$$CH$$
$$HOOC \quad COOH$$

γ-Carboxyglutamic acid

Factor IX

Factor IX (Christmas factor) is the agent deficient in the plasma of patients with the X-chromosome-linked recessive Christmas disease (hemophilia B). Both bovine and human factor IX are single-chain glycoproteins of molecular weight approximately 55,000 and with an amino-terminal tyrosine and a content of hexose, hexosamine and sialic acid (Fujikawa *et al.*, 1973; Østerud *et al.*, 1978). In common with the other vitamin K-dependent coagulation factors it contains eight to ten γ-carboxyglutamic acid residues (Fryklund *et al.*, 1976). Its concentration in the plasma is around 5 µg/ml (Thompson, 1977; Østerud *et al.*, 1978) and biosysnthesis takes place in the liver.

Activation of Factor IX

During the intrinsic coagulation process factor IX is activated by factor XIa in the presence of calcium ions with cleavage of two peptide bonds in the precursor factor IX molecule. Phospholipid is not required (Chuang *et al.*, 1974). The activation of bovine factor IX has been described by Fujikawa *et al.* (1974b). In the first stage factor IX is cleaved into a light chain (MW 16,000) with amino-terminal tyrosine and a heavy chain (MW 38,000) with amino-terminal alanine, the chains being linked by disulphide bridges: this protein lacks enzymatic activity. In the second stage, for which calcium ions are essential, the heavy chain is cleaved resulting in the formation of a new amino-terminal valine and the release of a peptide of molecular weight 9000 which contains about 50 per cent of the carbohydrate of the original molecule (Figure 3.5). The protein now possesses factor IXa activity; the active centre serine is located in the heavy chain. The activation of human factor IX by human factor XIa is very similar (Di Scipio *et al.*, 1978; Østerud *et al.*, 1978).

In addition to its activation by factor XIa in the classical intrinsic coagulation pathway, factor IX may be activated by other mechanisms. Factor Xa can catalyse the conversion of factor IX into its enzymatically active form, but only at about 20 per cent of the rate obtained with factor XIa (Kalousek *et al.*, 1975). The reaction product of tissue factor and factor VII is also capable of inducing the formation of a two-chain form of activated factor IX in the presence of calcium (Østerud and Rapaport, 1977).

Figure 3.5: The Activation of Factor IX by Factor XIa

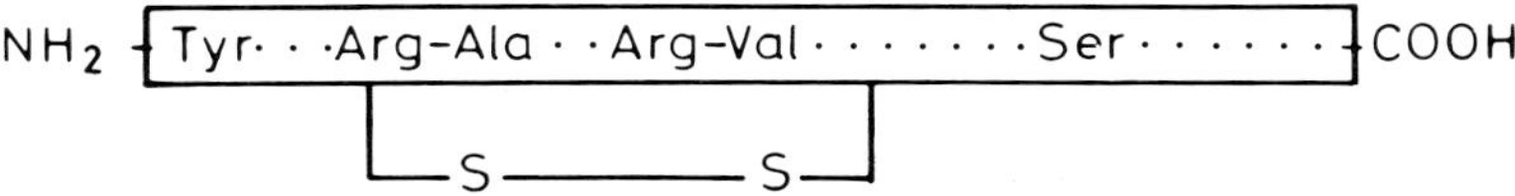

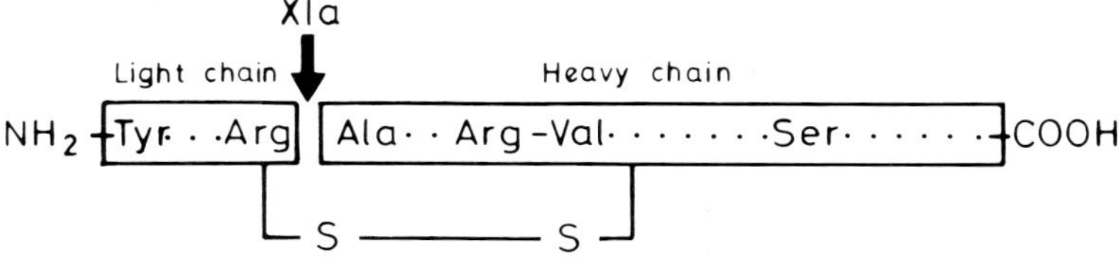

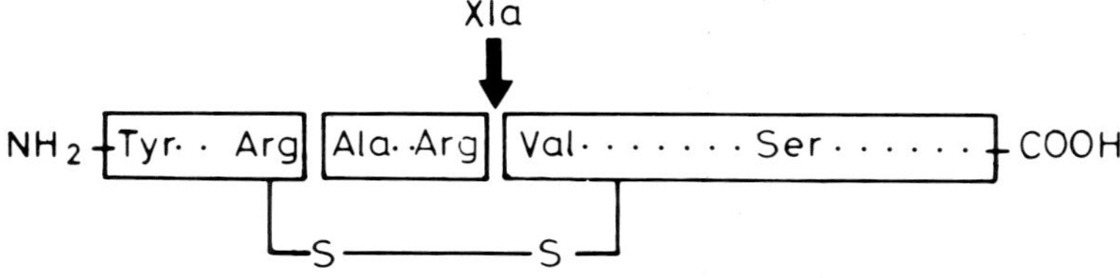

Factor VIII

The term factor VIII refers to a glycoprotein which exists in normal human plasma as a complex of two proteins, factor VIII coagulant factor (VIII:C; antihemophilic factor) and von Willebrand factor (vWF). The complex (factor VIII:C/vWF) is present in plasma at a concentration of about 5-10 µg/ml, most of it comprising the vWF component (Legaz *et al.*, 1973).

Structure

Factor VIII:C/vWF is a large molecule with a weight in excess of 0.86 × 10^6 which exists as a series of multimers: these may be reduced to a single subunit species of molecular weight 200,000 (Legaz *et al.*, 1973). The multimers appear to be built up from such subunits linked by disulphide bonds (Counts *et al.*, 1978). The amino acid composition of factor VIII:C/vWF has been reported (Legaz *et al.*, 1973; Shapiro *et al.*, 1973) and the molecule also contains carbohydrate with 154 ± 15 nmol of sialic acid/mg protein (Sodetz *et al.*, 1977).

The factor VIII:C and vWF molecules may be separated by gel filtration under conditions of high ionic strength (Weiss and Hoyer, 1973). Bovine factor VIII:C isolate freed from vWF by gel filtration and adsorption procedures appears as a triplet on urea-SDS-polyacrylamide gels with molecular weights of 93,000, 88,000 and 85,000 (Vehar and Davie, 1980); these may result from proteolysis of the native molecule during purification. Hoyer and Trabold (1981), using an immunoadsorbent technique, obtained a molecular weight of 285,000 for the human factor VIII:C protein. However, using the technique of electron irradiation, an estimated molecular weight of around 40,000 was obtained for circulating factor VIII:C (Harmon *et al.*, 1981). The vWF molecule free of VIII:C retains its large molecular weight with all multimers present (Fass *et al.*, 1978). Factors VIII:C and vWF have distinct immunological characteristics.

Synthesis and Function of von Willebrand Factor

Von Willebrand factor is synthesised by endothelial cells from which it can be released (Jaffe *et al.*, 1973, 1974; Jaffe and Nachman, 1975). It is also synthesised by megakaryocytes (Nachman *et al.*, 1977) and is present in platelets, both in the granules and on the membrane (Nachman and Jaffe, 1975), and can be released in response to ADP, collagen and thrombin (Koutts *et al.*, 1978). The platelet vWF has a multimeric structure similar to that of plasma

(Ruggeri and Zimmerman, 1980).

Von Willebrand factor is required to support aggregation by the antibiotic ristocetin and platelet retention in glass bead columns. Ristocetin-induced platelet aggregation needs an intact carbohydrate moiety on the vWF; removal of terminal galactose residues from asialo-vWF markedly reduces such platelet activity (Sodetz *et al.*, 1978).

The role of vWF in hemostasis is to promote adhesion of platelets to the subendothelium and thus lead to the formation of the platelet plug. Von Willebrand factor binds to the subendothelium (Sakariassen *et al.*, 1979) and also binds to platelets in the presence of ristocetin (Doucet de Bruine *et al.*, 1978). Whether it does so in the absence of ristocetin is not definitely established, but presumably it can in order to bind the platelet to the subendothelium. The platelet membrane receptor involved in the binding of platelets to vWF is probably glycoprotein I (see page 19).

Synthesis and Function of Factor VIII:C

Factor VIII:C is not synthesised by the same cells as vWF. Organ perfusion and transplantation studies in various mammalian species have suggested the involvement of the liver and spleen in the synthesis of VIII:C (Norman *et al.*, 1967; Webster *et al.*, 1967; Dodds, 1969; Dodds and Hoyer, 1974; Bowie *et al.*, 1975; Shaw *et al.*, 1979), but the hepatocyte does not appear to be the cell of origin since there is increased factor VIII:C activity in hepatocellular disease (Green and Ratnoff, 1974; Castillo *et al.*, 1977).

The importance of factor VIII:C for the hemostatic mechanism may be inferred from the severity of the bleeding state in its absence. Factor VIII:C is required as a cofactor to factor IX for the activation of factor X. It requires modification by thrombin to achieve its cofactor activity: this modification probably involves proteolytic cleavage (Vehar and Davie, 1980; Hoyer and Trabold, 1981). DFP-inactivation by thrombin-activated factor VIII:C has been demonstrated, suggesting that it may be a serine protease, but no direct enzymatic activity has been demonstrated. It is rapidly inactivated by the activated form of protein C.

Factor X

Factor X (Stuart factor; Prower factor) is the zymogen of the protease which catalyses the conversion of prothrombin into thrombin. Most of the initial characterisation of factor X and its activation was carried out on the bovine molecule, a glycoprotein with a molecular weight of around 55,000 (Fujikawa *et al.*, 1972; Jackson, 1972). The protein is composed of a heavy chain of molecular weight about 38,000 containing nearly all the carbohydrate, and a light chain of some 15,000 molecular weight. The chains are held together by one or more disulphide bonds.

Two forms of factor X, designated by X_1 and X_2, could be partly separated on DEAE-Sephadex (Jackson and Hanahan, 1968); the two forms were found to be identical in respect of molecular weight, amino acid composition and immunological properties (Fujikawa *et al.*, 1972), but had minor differences in carbohydrate composition (Jackson, 1972).

The complete amino acid sequences of both the light and heavy chain of bovine factor X have been elucidated (Enfield *et al.*, 1975; Titani *et al.*, 1975). The heavy chain has an amino-terminal tryptophan while the carboxyl-terminus is heterogeneous and ends with either proline or leucine. The heavy chain contains the active site serine residue while the light chain contains the γ-carboxyglutamic acid residues characteristic of the vitamin K-dependent clotting factors (Howard and Nelsestuen, 1975; Bucher *et al.*, 1976). The heavy chain of bovine factor X has considerable amino acid homology with the B chain of thrombin while the light chain has marked homology with fragment 1 of prothrombin.

Human factor X is also a glycoprotein with an apparent molecular weight of 59,000 and contains some 15 per cent carbohydrate (Di Scipio *et al.*, 1977). Its polypeptide chains have molecular weights of 49,000 and 17,000 as measured by SDS-polyacrylamide gel electrophoresis. The human factor X molecule differs from bovine factor X in having an amino-terminal serine in the heavy chain.

Synthesis of mammalian factor X takes place in the liver (Owen and Bowie, 1981), and its concentration in human plasma is about 10 µg/ml.

Activation of Factor X

The activation of factor X can be achieved physiologically through the extrinsic coagulation pathway by a complex of factor VII, tissue factor and calcium ions or through the intrinsic pathway involving the interaction of factor IXa, factor VIII, phospholipid and calcium ions. Non-physiological activation can be induced with trypsin or Russell's viper venom. All pathways of activation of factor X provide an identical product (Radcliffe and Barton, 1973).

During the conversion of bovine factor X into its activated form by Russell's viper venom a peptide of molecular weight 11,000 is released from the amino-terminal end of the heavy chain through the cleavage of a specific arginyl-isoleucine bond (Fujikawa *et al.*, 1972). Through this mechanism the enzymatically active serine protease Xa is formed (Figure 3.6).

The activation of factor X by the factor VII-tissue factor complex was found by Jesty and Nemerson (1974) to involve more than one peptide cleavage. The major pathway is initiated by the cleavage of an activation peptide from the amino-terminal portion of the heavy chain, the peptide released by Russell's viper venom in the studies of Fujikawa *et al.* (1972). The product, termed α-Xa, is converted into β-Xa by the Xa-catalysed cleavage of an arginyl-glycine bond to release a 17-residue glycoprotein from the carboxyl-terminus of the heavy chain of the molecule in a reaction which is dependent on lipid (Figure 3.6). The two forms of Xa have the same coagulant activity (Jesty *et al.*, 1974, 1975). An alternative pathway, favored at lower concentrations of the activating complex, is initiated by the action of factor Xa in splitting the same carboxyl-terminal peptide. The intermediate product can then be activated to factor βXa by the

Figure 3.6: The Activation of Factor X

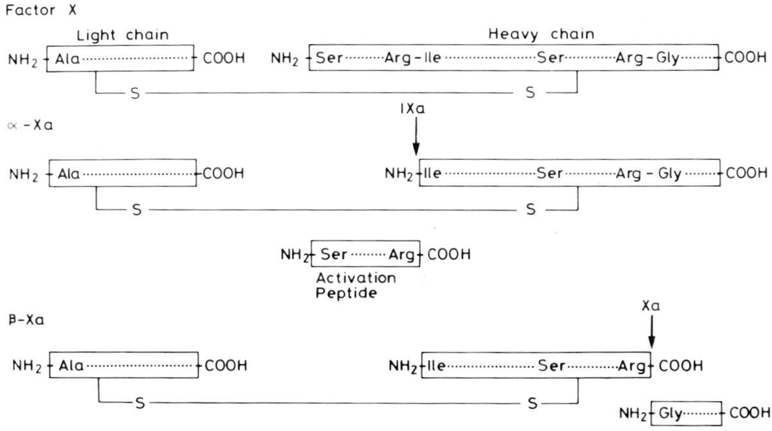

factor VII-tissue factor complex through removal of the amino-terminal activation peptide. Neither of the peptides released during the activation of factor X influence the coagulation rate by either the intrinsic or extrinsic pathways as assessed by the partial thromboplastin time or prothrombin time respectively (Jesty *et al.*, 1975). Essentially identical findings for the activation of bovine factor X by both the intrinsic and extrinsic pathways were reported by Fujikawa *et al.* (1974a).

Interaction Between Factors IXa, VIIIa and X

The activation of factor X via the intrinsic coagulation pathway requires factor IXa, thrombin-altered factor VIII, negatively charged phospholipid and calcium ions. These agents form a complex in which the enzymatic activity of the factor IXa is responsible for the activation of factor X (Hougie *et al.*, 1967; van Dieijen *et al.*, 1981). Factors IXa and X are bound to the phospholipid surface through calcium bridges, the γ-carboxyglutamic acid residues of the two vitamin K-dependent factors providing the calcium-binding sites. Factor X can be activated by factor IXa in the presence of phospholipid and calcium, but at a very slow rate (Suomela *et al.*, 1977; Hultin and Nemerson, 1978). The addition of thrombin-altered factor VIII has been found to increase the rate of factor Xa formation by more than 500-fold (Hultin and Nemerson, 1978), while van Dieijen and colleagues (1981) reported that the addition of factor VIIIa increased the V_{max} of factor X activation by some 200,000-fold. It appears likely that factor VIIIa functions in an analogous way to factor Va in the prothrombin-activating reaction, aligning the factor IXa and factor X on the phospholipid surface in optimal position for factor X activation to take place.

Tissue Factor

The extrinsic coagulation pathway is initiated by the interaction of factor VII with an agent present in many tissues. This tissue factor (tissue thromboplastin) has been shown to contain both protein and lipid (Chargaff *et al.*, 1944) and is now known to consist of a species of protein (apoprotein III) and a mixture of phospholipids (Nemerson and Pitlick, 1970).

74

A small amount of apoprotein III is probably present in all tissues, but especially rich sources are brain, thyroid, lungs and placenta. Apoprotein III can be synthesised by monocytes, the origin of their procoagulant activity (see page 204). The apoprotein portion of tissue factor has been purified from bovine brain and lung and found to have a molecular weight of some 56,000 (Liu and McCoy, 1975) while human brain has yielded a protein of similar size (Hvatum and Prydz, 1969; Bjørklid and Storm, 1977). A 142,000-fold purification was achieved from bovine brain by Bach and colleagues (1981): this preparation had an apparent molecular weight of 43,000 as assessed by SDS-polyacrylamide gel electrophoresis. It is a glycoprotein and all preparations have contained about ten per cent carbohydrate. The apoprotein has no proteolytic activity.

The apoprotein must be associated with phospholipid in order to possess biological activity. In experiments involving the reconstitution of coagulant activity of purified apoprotein by the addition of purified phospholipid Nemerson (1968) showed that phosphatidylethanolamine was the more effective, followed by phosphatidylcholine. Phosphatidylserine and phosphatidylinositol failed to restore tissue factor activity. Optimal activity was observed at phospholipid to protein ratios greater than 450:1 (Bach *et al.*, 1981).

Factor VII

The formation of a complex of tissue factor and factor VII initiates the extrinsic coagulation pathway. Factor VII, both bovine and human, is composed of a single polypeptide chain with an aminoterminal alanine. Bovine factor VII has been found to have an apparent molecular weight of around 53,000 (Kisiel and Davie, 1975; Radcliffe and Nemerson, 1975) or 45,500 (Kisiel *et al.*, 1977a). A human preparation with a molecular weight of 60,000 was described by Laake and Ellingsen (1974) while more recent preparations of human factor VII have had molecular weights close to 50,000 (Broze and Majerus, 1980; Bajaj *et al.*, 1981). Factor VII exhibits considerable amino acid homology with prothrombin and the light chain of factor X, but in contrast to these the single-chain zymogen of bovine factor VII can incorporate DFP, suggesting that it possesses proteolytic activity (Radcliffe and Nemerson, 1976). Human factor VII appears to be less sensitive to DFP inactivation than the bovine protein (Broze and Majerus, 1980).

Factor VII requires vitamin K for the γ-carboxylation of the glutamic acid residues in the final step of the biosynthesis of the functional protein. Both bovine and human factor VII contain approximately nine γ-carboxyglutamic acid residues per molecule (Di Scipio and Davie, 1979; Bajaj *et al.*, 1981). Synthesis takes place in the liver (Pool and Robinson, 1959). Its concentration in plasma is low, estimated at around 0.5 µg/ml (Bajaj *et al.*, 1981).

Activation of Factor VII

The coagulant activity of factor VII is greatly enhanced when it is converted into a two-chain molecule: this can be achieved by the action of a number of enzymes including factors IXa, Xa, XIIa and thrombin. The conversion of factor VII into its two-chain form involves the cleavage of a single specific arginyl-isoleucine bond between a disulphide bridge (Radcliffe and Nemerson, 1976; Kisiel *et al.*, 1977a). The activation of factor VII is represented diagrammatically in Figure 3.7. The molecular weight of both inactivated and activated forms of bovine factor VII were found to be identical (Kisiel *et al.*, 1977a), pointing to the absence of an activation peptide. The molecular weight of the heavy chain of the activated bovine molecule was found to be about 27,000 and to contain the active centre serine; the light chain had a molecular weight of 18,000 (Kisiel *et al.*, 1977a). Similar molecular weights have been reported for the two chains of activated

Figure 3.7: The Activation of Factor VII by Factor Xa

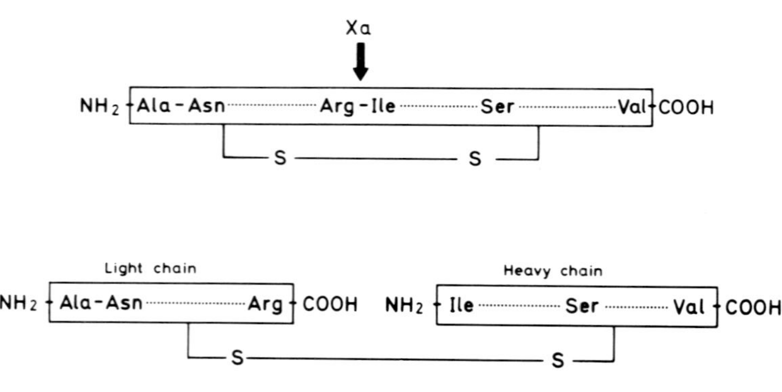

human factor VII (Bajaj *et al.*, 1981; Kisiel and McMullen, 1981).

Activation by Factor XIIa. In purified systems bovine factor VII can be activated to its two-chain form (VIIa) by factor XIIa or its active fragment (Kisiel *et al.*, 1977a; Radcliffe *et al.*, 1977). In this reaction additional cofactors are not required. Human factor VII has also been shown to be activated by the factor XIIa fragment in purified systems (Laake and Østerud, 1974; Seligsohn *et al.*, 1979; Broze and Majerus, 1980) and in plasma (Seligsohn *et al.*, 1979). It was originally suggested that kallikrein itself could increase factor VII activity (Gjønnaess, 1972), but its probable role lies in the activation of factor XII which can cleave factor VII directly.

Activation by Factor IXa. Direct activation of factor VII by factor IXa has been demonstrated (Laake and Østerud, 1974; Seligsohn *et al.*, 1979). This activation is possible at concentrations of factor IXa attainable in serum and the reaction does not appear to require platelets, added phospholipid or calcium (Seligsohn *et al.*, 1979).

Activation by Factor Xa and Thrombin. Both thrombin and factor Xa can activate factor VII in the absence of phospholipid with the formation of the two-chain form (Radcliffe and Nemerson, 1975). In the presence of phospholipid and calcium, however, the activation of factor VII by factor Xa is greatly enhanced, but phospholipid does not accelerate the rate of activation by thrombin. Factor Xa can also inactivate factor VIIa, but this occurs at a much slower rate than the process of activation. Inactivation is the result of cleavage at an arginyl-glycine bond in the heavy chain of the factor VII molecule with the release of a peptide of molecular weight 12,000 from the carboxyl end of the chain (Radcliffe and Nemerson, 1976).

Interaction of Tissue Factor with Factor VII

Purified factor VII, whether in the single or two-chain form, does not activate purified factor X in the absence of tissue factor. The mode of action of tissue factor and its constituent moieties is not yet established. It was proposed by Østerud and associates (1972) that tissue factor modifies factor VII with the formation of an activated form which is then able to activate factor X in the absence of tissue factor. Contrary to this view, Radcliffe and Nemerson (1975) found that both the single and the two-chain form of factor VII require tissue factor for their activity. It seems a tenable hypothesis that the apoprotein binds to factor VII resulting in its activation while the lipid portion binds to the substrate factor X, aligning the molecule optimally for activation by factor VIIa.

Interaction of Factor VII with Factor X

In the presence of tissue factor factor VII activates factor X. The two-chain form of factor VII is markedly more active in this reaction. A point of debate is whether normal plasma contains a low level of factor VII activity capable of activating a quantity of factor X on the appearance of tissue factor. The factor Xa would then activate factor VII in a positive feedback mechanism. This debate is analogous to the controversy over the initiation of factor XII activation.

It has been proposed recently that the activation of bovine factor VII to its two-chain form by factor Xa is a secondary mechanism: the initial activation is achieved by the rapid formation of a reversible stoichiometric complex between factor Xa at sub-nanomolar concentrations and one-chain factor VII in the presence of tissue factor (Morrison-Silverberg and Jesty, 1981). In this proposal it is the level of tissue factor which determines the quantity of the ternary reactive complex of tissue factor/factor VII/factor Xa formed, and it is this complex which is responsible for further generation of factor Xa.

Effect of Factor VIIa on Factor IX

In addition to its ability to activate factor X, factor VIIa has been shown to be able to activate human factor IX (Østerud and Rapaport, 1977). From the results of a study by Jesty and Silverberg (1979) it was predicted that, at normal bovine plasma concentrations of factors IX and X, the rate of factor X activation in the presence of tissue factor would exceed the rate of factor IX activation by a factor of 6.7. The kinetics of the reaction have been further examined in a purified system and the conclusion reached that the concentration of tissue factor regulates the ratio of the activation rates of factor IX and factor X by factor VIIa (Zur and Nemerson, 1980). The physiological relevance of this alternative route to the formation of factor Xa is presently undefined. Interactions of factor VII with other coagulation factors are illustrated in Figure 3.8.

Figure 3.8: Interactions Between Factor VII and Other Coagulation Factors

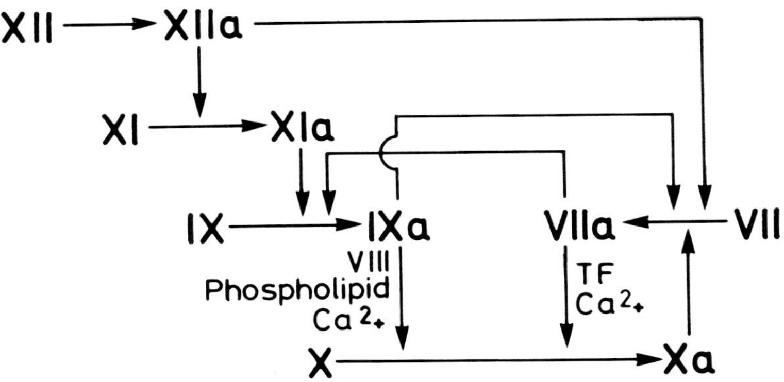

Factor V

Factor V (proaccelerin; accelerator globulin) is a high molecular weight protein found in plasma and platelets. Together with phospholipid, factor Xa and calcium ions factor V forms a complex which activates prothrombin. During the coagulation process it is consumed and is not present in serum. The instability of factor V has led to considerable difficulties with purification and reported molecular weights for the protein have varied widely from 38,000 (Philip *et al.*, 1970) to over a million (Saraswathi *et al.*, 1978). Nesheim and colleagues (1979) reported that their preparation of bovine factor V was a rod-like single-chain molecule with an amino-terminal sequence of Ala-Lys-Leu-Met and a molecular weight of 330,000, close to the value obtained by Esmon (1979). Human factor V isolated from plasma by Dåhlbäck (1980) was also a single-chain polypeptide with a molecular weight of 330,000. In contrast to these findings, the factor V in bovine blood obtained by venepuncture was reported to have a molecular weight of over one million (Saraswathi *et al.*, 1978): it was suggested that lower molecular weight preparations arise from proteolysis during blood collection. This high molecular weight for native factor V is in agreement with the data of Bartlett and associates (1980) who reported that both human and bovine blood collected by venepuncture provided factor V with an apparent molecular weight in excess of 800,000. Whether these differences in the molecular characteristics of the factor V preparations are the result of variations in blood handling or reflect the existence of aggregates of lower molecular forms is presently speculative.

Normal plasma has been found by a specific immunoassay to contain 4 to 14 µg/ml of factor V with a mean level of 7 µg/ml (Tracy *et al.*, 1982). There is good evidence that this factor is synthesised largely or exclusively in the liver: it was found to be elaborated by the perfused rat liver (Olson *et al.*, 1966; Owen and Bowie, 1977), while Giddings and colleagues (1975) identified factor V in human hepatic parenchymal cells using an indirect fluorescent antiglobulin technique.

Platelet Factor V

The association of factor V with bovine platelets was first described by Ware *et al.* (1948), and this was confirmed later for human platelets with the recognition that the platelet activity concerned with prothrombin conversion, termed platelet factor 1, was adsorbed factor V (Hjort *et al.*, 1955). It was believed that platelet factor V activity represented plasma factor V bound to the platelet surface, but there is now evidence that there is intracellular factor V (Breederveld *et al.*, 1975) which is secreted when platelets are stimulated by exposure to collagen (Østerud *et al.*, 1977). It is now known that factor V is present in the α-granules of platelets. The platelet-derived factor V is identical to plasma factor V in its immunochemical and procoagulant properties (Tracy *et al.*, 1979). Washed platelets lysed in Triton X-100 had 0.6 to 0.85 µg factor V per 2.5×10^8 platelets, representing 4400 to 6200 molecules of factor V per platelet (Tracy *et al.*, 1981).

Bovine platelets appear to possess two classes of binding sites for factor Va and a single class of factor V-binding site. Both the activated and unactivated factor V bind to the lower affinity sites whereas factor V does not bind to the high affinity factor Va-binding sites (Tracy *et al.*, 1979). It is likely that factor V, on release from the platelet, is activated and binds to the platelet surface and forms part of the binding site for factor Xa (Miletich *et al.*, 1977, 1978; Dählbäck and Stenflo, 1978).

Activation of Factor V

Factor V must be proteolytically altered in order to become functional as a cofactor in the activation of prothrombin by factor Xa: its activation by thrombin increases its affinity for phospholipid and allows it to bind prothrombin and factor Xa. Prolonged incubation of factor V with thrombin, however, results in its complete loss of activity. Esmon (1979) studied the thrombin-induced activation of bovine factor V and found that the activated form was composed of two polypeptide chains with molecular weights of 115,000 and 73,000: evidence was presented for the existence of an activation intermediate comprising two chains of molecular weights 210,000 and 115,000, the heavy chain of the intermediate being the precursor of the light chain of factor Va. Nesheim and Mann (1979) also concluded that the activation of factor V by thrombin proceeds through proteolytic cleavages with intermediate components of molecular weights 205,000 and 150,000: further rapid cleavages of the 150,000 chain provided two fragments of apparent molecular weights 94,000 and 74,000. The generation of factor Va activity coincided with the second cleavage. Activation of human factor V by thrombin produced two closely spaced doublets, one with a molecular weight of 110,000 and the other 72,000 (Dählbäck, 1980).

Prothrombin

A detailed review of historical aspects of prothrombin, its structure and its activation has been provided by Suttie and Jackson (1977). Both bovine and human prothrombin are glycoproteins consisting of a single polypeptide chain with an amino-terminal alanine and a molecular weight of 72,000. The primary structure of bovine prothrombin was described by Magnusson and co-investigators (1975), and the amino acid sequence of the closely similar human prothrombin has since been elucidated (Butkowski *et al.*, 1977; Walz *et al.*, 1977). The amino-terminal part of the molecule, which is not part of the thrombin structure, contains the ten γ-carboxyglutamic acid residues formed as a result of the action of vitamin K on the prothrombin precursor in the liver. The amino-terminal portion of prothrombin also contains two structures with a high degree of internal sequence homology and an identical disulphide bridge pattern. These are known as 'kringles' from their resemblance in shape to a type of Scandinavian cake when depicted in a two-dimensional drawing (Figure 3.9). The concentration of prothrombin in blood plasma is 70 to 150 µg/ml.

Figure 3.9: Kringle Structures in the Amino-Terminal Portion of the Prothrombin Molecule

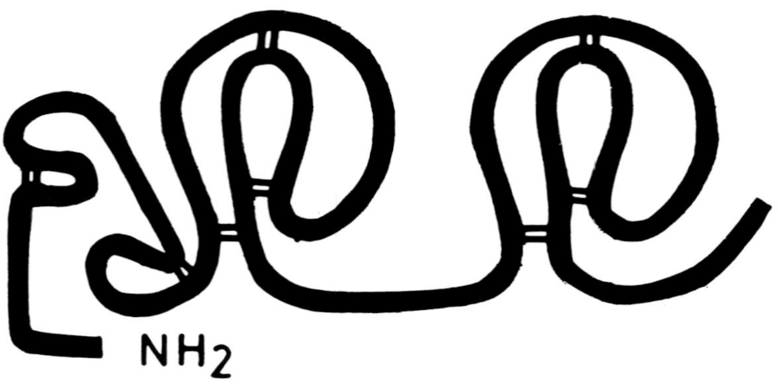

Activation of Prothrombin

Two peptide bonds are cleaved in the prothrombin molecule during its activation by factor Xa. Further bonds may be split by thrombin. The terminology used for these activation products is shown in Figure 3.10.

Under physiological conditions the conversion of prothrombin into thrombin is catalysed by factor Xa which is responsible for the essential peptide bond cleavages. Prothrombin activation by factor Xa involves the proteolytic scission of two peptide bonds: the first is an arginyl-threonine bond and this cleavage divides prothrombin into two halves (the amino-terminal fragment 1-2 and prothrombin 2) which remain associated through non-covalent forces. The second cleavage takes place at an arginyl-isoleucine bond to form the enzymatically active two-chain α-thrombin which then dissociates from the fragment 1-2 portion of the molecule. The two chains of thrombin are linked by a disulphide bond (Figure 3.11).

In purified systems thrombin can cleave fragment 1 from the prothrombin molecule (Owen *et al.*, 1974) and the rate of thrombin generation is markedly reduced if this fragment has been removed. However, the intact prothrombin fragment 1-2 is essentially the only amino-terminal polypeptide detected in freshly clotted human blood (Aronson *et al.*, 1977): the formation of prothrombin fragment 1 is, therefore, unlikely to be a physiological event. In the presence of large quantities of α-thrombin further degradation of the molecule takes place: a 13-residue peptide (fragment 3) is removed from the amino-terminus of the human thrombin A chain, and the fragment 1 is degraded into a disulphide-linked fragment 1α β which gives rise to fragment 1α and 1β on reduction.

Figure 3.10: Cleavage Sites and Terminology of Activation Products in Prothrombin

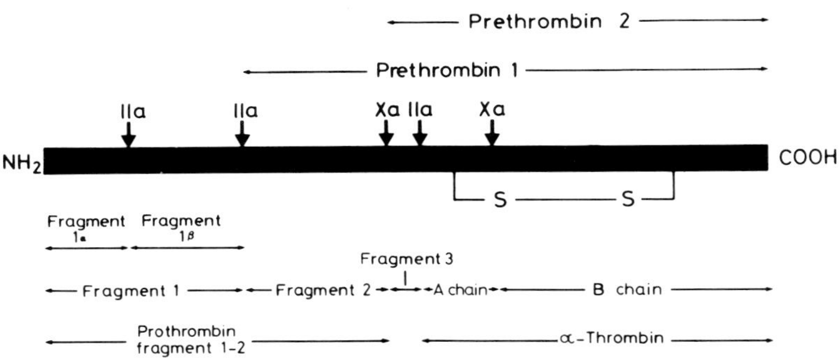

Figure 3.11: The Activation of Prothrombin by Factor Xa

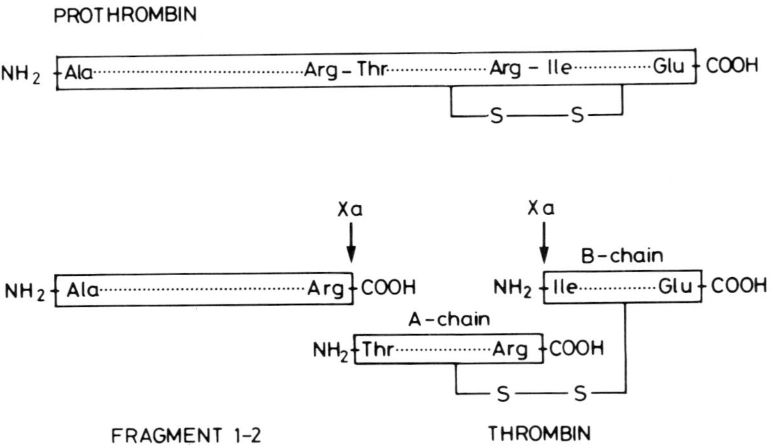

Properties of Prothrombin Fragments

It has been proposed that prothrombin fragment 1 may inhibit excess prothrombin conversion by competing with factor Xa (Prowse *et al.*, 1976). However, as indicated above, the quantity α-thrombin generated during clotting is probably insufficient to cleave the arginyl-serine bond between fragments 1 and 2.

Some of the possible biological actions of prothrombin fragments 1 (αβ) and 2 have been examined (Aronson *et al.*, 1980). In spite of similarities in the carboxyl-terminal sequences of fragment 1 (α β) and 2 with those of human C3a and C5a anaphylatoxins derived from complement components, they did not stimulate smooth muscle contraction, enhance vascular permeability, promote chemotaxis, or promote cellular proliferation of cultured embryo fibroblasts.

Interaction Between Factors Xa, Va and Prothrombin

Factor Va, phospholipid and calcium ions are required for maximal rates of prothrombin activation by factor Xa (Rosing *et al.*, 1980). Factor Xa and factor Va are organised into an enzymatic complex with calcium on a phospholipid surface, a complex which has been termed prothrombinase. It is probable that platelets are the normal source of the phospholipid and thereby serve to localise thrombin formation to sites of platelet deposition.

Prothrombin and factor Xa, like the other vitamin K-dependent coagulation factors, bind to negatively charged lipids such as phosphatidylserine, phosphatidylinositol and phosphatidic acid which are situated principally in the inner surface monolayer of the bilayer plasma membrane of the platelet. On platelet activation there is translocation of the negatively charged phospholipids to the outside of the platelet membrane with a consequential increase in binding sites for factor Xa and prothrombin. Factor Xa and prothrombin bind to the phospholipid surface via calcium, the γ-carboxyglutamic acid residues of fragment 1 of prothrombin providing its calcium-binding sites. Kinetic studies on the formation of thrombin from prothrombin by factor Xa in the presence of calcium ions and phospholipid suggest that the order of addition of reactants is calcium, phospholipid and the prothrombin substrate (Kosow and Orthner, 1979).

The platelet also contains binding sites for factor V: only factor Va binds to the high affinity sites while both factor V and Va bind to lower affinity sites (Tracy *et al.*, 1979). The sites on the membrane of the circulating platelet are believed to be largely saturated with factor V (Tracy *et al.*, 1980). In addition to the plasma source, factor V may also be secreted from the platelet α-granules and then activated by thrombin (Pifer *et al.*, 1977; Kane *et al.*, 1980; Chesney *et al.*, 1981).

Factor Va acts as the platelet receptor for factor Xa: in its absence there is defective binding of factor Xa to the platelet (Miletich *et al.*,

Figure 3.12: Diagram of the Role of the Platelet Surface and Factor V in the Activation of Prothrombin by Factor Xa

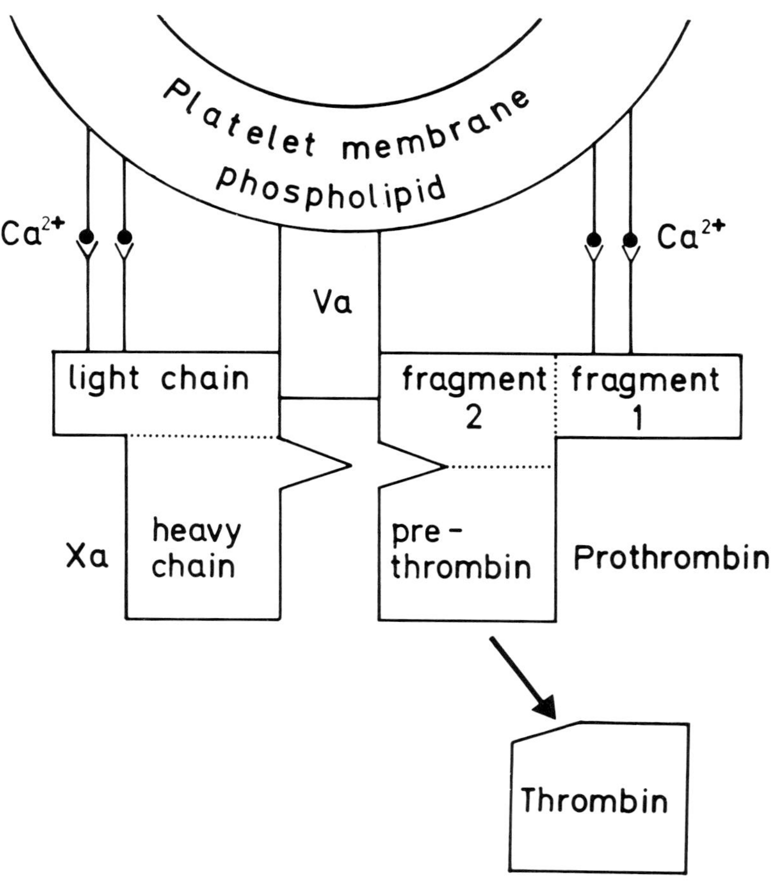

1978b). The phospholipid-bound factor Va and factor Xa form a stoichiometric enzymatic complex which assembles in the absence of its prothrombin substrate (Nesheim *et al.*, 1981; Tracy *et al.*, 1981). On binding to platelets the catalytic activity of factor Xa is increased some 300,000-fold whereas the substitution of platelets by phospholipid results in only a 50-fold increase in the factor Xa enzymatic activity (Miletich *et al.*, 1978a). Prothrombin also binds to factor Va through its fragment 2 region (Esmon and Jackson, 1974).

In the presence of factor Va the rate of thrombin formation is greatly enhanced and the main product of the reaction is thrombin: in the absence of factor V the main product formed in the initial stages of the reaction is prothrombin 2 which easily dissociates from factor X (Rosing *et al.*, 1980). It was suggested that reassociation with factor Xa and consequent conversion into thrombin is inhibited by an excess of prothrombin and that factor Va functions by preventing the dissociation of prothrombin 2 from the complex, thereby allowing thrombin formation.

It appears likely from the data available that factor V's role in the formation of thrombin is to align prothrombin and factor Xa in appropriate positions on the platelet surface for activation to thrombin to be achieved (Figure 3.12). The thrombin formed is released from the fragment 1-2 which remains attached to the phospholipid layer and from factor Va, and can diffuse out to act on the various substrates subserving a hemostatic function.

Thrombin

Prothrombin is converted into α-thrombin through the cleavage of a single arginyl-isoleucine bond with the formation of an A and a B chain. The cleavage causes a conformational change permitting expression of enzymatic activity. Human α-thrombin has a molecular weight of 36,000 and consists of a 36-residue A chain and a 259-residue B chain joined by a disulphide bridge between the mid-portions of the two chains (Fenton *et al.*, 1977). The B chain is structurally very similar to trypsin in its amino acid sequence, and it contains the specific catalytic site residues which are responsible for the proteolytic activity of thrombin. These amino acid residues are the active site serine which becomes acylated during the process in which thrombin cleaves a peptide bond, the active site histidine and aspartate which provide the charge relay system for the catalysis of peptide bond cleavage, and the aspartate residue responsible for the arginyl peptide bond specificity of thrombin.

Non-coagulant but enzymatically active forms arise from α-thrombin by proteolysis including three-chain β- and four-chain γ-thrombin. Whether these are formed *in vivo* is not known. The tentative structure of β-thrombin has been described (Lundblad *et al.*, 1979).

Thrombin is specifically and reversibly bound by fibrin: it is probable that there are both high and low-affinity binding sites. The bound thrombin is released following the degradation of fibrin by plasmin. Human thrombin has also been shown to bind reversibly and specifically to cultured human endothelial cells, a reaction which could have a role in the clearance of circulating thrombin (Awbrey *et al.*, 1979). In addition to the rapid initial binding, there is irreversible slow association of thrombin with the endothelium which involves the covalent linkage of thrombin to a 30,000 dalton protein (Lollar *et al.*, 1980). Thrombin has a number of effects on cultured endothelial cells including the release of prostacyclin, the release of ADP (Pearson and Gordon, 1979) and the inactivation of endothelial plasminogen activator (Loskutoff, 1979).

Fibrinogen

Structure

Fibrinogen, the precursor of insoluble fibrin, has a molecular weight of 340,000 (McDonagh *et al.*, 1972). It contains 4 to 5 per cent carbohydrate composed of mannose, galactose, glucosamine and sialic acid. The molecule is a dimer with two identical halves consisting of three polypeptide chains Aα, Bβ and γ with molecular weights of 64,000, 57,000 and 48,000 respectively. The chains are connected by disulphide bridges, and disulphide bonds also join the two half molecules through their Aα and γ-chains near the amino-terminus. The area of high intra- and inter-chain disulphide bonding, making up the amino-terminal regions of both halves of the dimer, has been termed the *N*-terminal disulphide knot (Blombäck *et al.*, 1968).

Based on electron microscopy studies a variety of models have been proposed for the three-dimensional structure of the fibrinogen molecule. The most generally accepted model is that of Hall and Slayter (1959) in which fibrinogen is perceived as a trinodular rod about 47 nm in length with two identical outer nodules and a smaller central nodule (Figure 3.13). Others have obtained similar appearances by different techniques (Fowler and Erickson, 1979). Alternative proposals on the structure of fibrinogen have included a spherical-like molecule 20 to 24 nm in diameter (Pouit *et al.*, 1972), a disc of around 14 nm in diameter and 5 nm wide (Mosesson *et al.*, 1979), and a cylinder 45 nm long (Bachmann *et al.*, 1975). Support for a trinodular structure comes from physicochemical studies of fibrinogen fragments (Budzynski, 1971; Donovan and Mihalyi, 1974). From electron microscopy studies on fibrinogen fragments produced by plasmin (see page 147) Fowler and colleagues (1980) have concluded that the outer nodules of the trinodular fibrinogen molecule are the fragment D-containing regions while the central nodule is the fragment E-containing region.

The fibrinogen molecule contains 2964 amino acids, 610 in each α-chain, 461 in each β-chain and 411 in the γ-chains (Lottspeich and Henschen, 1977a, b; Henschen and Lottspeich, 1977; Watt *et al.*, 1978; Doolittle *et al.*, 1979). Carbohydrate is present in the Bβ and γ-chains (Gaffney, 1972); it is probable that there are two carbohydrate sites of attachment on the Bβ chain and one on the γ-chain (Pepper *et al.*, 1974). Each molecule contains 29 disulphide bonds; three of these hold the two halves of the dimeric molecule together (Blombäck *et al.*, 1976). The arrangement of a number of the 29 disulphide bonds has been detailed by Bouma and colleagues (1978).

Figure 3.13: Schematic Representation of the Trinodular Structure of the Fibrinogen Molecule

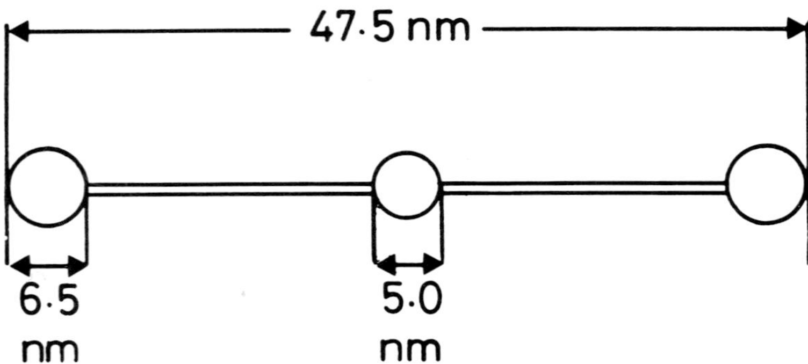

Through the endeavours of a number of laboratories the amino acid sequencing of fibrinogen is now virtually complete (Cottrell *et al.*, 1979; Hassel *et al.*, 1979; Strong *et al.*, 1979; Watt *et al.*, 1979).

Each chain of fibrinogen incorporates in its amino-terminal portion two sequences of two cysteines bounding three other amino acids and separated by 111 or 112 amino acids. These form the disulphide rings which connect the chains together. Between the disulphide rings the chains have an α-helical structure which probably represents the inter-nodular regions in the fibrinogen model of Hall and Slayter (Doolittle *et al.*, 1977, 1978). The carboxyl-terminal portions of the β- and γ-chains are folded and compact and are represented by the peripheral nodules of the molecule as seen on electron micrographs.

Formation of Fibrin

The transformation of fibrinogen into fibrin involves the thrombin-catalysed two-stage removal of fibrinopeptides from the amino-terminal end of the α and β-chains (Bettelheim and Bailey, 1952; Blombäck and Vestermark, 1958; Bilezikian *et al.*, 1975). Thrombin first attacks the Aα chain of fibrinogen, cleaving the bond between arginine and glycine (Arg_{16}-Gly_{17}) to release fibrinopeptide A. The remainder of the molecule has been termed fibrin I. Fibrin I polymerises to form a visible clot composed of thin strands believed to represent end-to-end polymers. Following the polymerisation of fibrin I, thrombin splits the bond between arginine and glycine (Arg_{14}-Gly_{15}) in the Bβ-chain to produce fibrinopeptide B and fibrin II (Figure 3.14). The strands of fibrin II are thicker than those of fibrin I and are thought to result from the uncovering of a second polymerisation site and consequent lateral aggregation of fibrin molecules in addition to the end-to-end aggregation (Ferry, 1952; Laurent and Blombäck, 1958; Laki and Gladner, 1964; Blombäck *et al.*, 1978).

The precise mechanism of polymerisation is not yet established with certainty. There is evidence that one major polymerisation site is located in each fragment D comprising the terminal domains of the molecule with the carboxyl-terminal parts of the six chains (Doolittle *et al.*, 1972; Kudryk *et al.*, 1974). More recently, Olexa and Budzynski (1981) have isolated a peptide from the γ-chain remnant of fragment D which has the ability to bind to the amino-terminal region of fibrinogen following removal of fibrinopeptide A: the site was localised to the γ 373-410 region of fibrinogen. The other binding site has been located in the *N*-terminal disulphide knot (Kudryk *et al.*,

Figure 3.14: The Action of Thrombin on Fibrinogen

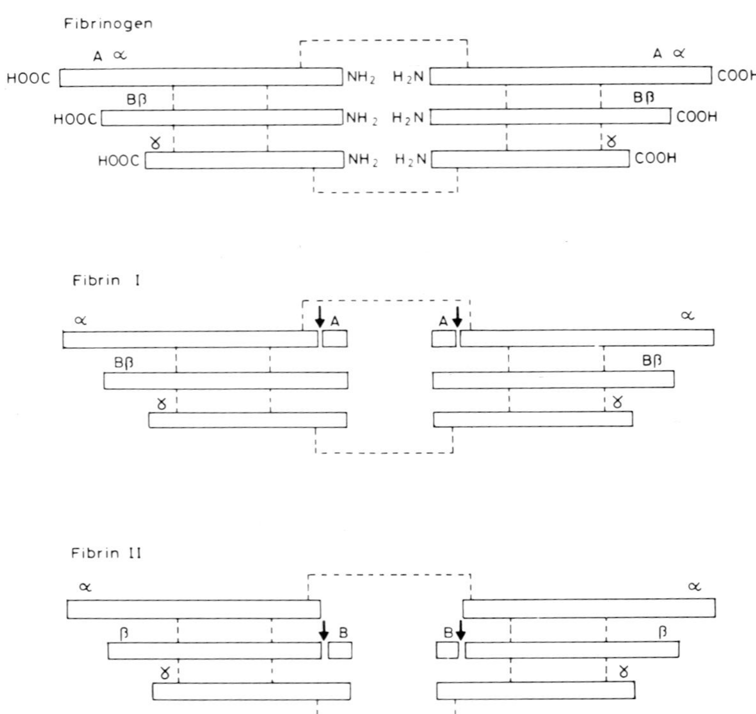

1974), while Laudano and Doolittle (1978, 1980) have provided data to indicate that the tripeptide Gly-Pro-Arg which is contiguous with fibrinopeptide A binds to fibrinogen; this suggests that the amino-terminal binding site is on the Aα-chain of the molecule immediately following fibrinopeptide A.

From the available information a hypothesis to explain the mode of polymerisation has been advanced (Doolittle, 1981), represented diagrammatically in Figure 3.15. Cleavage of the fibrinopeptides removes an excess of negative charge from the central portion of the molecule while the terminal portions retain a net negative charge. Each positively charged central portion has a complementary electrostatic interaction with a negatively charged terminal portion: this gives rise to a staggered, overlapping aggregation of fibrin monomers. The initial polymerisation depends on binding of the exposed

Figure 3.15: Diagrammatic Representation of the Polymerisation of Fibrin

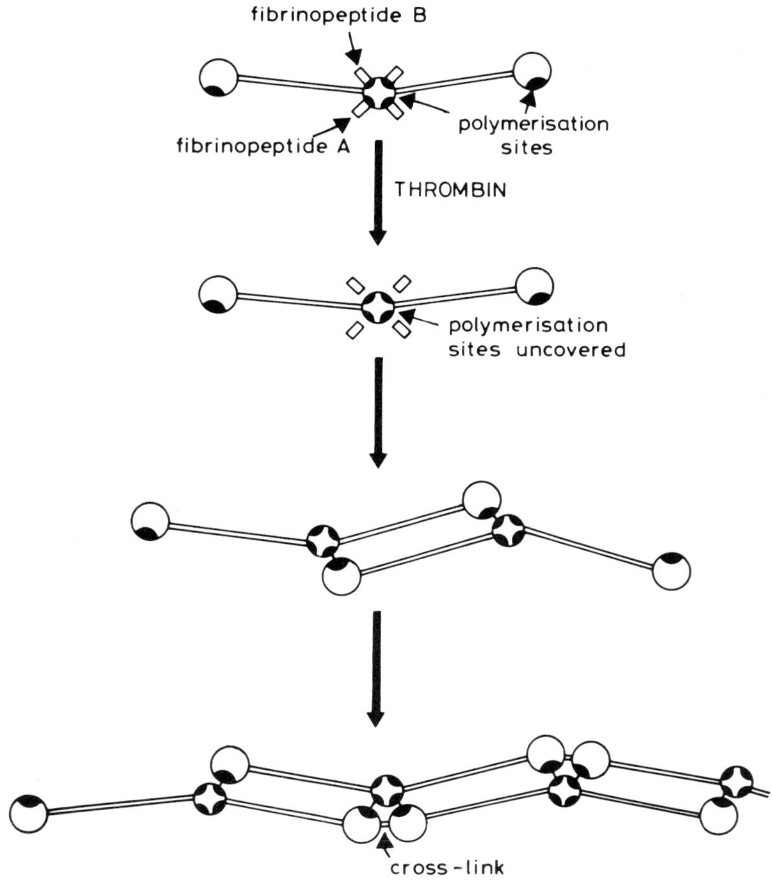

amino-terminal end of the α-chain to a specific portion of the carboxyl end of the γ-chain of another fibrinogen molecule. Lateral growth may be mediated by the sites exposed on removal of fibrinopeptide B. Finally, stabilisation of the fibrin is achieved by the formation of covalent bonds between adjacent monomers under the influence of activated factor XIII (see page 97).

Synthesis

Fibrinogen is synthesised in the liver (Miller *et al.*, 1951; Miller and Bale, 1954; Barnhart and Anderson, 1962). The plasma fibrinogen level is maintained in health within relatively narrow limits, but the precise mechanism of the control of fibrinogen synthesis is not established. A number of agents have been incriminated as stimulators of fibrinogen synthesis; those of possible physiological relevance include fibrinogen degradation products, thrombin, adrenocortico-trophic hormone, growth hormone and prostaglandins.

Plasma Fibrinogen Level. Experiments to determine the influence of the circulating fibrinogen concentration on its synthesis rate, based on the infusion of purified fibrinogen into rabbits, have provided discordant conclusions. An increase in fibrinogen synthesis after fibrinogen infusion has been reported by some investigators (Kropatkin and Izak, 1968; Regoeczi, 1970; Bocci and Pacini, 1973): others have found no change in synthesis when increased fibrinogen levels have been maintained in rabbits (Atencio *et al.*, 1969a; Alving *et al.*, 1977a). Reduced fibrinogen levels *per se* do not appear to be responsible for increased fibrinogen synthesis (Alving *et al.*, 1977a).

Fibrin(ogen) Degradation Products. The possibility that the degradation products of fibrin or fibrinogen might have a feed-back role in stimulating synthesis of fibrinogen has received considerable attention. The conclusions drawn from the experimental data obtained have been, however, variable. Some investigators have been unable to demonstrate increased synthesis of fibrinogen in response to infusions of fibrinogen degradation products (Kropatkin and Izak, 1968; Otis and Rapaport, 1973). In contrast, accelerated fibrinogen synthesis was observed by others (Barnhart *et al.*, 1970; Young and Kolmen, 1970; Bocci *et al.*, 1974). Using the rate of appearance of [75Se]selenomethionine in circulating fibrinogen to assess the synthesis rate in rabbits Kessler and Bell (1979b) found that plasmin-cleaved fibrinogen degradation products, particularly fragments D and E, accelerated fibrinogen synthesis, and they proposed that thrombin-induced increases in fibrinogen synthesis are mediated by fibrinogen degradation products. However, the quantity of fibrinogen degradation products required to stimulate fibrinogen synthesis were much higher than would be found under physiological circumstances. Franks and colleagues (1981) reported that purified homologous fragment D increased fibrinogen synthesis in rats whereas fragment E had no effect. In further studies by Kessler and Bell (1979a) neither fibrin degradation products nor fibrinopeptides A and B increased the synthesis of fibrinogen. The absence of an increase in the fibrinogen synthesis rate following the infusion of fibrin degradation products was confirmed by Ittyerah *et al.* (1979).

Thrombin. Intravenous thrombin causes an increase in fibrinogen synthesis in rabbits (Pickart and Pilgeram, 1967; Alving *et al.*, 1977b), even in afibrinogenemic animals (Alving *et al.*, 1977b). The mechanism of the stimulatory effect of thrombin is disputed. It does not appear to result from an alteration in the fibrinogen level (Alving *et al.*, 1977a). One proposal is that fibrinogen synthesis is mediated by the fibrin(ogen) degradation products generated by thrombin (Kropatkin and Izak, 1968). On the basis of the relationship found between the titre of fibrin(ogen) degradation products produced by thrombin infusion and the degree of fibrinogen synthesis, and the inhibitory effect of epsilon-aminocaproic acid on thrombin-stimulated fibrinogen synthesis and fibrin(ogen) degradation product production, Kessler and Bell (1979b) suggested that thrombin-augmented fibrinogen synthesis is mediated by fibrinogen degradation products. Ittyerah and associates (1979) reported that the infusion of small quantities of thrombin increased the fibrinogen synthesis rate without decreasing the plasma fibrinogen level or increasing fibrin(ogen) degradation products, while the infusion of thrombin with antithrombin III did not affect fibrinogen synthesis: they concluded that thrombin stimulates fibrinogen synthesis by an enzymatic mechanism independent of fibrinogenolysis or significant disseminated intravascular coagulation. Pickart and Thaler (1976) proposed that free fatty acid, mobilised by thrombin, mediates increased fibrinogen synthesis: they suggested a sequence of formation of fibrin degradation products which potentiate the epinephrine effect on β-adrenergic receptors with consequent mobilisation of free fatty acids; these in turn directly stimulate fibrinogen synthesis.

Adrenocorticotrophic Hormone. Injections of adrenocorticotrophic hormone (ACTH) have been observed to increase the plasma fibrinogen concentration in rabbits without altering the catabolism of fibrinogen (Atencio *et al.*, 1969b). This effect did not appear to be mediated by cortisol (Atencio and Lorand, 1970), and the influence of ACTH on the fibrinogen level was not altered by adrenalectomy (Seligsohn *et al.*, 1973). It is not known whether ACTH stimulates fibrinogen synthesis by the hepatic cell directly or through an indirect extra-adrenal mechanism. In any event, the high ACTH levels required to stmulate increased fibrinogen synthesis make it improbable that it acts as a normal stimulator (Chen *et al.*, 1974).

Growth Hormone. The administration of growth hormone to dogs causes an early and marked increase in plasma fibrinogen (Campbell *et al.*, 1953), later found to be the result of increased fibrinogen synthesis (Jeejeebhoy *et al.*, 1970). In further studies Jeejeebhoy and associates (1975) found that fibrinogen synthesis by isolated rat hepatocyte suspensions was enhanced by the addition of a hormone mixture comprising glucagon, cortisol, tri-iodothyronine and growth hormone, but the hormones were not tested individually.

Insulin. Fibrinogen synthesis by cultured chick embryo hepatocytes could be induced by prolonged exposure to insulin (Liang and Grieninger, 1981). The relationship of this finding, and the other observations on the influence of hormones on fibrinogen synthesis, to the physiological regulation of the circulating fibrinogen level is obscure.

Inflammation. A rise in the plasma fibrinogen concentration is characteristic of inflammation, but the mechanism of the increase is not established. Prostaglandins act as mediators of inflammation and prostaglandin E_1 has been shown to be an effective stimulator of fibrinogen synthesis (Carlson *et al.*, 1977). Other factors may be involved in the mediation of increased fibrinogen synthesis during inflammation or injury. A number of investigators have reported that substances released from leukocytes (crude leukocyte extract, leukocytic endogenous mediator, hepatocyte-stimulating factor) increase the fibrinogen concentration when injected into animals (Kampschmidt and Upchurch, 1974; Bocci *et al.*, 1976), although this has been denied (Seligsohn and Klein, 1973). Weidner and colleagues (1979) claimed that crude leukocyte extract harvested from inflamed rabbit peritoneal cavities increased fibrinogen synthesis whereas the pyrogenic extract derived from blood leukocytes had no effect. Hepatocyte-stimulating factor has been shown also to increase fibrinogen production by monolayer cultures of primary hepatocytes (Rupp and Fuller, 1979), and Fuller and Ritchie (1981) propose that plasmin-derived fragments of fibrinogen or fibrin, produced at the site of injury, stimulate the production of hepatocyte-stimulating factor by monocytes which is then carried in the blood to the liver to promote increased synthesis of fibrinogen. Leukocytic endogenous mediator prepared from rabbit peritoneal leukocytes has been found to be a heat-labile protein of molecular weight 13,000 to 16,000: it has both pyrogenic and fibrinogen-increasing activity (Kampschmidt *et al.*, 1981).

Fetal Fibrinogen

The plasma thrombin time of neonates is prolonged (Larrieu *et al.*, 1952). This was attributed by some to the presence of fibrin(ogen) degradation products (Von Felton and Straub, 1969), but prolonged thrombin clotting times have been found in the absence of detectable degradation products (Aguercif *et al.*, 1973). Others have ascribed the thrombin time abnormality to the presence of a molecular variant of fibrinogen during the first few days of life.

Witt and her colleagues have marshalled the evidence for a structural difference between adult and fetal fibrinogen in a series of reports. They have shown a greater negative charge on fetal fibrinogen, detected by chromatography on DEAE-cellulose, and differing pH-dependency of the thrombin clotting time (Witt *et al.*, 1969): this latter finding has been disputed by Teger-Nilsson and Ekelund (1974) who found prolonged thrombin times in cord plasma, but the thrombin clotting time of purified cord and adult fibrinogen was the same and showed no pH-dependency. Fetal fibrinogen was found to have a greater phosphorus content, but an equivalent amount of hexose (Witt and Müller, 1970): the phosphorus moiety of the fibrinogen molecule was noted later not to influence either the fibrinogen-fibrin conversion rate or the affinity between fibrinogen and thrombin (Witt and Hasler, 1972). Tryptic peptide maps of fetal and adult fibrinogen have been claimed to show identity of the Bβ and γ-chains, but differing Aα-chains (Witt and Tesch, 1979). Fetal and adult fibrin have also a differing electron microscopy appearance with the fetal fibrin having a more compact network and a lesser fiber thickness (Tesch *et al.*, 1979). In addition to the prolonged thrombin time, fetal fibrinogen shows delayed fibrin monomer aggregation (Guillin and Menache, 1973). The delayed fibrin aggregation of cord fibrinogen has been found to be corrected by partial removal of sialic acid, suggesting that the phenomenon is due to the relatively high sialic acid content of fetal fibrinogen (Galanakis and Mosessen, 1977, 1979).

In spite of the accumulating suggestive evidence it is not firmly established at the present time that a distinct molecular form of fibrinogen exists in fetal life.

Factor XIII

The final stage of the blood coagulation mechanism involves the formation of covalent bonds between adjacent fibrin monomers through the action of an enzyme termed factor XIII or fibrin-stabilising factor: such bonds produce stabilised fibrin of high tensile strength.

Factor XIII circulates in plasma in the form of an inactive zymogen: the molecule consists of separate a and b subunits held together as a tetramer a_2b_2 by non-covalent bonds. The active enzymatic site consists of a cysteine-SH group sited in the a subunit. The molecular weight of the tetrameric complex is approximately 320,000 daltons; each a subunit has a molecular weight of 75,000, and each b subunit a weight of 88,000 (Schwartz *et al.*, 1973).

Factor XIII is also found in platelets where it exists as an a_2 dimer. The platelet zymogen and the plasma a subunits are electrophoretically identical (Schwartz *et al.*, 1973). The platelet content of factor XIII is relatively large and contributes significantly to the total plasma factor XIII content (Rider *et al.*, 1978). Factor XIII is also found in placental tissue: the plasma, platelet and placental a subunits have been shown to be immunologically identical (Bohn, 1972).

The plasma concentration of factor XIII is around 1 mg/100 ml (Bohn *et al.*, 1974). The liver is probably the major site of synthesis of the plasma zymogen (Losowsky and Walls, 1971; Lee and Chung, 1976). The platelet a subunit is synthesised in megakaryocytes (Kiesselbach and Wagner, 1972).

It has been demonstrated that the a_2 subunits of platelet and plasma factor XIII zymogen bind specifically and with high affinity to fibrinogen (Greenberg and Shulman, 1982): such binding would provide an efficient mechanism for the localisation of factor XIII to the fibrin clot.

Activation of Factor XIII

The activation of the factor XIII zymogen takes place in stages. In the first stage the thrombin cleaves an arginyl-glycine bond near the amino-terminal end of the a subunit resulting in the release of a 36-

residue activation peptide with a molecular weight of approximately 4000 (Schwartz *et al.*, 1973; Takagi and Doolittle, 1974). In the second stage, induced by the presence of calcium ions, there is dissociation of the a_2b_2 complex and unmasking of the active center cysteine. The concentration of calcium ion required to dissociate the complex is greater than that found in plasma so that a mechanism must exist to reduce the calcium requirement. This is provided by one of the functional domains of fibrinogen: the interaction of a domain in the α-chain of fibrinogen with thrombin-modified factor XIII results in a reduction in the concentration of calcium ion required for the dissociation of the complex to that available in plasma (Credo *et al.*, 1978, 1981).

Cross-linking Function of Factor XIIIa

The activated factor XIII (fibrinoligase) is rapidly adsorbed onto fibrin. Through the specific transamidase action of factor XIIIa γ-glutamyl-ϵ-lysine bridges are formed between adjacent fibrin monomers (Figure 3.16). It is these covalent bonds that endow the polymerised fibrin with increased mechanical stability. Six moles of the γ-glutamyl-ϵ-lysine bonds are normally formed per mole of fibrin; two moles are involved in the cross-linking of the γ-chains of the fibrin monomers while four moles are formed between the α-chains (Pisano *et al.*, 1971). The γ-γ dimers are formed very rapidly, the α-α linkages take much longer to form. The cross-linking sites on the fibrinogen molecule are located in the carboxyl-terminal portions of the γ-chains and in the mid-sections of the Aα-chains (Fretto *et al.*, 1978): in particular, the acceptor sites have been pinpointed to the glutamyl residues α_{328} and α_{366} (Cottrell *et al.*, 1979).

Activated factor XIII has also been shown to catalyse the cross-linking of actin and fibrin (Mui and Ganguly, 1977), and has been implicated in the cross-linking of platelet actin, an action prevented by ATP (Cohen *et al.*, 1979). It has been suggested that depletion of platelet ATP after aggregation and secretion may stimulate cross-linking processes in the final stages of platelet activation.

Cross-linked fibrin is less susceptible to plasmin degradation than non-cross-linked fibrin. This resistance is dependent on the α-chain cross-links and not on the γ-chain cross-links (Schwartz *et al.*, 1973).

Coagulation Inhibitors

Heparin

Interest in heparin has stemmed principally from its therapeutic potential as an anticoagulant drug for the prevention of thrombosis. It is present in tissue mast cells in relatively high concentration and has been reported to be present within platelets and to be released on adhesion to a surface (Vannucchi *et al.*, 1982). There is, however, no firm evidence that endogenous heparin circulates in human blood and its physiological role is uncertain.

Heparin belongs to a group of sulphated carbohydrates termed glycosaminoglycans which are widely distributed in human and animal tissues. A further compound of this group with heparin-like properties is heparan sulphate which is the major surface glycosaminoglycan of enthothelial cells (Bounassisi, 1973; Kraemer, 1977; Wasteson *et al.*, 1977). The heparins are straight-chain anionic glycosaminoglycuronan sulphate esters of variable molecular weight ranging from some 4000 to 40,000 (Johnson and Mulloy, 1976). The glycosaminoglycuronan chains consist of alternating uronic acids and glycosamines, both in pyranose form.

Heparin acts as an anticoagulant by forming complexes with antithrombin III and thereby accelerating the rate at which the inhibitor can inactivate the various enzymes of the coagulation mechanism (see page 102). Heparin has a similar catalytic effect on the reaction between antithrombin III and thrombin (Hatton *et al.*, 1978; Cofranesco *et al.*, 1979). The precise physicochemical characteristics of heparin responsible for its anticoagulant properties are not yet established, but a number of differences in the composition of heparin molecules that bind strongly to antithrombin III and those that do not have been identified. Rosenberg and Lam (1979) suggested that a unique tetrasaccharide in heparin is required for binding antithrombin III, while Lindahl and associates found the high affinity species of heparin to contain a relatively high proportion of non-sulphated iduronic acid and they isolated fragments of 12 to 14 saccharide units with high biological activity. Some of the units within the sequence appear to be essential for antithrombin III binding and an octasaccharide with high affinity for antithrombin has been described (Thunberg *et al.*, 1980). Anionic density may also be an important factor with a decrease in the anticoagulant activity as the degree of sulphation decreases (Hurst *et al.*, 1979; Ayotte *et al.*, 1981).

Figure 3.16: Formation of γ-Glutamyl-ε-Lysine Bridges

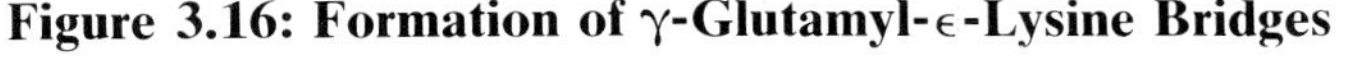

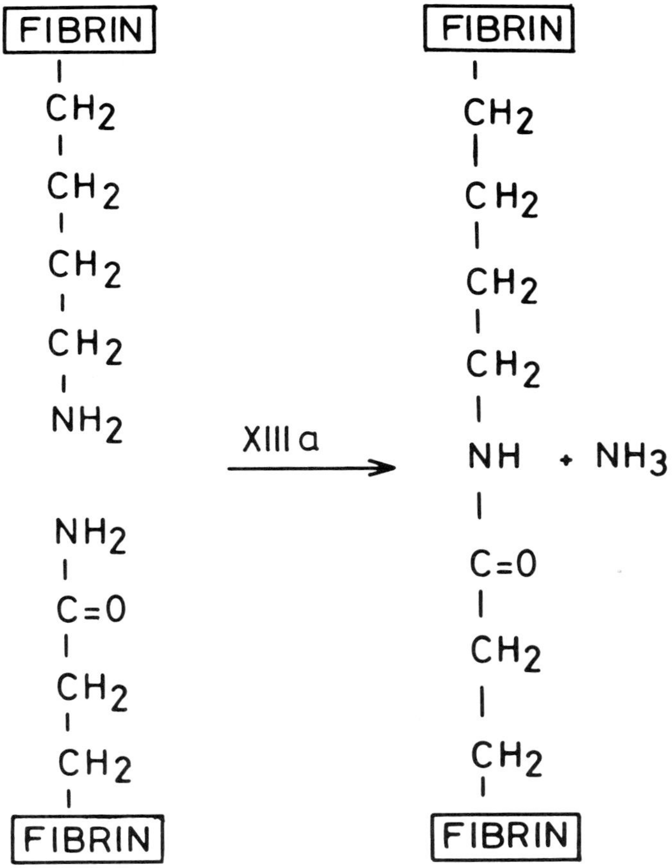

When heparins are fractionated on the basis of molecular weight the larger molecules are found to possess greater anticoagulant activity and to promote stronger inhibition of thrombin by antithrombin III, whereas lower molecular weight fractions are more effective in potentiating the inhibition of factor Xa (Andersson *et al.*, 1976, 1979; Laurent *et al.*, 1978).

Antithrombin III

In their classification of plasma mechanisms for the inactivation of
thrombin Seegers and his colleagues (1954) applied the term antithrombin III to the agent causing progressive inhibition of thrombin
in plasma. There is now overwhelming evidence that antithrombin III
is identical to heparin cofactor, the agent which in association with
heparin inhibits the action of thrombin (Abildgaard, 1968; Rosenberg and Damus, 1973). Antithrombin III is the major natural inhibitor of the blood coagulation system.

Chemistry

Antithrombin III is a single-chain glycoprotein with a molecular
weight of approximately 60,000, an amino-terminal histidine and a
carbohydrate content of 10 to 17 per cent (Abildgaard, 1967; Miller-
Andersson *et al.*, 1974; Kurachi *et al.*, 1976a; Nordenman *et al.*,
1977; Koide, 1979). The primary structure has been determined and
the molecule found to contain 425 amino acid residues (Petersen *et
al.*, 1979): the sequence has a degree of homology with that of
α_1-antitrypsin.

Antithrombin III contains covalently-linked *N*-acetylgluco-
samine, mannose, galactose and sialic acid (Danishefsky *et al.*, 1978):
the sialic acid moiety does not appear to be required for the inhibitory
action of antithrombin III on thrombin. Each molecule has been
shown to contain four identical *N*-glycosidally-linked carbohydrate
chains (Franzen *et al.*, 1980). The preparations of Miller-Andersson
and her co-workers (1974) were free from cystine and cysteine and
they inferred that antithrombin III has no disulphide bridges. In
contrast, Nordenman *et al.* (1977) concluded that each molecule
contains three disulphide bonds.

Inhibition of Coagulation by Antithrombin III

Antithrombin III appears to be capable of forming complexes with all
serine proteases. It is established that it is the principal physiological
inhibitor of thrombin, forming a 1:1 stoichiometric complex which is
enzymatically inactive (Abildgaard, 1969; Rosenberg and Damus,
1973). Complex formation is markedly accelerated in the presence of
heparin resulting in much more rapid neutralisation of thrombin.

The plasma inhibitory activity against factor Xa was identified as
residing in antithrombin III by Biggs and associates (1970) and,
independently, by Yin and co-workers (1971). The formation of a

100

1:1 complex between factor Xa and the inhibitor was demonstrated by the use of SDS-polyacrylamide electrophoresis (Kurachi *et al.*, 1976b). In the presence of physiological concentrations of factor V, phospholipid and calcium the affinity of factor Xa for antithrombin III is markedly decreased (Marciniak, 1973), raising doubts on the *in vivo* importance of antithrombin III as a factor Xa inhibitor.

The procoagulant activity of both high and low molecular weight forms of factor XIIa are slowly neutralised by antithrombin III following the formation of a 1:1 stoichiometric complex of activated factor XII and inhibitor (Stead *et al.*, 1976): the reaction is very substantially accelerated in the presence of heparin. Inhibition of the prekallikrein-activating activity of factor XIIa by antithrombin III has also been recorded (Chan *et al.*, 1977). The heparin-potentiated inhibitory activity of antithrombin III on factor XIa was demonstrated by Damus and colleagues (1973).

Predictably, antithrombin III also inactivates factor IXa, a further serine protease (Rosenberg *et al.*, 1975; Østerud *et al.*, 1976). Again the formation of a 1:1 enzyme-inhibitor complex takes place (Rosenberg *et al.*, 1975; Kurachi *et al.*, 1976b). The rate of factor IXa activity neutralisation was noted to be slow in the absence of heparin. Recently, however, it was found that therapeutic concentrations of heparin did not enhance the inactivation rate of factor XIa by purified antithrombin III (Scott *et al.*, 1982a).

The effect of antithrombin III on factor VII has been disputed. Godal and co-workers (1974) reported progressive inactivation of factor VIIa in the presence of both heparin and antithrombin III at 2°C and 25°C, but inactivation of factor VII activity could not be demonstrated at 37°C. However, using purified human preparations Østerud and associates (1976) found no inhibition of factor VII by antithrombin III in the absence of heparin. Human factor VII was reported by others to react slowly with antithrombin III in the presence of heparin (Broze and Majerus, 1980), factor VIIa being inhibited at a rate about 25 times that of the unactivated factor VII. In contrast to this finding with human factor VII, bovine factor VII was found to be stable when mixed with antithrombin III in the presence or absence of heparin (Jesty, 1978; Zur and Nemerson, 1978).

Mechanism of Action of Antithrombin III

The initial proposal for the mechanism of neutralisation of thrombin by antithrombin III was that a 1:1 stoichiometric complex of enzyme and inhibitor is formed through the interaction of the reactive site

101

arginine of antithrombin III with the active center serine of thrombin (Rosenberg and Damus, 1973). Fish and Björk (1979a) provided evidence that the interaction is between the active site of thrombin and a specific arginyl-serine bond in the carboxyl-terminal region of the inhibitor. They also showed that the stable inactive complex between antithrombin and thrombin, when dissociated by hydroxylamine or ammonia, releases a proteolytically modified two-chain form of antithrombin which is indistinguishable from that produced free in solution during the reaction between antithrombin and thrombin (Fish and Björk, 1979b): these observations suggest that limited proteolysis of antithrombin III may be involved in the formation of the inactive complex with thrombin. A similar cleavage of the arginyl-serine bond in antithrombin III has been found in the reaction between equimolar amounts of the inhibitor and factors IXa or Xa with the formation of the two-chain form of the inhibitor (Björk *et al.*, 1981). This is in keeping with the finding that complexes of antithrombin III and thrombin or factor Xa show a two-chain form of antithrombin III when dissociated at pH 9 and studied on SDS-polyacrylamide gels (Jesty, 1979).

Villanueva and Daneshefsky (1979) studied possible conformational changes accompanying the binding of antithrombin III to thrombin by difference spectroscopy, circular dichroism and optical rotatory dispersion. Their results suggested that conformational changes take place on the formation of the complex between thrombin and antithrombin with an increase in the α-helix content from 5-7 to 12-15 per cent and a corresponding loss in the β structure.

It has been suggested that antithrombin III itself is regulated by a specific binding protein in plasma which has been partially purified and designated anti-XaI (Yin *et al.*, 1979). An observation which has a possible similar significance is that of MacGregor and colleagues (1979): they found that normal human plasma contains a component which inhibits the heparin and antithrombin neutralisation of factor Xa.

Mechanism of the Potentiating Action of Heparin on the Antithrombin III-Enzyme Reaction

Heparin markedly accelerates the rate at which antithrombin III reacts with thrombin and other coagulation system enzymes, but the mechanism by which it achieves this effect is not finally established. Heparin is capable of binding to antithrombin III although the binding sites on the antithrombin are not defined: lysine residues (Rosenberg and Damus, 1973) or tryptophan residues (Einarsson and Andersson, 1977; Björk *et al.*, 1979) may be involved. The number of binding sites for heparin on the antithrombin III molecule has also been disputed. Danielsson and Björk (1978) proposed that there is a single site while Piepkorn and colleagues (1978), on the basis of masurements of enhancement of tryptophan fluorescence, suggested that two sites on antithrombin III can accommodate one large heparin molecule with high affinity or two smaller molecules with low affinity. The conformational changes seen on the formation of complexes of thrombin and antithrombin appear to be different when the complex is formed in the presence of heparin (Villanueva and Danishefsky, 1979).

Rosenberg and Damus (1973) proposed that heparin binds to the lysyl residues of antithrombin III to produce a conformational alteration in the inhibitor which renders the reactive site arginine more accessible to the active center serine; in this way the interaction between inhibitor and enzyme is accelerated. In later studies Jordan and associates (1979) provided kinetic evidence that heparin acts as a catalyst in the thrombin-antithrombin interaction and is able to initiate multiple rounds of enzyme-inhibitor complex formation. At high heparin concentrations complexes of thrombin and heparin are formed.

A further proposal, also based on heparin-antithrombin complex formation, is that antithrombin III is transformed into a more reactive inhibitor after binding with heparin (Andersson *et al.*, 1977; Hook *et al.*, 1976). Holmer and colleagues (1979) suggested that heparin enhances the thrombin-antithrombin III reaction by binding first to antithrombin III and then by linkage of thrombin to both the bound heparin and the antithrombin.

All the preceding views on the actin of heparin have assumed that the binding of heparin to antithrombin III is the primary event in the reaction. The contrasting hypothesis is that heparin reacts with thrombin to produce changes which make the enzyme more susceptible to attack by the inhibitor (Machovich, 1975; Machovich *et al.*, 1975; Smith and Craft, 1976; Smith, 1977). The evidence for the view that thrombin neutralisation by antithrombin III and heparin proceeds through the formation of a heparin-thrombin complex has been marshalled by Smith and Sundboom (1981a, b).

Synthesis

The site of synthesis of antithrombin III is not finally established.

The Coagulation System

Immunoreactive antithrombin III may be demonstrated in fresh endothelial cells from human umbilical cord (Chan and Chan, 1979), and after culture antithrombin III was detected in the culture medium suggesting that it had been synthesised in the endothelial cells. Based on the finding of antithrombin III in hepatocytes of rat liver by an immunofluorescent technique Watada and colleagues (1981) concluded that the liver is the major site of antithrombin production. In contrast, Lee and co-workers (1979) found only a weak or absent reaction in human hepatocytes to an antithrombin III antibody.

Plasma Level and Turnover

The plasma concentration of antithrombin III in man has been reorted to be 19.6 ± 2.3 mg/100 ml (Collen *et al.*, 1977): very similar plasma levels were obtained by Chan and associates (1979) using a specific radioimmunoassay. From studies with radiolabelled antithrombin III it was estimated that it has a half-life of 2.83 ± 0.36 days (Collen *et al.*, 1977).

Antithrombin III-like Protein

Koide and co-workers (1982) have described the isolation of a protein from human plasma which may be identical to histidine-rich glycoprotein. It had an amino-terminal sequence which is homologous to that of antithrombin III, and had strong heparin-binding properties although it did not inhibit thrombin or show heparin cofactor activity.

Alpha$_2$-Macroglobulin

Alpha$_2$-macroglobulin is a glycoprotein with an apparent molecular weight of 725,000 which is capable of forming complexes with a number of serine proteases including thrombin, plasmin and kallikrein. It possesses a dimeric structure, consisting of two half-molecules joined by non-covalent bonds (Jones *et al.*, 1972). The subunits have a molecular weight of around 180,000 and each half-molecule consists of two subunit chains held together by disulphide bridges (Harpel, 1973). It is present in the plasma of healthy adults at a concentration of 130 to 335 mg/100 ml (Wilding *et al.*, 1967).

Alpha$_2$-macroglobulin is capable of inhibiting the coagulant activity of thrombin on fibrinogen in purified systems, while its esterase activity is unaffected (Lanchantin *et al.*, 1966; Steinbuch *et al.*,

1967). The binding of thrombin by α_2-macroglobulin is slow (Iwamota and Abiko, 1970). When α_2-macroglobulin is incubated with thrombin there is proteolytic modification of the subunit chain and the production of a 85,000 dalton derivative which remains bound to the parent molecule by disulphide bonds (Harpel, 1973). The proteolytic cleavage of α_2-macroglobulin, associated with complex formation, takes place at an arginyl-leucine bond in the 'middle' segments of the polypeptide chains (Sottrup-Jensen *et al.*, 1982). The α_2-macroglobulin-thrombin complex may possess some biological activity: α_2-macroglobulin purified from serum and presumably complexed with thrombin is able to clot fibrinogen (Rinderknecht and Geokas, 1973).

Alpha$_2$-macroglobulin also forms complexes with kallikrein producing inhibition of its proteolytic activity, but only partial inhibition of its esterase properties (Harpel, 1970). Inhibition of the chemotactic activity of kallikrein by α_2-macroglobulin has also been reported (Gallin and Kaplan, 1974). Other coagulation factors examined do not appear to be inhibited by α_2-macroglobulin: these include factor Xa (Østerud *et al.*, 1976) and factor XIIa (Chan *et al.*, 1977).

Alpha$_2$-macroglobulin-protease complexes are removed rapidly from the circulation. Rabbit alveolar and peritoneal macrophages have been shown to be capable of the uptake of such complexes (Debanne *et al.*, 1975, 1976).

Alpha$_1$-Antitrypsin

Alpha$_1$-antitrypsin is a single-chain glycoprotein with a molecular weight of around 55,000. Its structure has been reviewed recently (Carrell *et al.*, 1982). Alpha$_1$-antitrypsin has the capacity to inhibit trypsin and a number of other proteases, including chymotrypsin, plasmin and elastase in a stoichiometric and irreversible reaction (Bundy and Mehl, 1959; Shamash and Rimon, 1966; Crawford, 1973). Multiple generic types of the protein in human serum have been shown to be inherited by a series of autosomal co-dominant alleles that determine the concentration and electrophoretic mobility. Its concentration in normal adult plasma is in the range 200 to 400 mg/100 ml (36 to 72 μM), and its half-life has been estimated at 6 days (Kueppers and Fallat, 1969).

There has been a conflict of view over the effect of α_1-antitrypsin on thrombin. On the basis of a reduction in fast antithrombin activity in patients with genetic α_1-antitrypsin deficiency it was claimed that thrombin is inhibited by α_1-antitrypsin (Gans and Tan, 1967). Rimon and colleagues (1966) found thrombin to be inhibited by a purified α_1-protease inhibitor considered to be identical to α_1-antitrypsin, while Matheson and Travis (1976) reported that the clotting and esterase activities of purified thrombin were inhibited by purified α_1-antitrypsin in a time- and temperature-dependent reaction, the inhibitor forming a 1:1 complex with thrombin. Lane and her colleagues (1975), using the technique of measuring the residual antithrombin activity of plasma after precipitation with specific antibodies, concluded that α_1-antitrypsin is responsible for 25 per cent of the total progressive plasma antithrombin activity. In contrast to these various findings, Heimburger and associates (1971) were unable to detect inhibition of thrombin with purified α_1-antitrypsin. In addition, Learned and co-workers (1976) found that their partially purified preparation of α_1-antitrypsin did not inhibit the conversion of fibrinogen into fibrin by bovine thrombin when used at a concentration within the normal plasma level range; they concluded that α_1-antitrypsin does not contribute significantly to the progressive antithrombin activity of plasma under normal conditions.

In an examination of the plasma inhibitors of activated factor XI Heck and Kaplan (1974) identified α_1-antitrypsin as a major inhibitor of factor XIa; the reduction in factor XIa-inhibitory activity in fractionated α_1-antitrypsin-deficient plasma provides confirmatory evidence of this action. Alpha$_1$-antitrypsin has been claimed to be capable of inhibiting kallikrein (McConnell, 1972), but this could not be confirmed by others (Gallimore *et al.*, 1979). Alpha$_1$-antitrypsin has no effect on factor XIIa (Chan *et al.*, 1977), but has been shown to inhibit the amidolytic activity of factor Xa (Scully *et al.*, 1981).

Cl Inactivator

Cl inactivator is a neuraminoglycoprotein with a molecular weight of approximately 105,000 which was recognised originally as an inhibitor of the esterase activity of the first component of complement. Its inhibitory action on the coagulation system is directed at the initial stages of the intrinsic pathway. Inhibition of the procoagulant activity of the activated forms of both factor XII and factor XI by purified Cl inactivator was demonstrated by Forbes and associates (1970). The inhibition of factor XIIa has been confirmed (Chan *et al.*, 1977), and others have shown the inhibition of factor XII fragments (Schreiber *et al.*, 1973). Cl inactivator is a major inhibitor of plasma kallikrein (Ratnoff *et al.*, 1969; Gigli *et al.*, 1970; McConnell, 1972), and has been estimated to account for around half the inhibitory activity against kallikrein in human plasma (Scapira *et al.*, 1981).

Protein C

In 1976 Stenflo described the isolation and some properties of a previously unrecognised vitamin K-dependent protein from bovine plasma. Its name was derived from its presence in pool C of a chromatographic separation on DEAE-Sephadex. Protein C was shown to be composed of two polypeptide chains, the light chain having an amino-terminal sequence containing γ-carboxyglutamic acid residues and showing considerable homology with the other vitamin K-dependent proteins. Its molecular weight was found to be approximately 56,000 (Stenflo, 1976). Protein C is the zymogen of a serine protease, the active center serine being located in the heavy chain (Esmon *et al.*, 1976). Activation of the bovine zymogen results from the cleavage of an arginyl-isoleucine bond in the heavy chain which can be achieved by trypsin or a protease from Russell's viper venom (Kisiel *et al.*, 1976). Bovine α-thrombin was found later to be also capable of activating protein C (Kisiel *et al.*, 1977b).

Human protein C has a molecular weight of 62,000 composed of a light chain of 21,000 and a heavy chain of 41,000 held together by one or more disulphide bonds. On activation with α-thrombin the apparent molecular weight of the heavy chain decreases to 40,000 (Kisiel, 1979); the bond cleaved is probably the arginyl-leucine bond between residues 12 and 13 in the amino-terminal region of the heavy chain (Figure 3.17). The amidolytic and anticoagulant actions of activated protein C are inhibited by DFP, indicating that both activities are dependent on the active site serine residue (Kisiel, 1979). The thrombin-catalysed activation of protein C has been reported to be accelerated by a human endothelial cell surface cofactor which represents a thrombin receptor site (Owen and Esmon, 1981): Esmon *et al.* (1982) have isolated a plasma membrane from rabbit lung which acted as a Ca^{2+}-dependent cofactor for the thrombin-catalysed activation of protein C. It has been shown that low levels of thrombin *in vivo* can activate protein C (Comp *et al.*, 1982).

Figure 3.17: Structure and Activation of Protein C

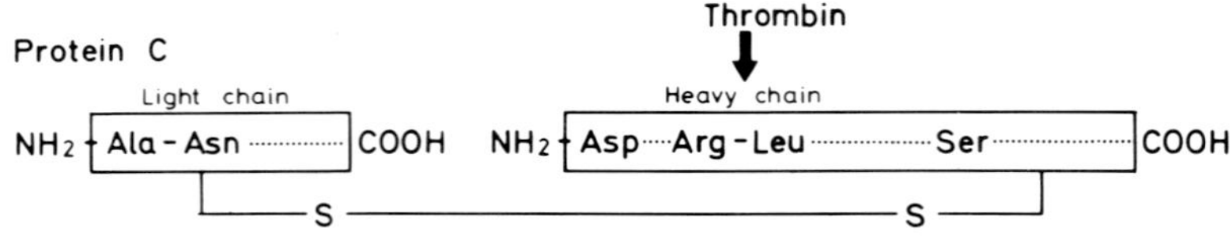

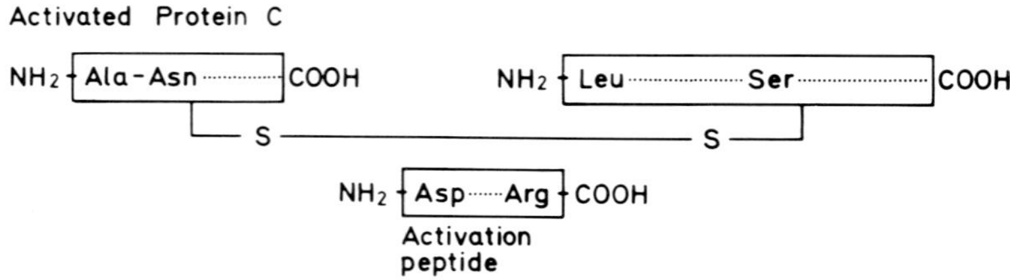

Initially no function could be ascribed to the newly discovered protein, but Kisiel and colleagues (1977) found that the activated bovine protein inactivated bovine factor V in a reaction which required calcium and phospholipid and was time- and temperature-dependent. This was confirmed by Walker and associates (1979). Nesheim and co-workers (1982) have provided evidence that factor Xa and activated protein C compete for factor Va, and that factor Xa protects factor Va from inactivation by activated protein C. The inactivation of factor Va involves the proteolytic cleavage of the factor Va molecule (Canfield *et al.*, 1978). It is now known that activated protein C also rapidly inactivates the coagulant activity of factor VIII in the presence of calcium and phospholipid (Vehar and Davie, 1980): the rate of inactivation of thrombin-activated factor VIII by protein C was found to be significantly faster than that of the unactivated form. In addition to its ability to inhibit plasma factor Va activity protein C can inhibit platelet prothrombin-converting activity, presumably through its effect on platelet factor Va acting as a factor Xa receptor (Comp and Esmon, 1979; Dahlbäck and Stenflo, 1980).

Intravenous injection of activated protein C in dogs has been found to result in an elevation of the circulating plasminogen activator, probably due to release from the vascular endothelium (Comp and Esmon, 1981).

Protein C has been shown to be identical to the autoprothrombin II-A described as a factor Xa inhibitor with fibrinolytic activity by Seegers and colleagues (Murano *et al.*, 1974; Seegers *et al.*, 1976). It is probably responsible for the anticoagulant activity in thrombin-treated prothrombin complex reported by Marciniak (1970, 1972).

The complexity of the relationship between protein C and clotting factors may be yet greater as it has been reported that normal human plasma can inhibit the amidolytic activity of activated protein C (Marlar and Griffin, 1980).

While the physiological function of protein C is not established, its ability to inactivate the activated forms of factors V and VIII suggest that it may make an important contribution to the anticoagulant activity of blood and thereby play a role in the regulation of coagulation. The importance of its influence on fibrinolytic activity also requires clarification. The potentiation of the activation of protein C by a cofactor on the endothelial wall points to a further contribution of the endothelium to the maintenance of blood fluidity.

Protein S

Protein S is a single-chain glycoprotein with an amino-terminal alanine and a molecular weight of 69,000. It derives its name from its isolation and characterisation in Seattle (Di Scipio *et al.*, 1977). Its amino acid sequence homology with other vitamin K-dependent proteins suggests that it also belongs to this group, and in further studies Di Scipio and Davie (1979) found that it contains 10 residues of γ-carboxyglutamic acid per mol and 7.8 per cent carbohydrate. It is likely that protein S is identical to a factor isolated from bovine plasma which enhanced the rate of inactivation of factor Va by activated protein C (Walker, 1980). The stimulatory effect of protein S on the inactivation of factor Va by activated protein C appears to be due, at least in part, to the enhancement of the binding of activated protein C to phospholipid (Walker, 1981a). The anticoagulant activity of protein C is species specific (Kisiel *et al.*, 1977b): evidence has been provided that this results from species specificity of protein S cofactor activity (Walker, 1981b).

Protein Z

A further γ-carboxyglutamic acid-containing protein, originally described as a single-chain variant of factor X (Mattock and Esnouf, 1973), has been isolated from bovine plasma (Petersen *et al.*, 1980). The amino-terminal sequence shows considerable homology with the other vitamin K-dependent plasma proteins; residues 1-43 have been isolated and 13 γ-carboxyglutamic acid residues identified (Højrup *et al.*, 1982). Its function is unknown.

Lipoprotein Factor Xa Inhibitor

An inhibitor of factor Xa, separate from antithrombin III, has been recognised in the high molecular weight fractions of plasma (Marciniak and Tsukamura, 1972). Low-density lipoproteins have been identified as carriers of anti-Xa activity (Marciniak, 1974; Barrowcliffe and Eggleton, 1981). The macromolecular inhibitor of thrombin described by Miller-Andersson and colleagues (1973) may represent the same inhibitor. While it is likely that low-density lipoprotein contains binding sites for factor Xa, its importance in the regulation of the coagulation system is uncertain.

Inhibition of Thrombin

The central place of thrombin in hemostasis underscores the potential importance of thrombin-inhibitory mechanisms. While a number of mechanisms make a contribution to the neutralisation of thrombin in blood, including adsorption of thrombin by fibrin, there is now no doubt that the principal thrombin-inhibitory activity resides in the plasma protease inhibitors. There is an overwhelming evidence that antithrombin III is the major functional thrombin inhibitor in plasma although other protease inhibitors, notably α_2-macroglobulin and α_1-antitrypsin may contribute. From the results of experiments involving the precipitation of antithrombin activity by specific antibodies and the subsequent assay of the residual antithrombin activity by both a clotting method and a thrombin-agarose gel diffusion method, Lane and associates (1975) concluded that antithrombin III accounts for around 50 per cent of the total progressive antithrombin activity while α_2-macroglobulin and α_1-antitrypsin each contribute about 25 per cent. Others have found that antithrombin III accounts for 70 to 80 per cent of the total thrombin-inhibiting activity of plasma, α_2-macroglobulin contributing the bulk of the remainder (Abildgaard, 1967; Shapiro and Anderson, 1977). However, using a radioimmunoassay of the thrombin-antithrombin complex, Lau and Rosenberg (1980) demonstrated that antithrombin III is responsible for the neutralisation of virtually all added thrombin under conditions approaching the *in vivo* situation. The molar concentrations of antithrombin III and α_2-macroglobulin in normal plasma are approximately equal: the greater contribution of antithrombin III to the total plasma thrombin neutralisation activity reflects the more rapid enzyme-inhibitor reaction. When large amounts of thrombin are incubated with plasma even the α_2-macroglobulin may be saturated.

The conclusions drawn from *in vitro* studies on the importance of antithrombin III in the neutralisation of thrombin are reinforced by the finding that kindred congenitally deficient in antithrombin III suffer from recurrent venous thromboembolic disease at a relatively early age (Egeberg, 1965; Marciniack *et al.*, 1974; Mackie *et al.*, 1978; Matsuo *et al.*, 1979). In contrast, a familial deficiency of α_2-macroglobulin was not associated with a thrombotic tendency (Bergqvist and Nilsson, 1979), while patients with congenital α_1-antitrypsin deficiency have a liability to pulmonary emphysema or hepatic disease rather than thrombosis.

Inhibition of Factor Xa

The pivotal position of factor X in the coagulation pathways suggests that its inhibition would have an important controlling influence on fibrin formation. It is clear that factor Xa is inhibited by antithrombin III in purified systems, although more slowly than thrombin (Ødegård *et al.*, 1977). However, the physiological role of antithrombin III in factor Xa inhibition is speculative in view of the protection of factor Xa from inactivation when associated with platelets or phospholipid (Marciniak, 1973). It is probable that the affinity of factor Xa for antithrombin III becomes negligible when complexed with phospholipid and factor V compared to its affinity towards prothrombin (Josso and Beguin, 1981). Such a mechanism would ensure that factor Xa bound to a phospholipid surface is able to convert prothrombin into thrombin whereas factor Xa in the cir-

111

culating plasma will be rapidly neutralised.

The contribution of other protease inhibitors to the inactivation of factor Xa is not established. In one study (Scully *et al.*, 1981) the inhibitory actions of α_1-antitrypsin, antithrombin III and α_2-macroglobulin were compared. The relative order of effectiveness at normal plasma concentrations and in the absence of heparin was α_1-antitrypsin $>$ antithrombin III $>$ α_2-macroglobulin in the proportions 4.64:3.08:1.00.

Inhibition of the Contact Activation System

Control of the contact activation system may be achieved by prevention of the initial activation or by the inhibition of factor XIIa, factor XIa or kallikrein. Plasma appears to contain agents which interfere with the adsorption of factor XII to a surface and thereby prevent its activation (Saito *et al.*, 1974), but their characterisation is incomplete and their physiological importance obscure.

Cl inactivator, antithrombin III and α_2-antiplasmin have been shown to be able to neutralise the biological actions of activated factor XII (Forbes *et al.*, 1970; Stead *et al.*, 1976; Chan *et al.*, 1977; Saito *et al.*, 1979). A further plasma protein, presently uncharacterised and previously described as an inhibitor of plasminogen activation, has been reported to be able to inhibit factor XIIa (Hedner and Martinsson, 1978; Hedner, 1979). The relative contribution of these various inhibitors of factor XIIa under physiological conditions is unknown.

In purified systems the clot-promoting activity of factor XI has been shown to be inhibited by Cl inactivator (Forbes *et al.*, 1970), antithrombin III (Damus *et al.*, 1973), α_1-antitrypsin (Heck and Kaplan, 1974) and α_2-antiplasmin (Saito *et al.*, 1979). An additional factor XI inhibitor has been described (Nossel and Niemetz, 1965; Ratnoff *et al.*, 1972), but its nature and distinction from the other protease inhibitors requires clarification. The contributions of α_1-antitrypsin, antithrombin III, Cl inactivator and α_2-antiplasmin towards the inactivation rate of factor XIa by plasma has been found to be 68 per cent, 16 per cent, 8 per cent and 8 per cent respectively (Scott *et al.*, 1982b).

Kallikrein has also been shown to be inhibited in purified systems by a number of the plasma protease inhibitors: those shown to have this property are Cl inactivator (Ratnoff *et al.*, 1969; Gallimore *et al.*,

1979), α_2-macroglobulin (Harpel, 1970), antithrombin III (Burrowes *et al.*, 1975; Vennerod *et al.*, 1976) and α_2-antiplasmin (Saito *et al.*, 1979). Most have concluded that Cī inactivator is the major plasma inhibitor of plasma kallikrein. The relative importance of Cī inactivator is supported by the experiments of Gallimore *et al.* (1979): using a chromogenic substrate to assay residual kallikrein activity they found that α_1-antitrypsin, antithrombin III and α_2-antiplasmin produced little or no inhibition of plasma kallikrein whereas the inhibitory activity in plasma correlated with that of Cī inactivator. They found a further inhibitor of kallikrein, distinct from Cī inactivator and α_2-macroglobulin, but further information on this agent is not available. Schapira and colleagues (1981b, 1982) demonstrated that HMW kininogen decreases the inactivation of kallikrein by all inhibitors except α_2-macroglobulin and calculated from their data on both purified systems and plasma that Cī inactivator and α_2-macroglobulin were each responsible for about half the inhibition of kallikrein and all the other inhibitors together contributed less than 2 per cent.

Anticoagulant Effect of Fibrinogen Degradation Products

The anticoagulant effect of plasmin digests of fibrinogen were first reported twenty-five years ago (Niewiarowski and Latallo, 1957; Niewiarowski and Kowalski, 1958; Triantaphyllopoulos, 1958). The mechanism of this anticoagulant effect has been controversial. Evidence has been provided to indicate an antithrombin action (Triantaphyllopoulos, 1959; Triantaphyllopoulos and Triantaphyllopoulos, 1966). Others ascribed the anticoagulant effect to interference with the polymerisation of fibrin monomers (Alkjaersig *et al.*, 1962), while Latallo *et al.* (1964) showed that fibrinogen degradation products could produce both effects.

The identification of specific digestion fragments of fibrinogen has aided understanding of the different mechanisms of anticoagulant action (see page 147). The high molecular fragments X and Y, particularly Y, were found to be more potent anticoagulants than the end-stage fragments D and E (Marder and Shulman, 1969): this was in agreement with earlier observations suggesting that degradation products present early in the course of plasmin digestion had greater anticoagulant activity than those present later (Niewiarowski and Kowalski, 1958; Triantaphyllopoulos, 1958). Although the mechanism of the anticoagulant action was not fully elucidated by Marder and Shulman, the non-clottable complexes which form between fibrin monomers and degradation products appeared to be involved.

The core fragments D and E have been studied by Larrieu and associates (1972). Fragment D was found to act primarily by inhibiting fibrin monomer polymerisation. This resulted from the competition of fragment D molecules for the fibrin monomer linkages leading to the formation of non-clottable complexes. The maximal inhibition of polymerisation was obtained by the addition of D dimer from cross-linked fibrin (Larrieu *et al.*, 1975). Fragment E acted as a competitive inhibitor of the clotting action of thrombin on fibrinogen; fragment D had no specific antithrombin effect.

The action of fibrin(ogen) degradation products on clotting may contribute to the hemorrhagic tendency in patients with plasma proteolytic states, but the concentration required to produce the anticoagulant effects do not suggest that they have any physiological role in the control of coagulation.

References

Albildgaard, U. (1967) 'Purification of two progressive antithrombins of human plasma', *Scandinavian Journal of Clinical and Laboratory Investigation, 19,* 190-5

Abildgaard, U. (1968) 'Highly purified antithrombin III with heparin cofactor activity prepared by disc electrophoresis', *Scandinavian Journal of Clinical and Laboratory Investigation, 21,* 89-91

Abildgaard, U. (1969) 'Binding of thrombin to antithrombin III', *Scandinavian Journal of Clinical and Laboratory Investigation, 24,* 23-7

Aguercif, M., Giacometti, N., Nigg, O.M., Lacourt, G. and Bouvier, C.A. (1973) 'Existe-t-il un fibrinogene foetal? Explanation statistique de la fibrinoformation dans une population de nouveau-nés prematures, matures et post-matures', *Pediatrie, 28,* 381-99

Alkjaersig, N., Fletcher, A.P. and Sherry, S. (1962) 'Pathogenesis of the coagulation defect developing during pathological plasma proteolytic ("fibrinolytic") states. II. The significance, mechanism and consequences of defective fibrin polymerisation', *Journal of Clinical Investigation, 41,* 917-34

Alving, B.M., Bell, W.R. and Evatt, B.L. (1977a) 'Fibrinogen synthesis in rabbits: effects of altered levels of circulating fibrinogen', *American Journal of Physiology, 232,* H478-84

Alving, B.M., Evatt, B.L. and Bell, W.R. (1977b) 'Stimulation of fibrinogen synthesis in rabbits with ancrod-induced afibrinogenemia', *American Journal of Physiology, 233,* H562-7

Andersson, L.-O., Barrowcliffe, T.W., Holmer, E., Johnson, E.A. and Sims, G.E.C. (1976) 'Anticoagulant properties of heparin fractionated by affinity chromatography on matrix-bound antithrombin-III and by gel filtration', *Thrombosis Research, 9,* 575-83

Andersson, L.-O., Engman, L. and Henningsson, E. (1977) 'Crossed immunoelectrophoresis as applied to studies on complex formation. The binding of heparin to antithrombin III and the antithrombin-thrombin complex', *Journal of Immunological Methods, 14*, 271-81

Andersson, L.-O., Barrowcliffe, T.W., Holmer, E., Johnson, E.A. and Söderström, G. (1979) 'Molecular weight dependency of the heparin potentiated inhibition of thrombin and activated factor X. Effect of heparin neutralization in plasma', *Thrombosis Research, 15*, 531-41

Aronson, D.L., Stevan, L., Ball, A.P., Franza, B.R. and Finlayson, J.S. (1977) 'Generation of the combined prothrombin activation peptide (F 1.2) during the clotting of blood and plasma', *Journal of Clinical Investigation, 60*, 1410-8

Aronson, D.L., Ball, A.P., Franza, R.B., Hugli, T.E. and Fenton, J.W. II (1980) 'Human prothrombin fragments in F 1 ($\alpha\beta$) and F 2: preparation and characterization of structural and biological properties', *Thrombosis Research, 20*, 239-53

Atencio, A.C. and Lorand, L. (1970) 'Effect of ACTH on biosynthesis of fibrinogen in the rabbit', *American Journal of Physiology, 219*, 1161-5

Atencio, A.C., Joiner, K. and Reeve, E.B. (1969a) 'Experimental and control systems studies of plasma fibrinogen regulation in rabbits', *American Journal of Physiology, 216*, 764-72

Atencio, A.C., Chao, P.-Y., Chen, A. and Reeve, E.B. (1969b) 'The fibrinogen response to corticotropin preparations in rabbits', *American Journal of Physiology, 216*, 773-80

Awbrey, B.J., Hoak, T.C. and Owen, W.G. (1979) 'Binding of human thrombin to cultured human endothelial cells', *Journal of Biological Chemistry, 254*, 4092-5

Ayotte, L., Lorimeau, J.C. and Perlin, A.S. (1981) 'Influence of variations in the chemical structure of heparin on its anticoagulant and anti-factor Xa activities', *Thrombosis Research, 22*, 97-102

Bach, R., Nemerson, Y. and Konigsberg, W. (1981) 'Purification and characterization of bovine tissue factor', *Journal of Biological Chemistry, 256*, 8324-31

Bachmann, L., Schmitt-Fumian, W., Hammel, R. and Lederer, K. (1975) 'Size and shape of fibrinogen. I. Electron microscopy of the hydrated molecule', *Die Makromokulare Chemie, 176*, 2603-18

Bagdasarian, A., Lahiri, B., Talamo, R.G., Wong, P. and Colman, R.W. (1974) 'Immunochemical studies of plasma kallikrein', *Journal of Clinical Investigation, 54*, 1444-54

Bajaj, S.P., Rapaport, S.I. and Brown, S.F. (1981) 'Isolation and characterization of human factor VII. Activation of factor VII by factor Xa', *Journal of Biological Chemistry, 256*, 253-9

Barnhart, M.I. and Anderson, G.F. (1962) 'Intracellular localization of fibrinogen', *Proceedings of the Society for Experimental Biology and Medicine, 110*, 734-7

Barnhart, M.I., Cress, D.C., Noonan, S.M. and Walsh, R.T. (1970) 'Influence of fibrinolytic products on hepatic release and synthesis of fibrinogen', *Thrombosis et Diathesis Haemorrhagica*, Suppl. 39, 143-59

Barrowcliffe, T.W. and Eggleton, C.A. (1981) 'The anti-Xa activity of lipoproteins', *Thrombosis and Haemostasis, 46*, 351

Bartlett, S., Latson, P. and Hanahan, D.J. (1980) 'High molecular weight factor V of bovine and human plasma', *Biochemistry, 19*, 273-7

Bergqvist, D. and Nilsson, I.M. (1979) 'A family with α_2-macroglobulin deficiency', *Thrombosis and Haemostasis, 42*, 219

Bettelheim, F.R. and Bailey, K. (1952) 'The products of the action of thrombin on fibrinogen', *Biochimica et Biophysica Acta, 9*, 578-9

Biggs, R., Denson, K.W.E., Akman, N., Barrett, R. and Hadden, M. (1970)

'Antithrombin III, anti-factor Xa and heparin', *British Journal of Haematology, 19*, 283-305

Bilezikian, S.B., Nossel, H.L., Butler, V.P. and Canfield, R.E. (1975) 'Radioimmunoassay of human fibrinopeptide B and kinetics of fibrinopeptide cleavage by different enzymes', *Journal of Clinical Investigation, 56*, 438-45

Björk, I., Danielsson, Å., Fish, W.W., Larsson, K., Lieden, K. and Nordenman, B. (1979) 'The interaction between heparin and antithrombin', in D. Collen, B. Wiman and M. Verstraete (eds.) *The Physiological Inhibitors of Blood Coagulation and Fibrinolysis*, Elsevier/North Holland, Amsterdam, pp. 67-84

Björk, I., Jackson, C.M., Jornvall, H., Lavine, K.K., Nordling, K. and Salsgiver, W.J. (1981) 'The active site of antithrombin. Release of the same proteolytically cleaved form of the inhibitor from complexes with factor IXa, factor Xa and thrombin', *Thrombosis and Haemostasis, 46*, 286

Bjørklid, E. and Storm, E. (1977) 'Purification and some properties of the protein component of tissue thromboplastin from human brain', *Biochemical Journal, 165*, 89-96

Blombäck, B. and Vestermark, A. (1958) 'Isolation of fibrinopeptides by chromatography', *Arkiv för Kemi, 12*, 173-83

Blombäck, B., Blombäck, M. Henschen, A., Hessel, B., Iwanaga, S. and Woods, K.R. '*N*-terminal disulphide knot of human fibrinogen', *Nature, 218*, 130-4

Blombäck, B., Hessel, B. and Hogg, D. (1976) 'Disulphide bridges in NH_2-terminal part of human fibrinogen', *Thrombosis Research, 8*, 639-58

Blombäck, B., Hessel, B., Hogg, D. and Therkildsen, L. (1978) 'A two-step fibrinogen-fibrin transition in blood coagulation', *Nature, 275*, 501-5

Bocci, V. and Pacini, A. (1973) 'Factors regulating plasma protein synthesis. II. Influence of fibrinogenolytic products on plasma fibrinogen concentration', *Thrombosis et Diathesis Haemorrhagica, 29*, 63-75

Bocci, M., Conti, T., Muscellota, M., Pacini, A. and Pessina, G.P. (1974) 'Factors regulating plasma protein synthesis. IV. Influence of fragments D and E on plasma fibrinogen concentrations', *Thrombosis et Diathesis Haemorrhagica, 31*, 395-402

Bocci, M., Muscettoia, M., Pacini, A. and Pessina, G.P. (1976) 'The effect of leucocyte pyrogen on rabbit fibrinogen. A third view', *Thrombosis Research, 9*, 505-12

Bohn, H. (1972) 'Comparative studies on the fibrin-stabilizing factors from human plasma, platelets and placentas', *Annals of the New York Academy of Sciences, 202*, 256-72

Bohn, H. (1974) 'Immunologische Untersuchungen über Faktor XIII und das PSF-bindende Globulin', *Blut, 28*, 81-90

Bouma, B.N. and Griffin, J.H. (1977) 'Human blood coagulation factor XI. Purification, properties, and mechanism of activation by activated factor XII', *Journal of Biological Chemistry, 252*, 6432-7

Bouma, B.N., Miles, L.A., Beretta, G. and Griffin, J.H. (1980) 'Human plasma prekallikrein. Studies of its activation by activated factor XII and of its inactivation by diisopropyl phosphofluoridate', *Biochemistry, 19*, 1151-60

Bouma, H., Takagi, T. and Doolittle, R.F. (1978) 'The arrangement of disulphide bonds in fragment D from human fibrinogen', *Thrombosis Research, 13*, 557-62

Bounassisi, V. (1973) 'Sulfated mucopolysaccharide synthesis and secretion in endothelial cell cultures', *Experimental Cell Research, 76*, 363-8

Bowie, E.J.W., Woods, J.E., Fass, D.N., Zollman, P.E. and Owen, C.A. (1975) 'Liver transplantation in pigs with von Willebrand's disease', *British Journal of Haematology, 31*, 37-44

Breederveld, K., Giddngs, J.C., ten Cate, J.W. and Bloom, A.L. (1975) 'The localisation of factor V within normal human platelets and the demonstration of a platelet-factor V antigen in congenital factor V deficiency', *British Journal of Haematology, 29*, 405-12

Broze, G.J. and Majerus, P.W. (1980) 'Purification and properties of human coagulation factor VII', *Journal of Biological Chemistry, 255*, 1242-7

Bucher, D., Nebelin, E., Thomsen, J. and Stenflo, J. (1976) 'Identification of γ-carboxyglutamic acid residues in bovine factors IX and X, and in a new vitamin K-dependent protein', *FEBS Letters, 68*, 293-6

Budzynski, A. (1971) 'Difference in conformation of fibrinogen degradation products as revealed by hydrogen exchange and spectropolarimetry', *Biochimica et Biophysica Acta, 229*, 663-71

Bundy, H.F. and Mehl, J.W. (1959) 'Trypsin inhibitors of human serum', *Journal of Biological Chemistry, 234*, 1124-8

Burrowes, C.E., Habal, F.M. and Movat, H.Z. (1975) 'The inhibition of human plasma kallikrein by antithrombin III', *Thrombosis Research, 7*, 175-83

Butkowski, R.J., Elion, J., Downing, M.R. and Mann, K.G. (1977) 'Primary structure of human prethrombin 2 and α-thrombin', *Journal of Biological Chemistry, 252*, 4942-57

Campbell, J., Hausler, H.E., Munroe, J.S. and Davidson, I.W.F. (1953) 'Effect of growth hormone in dogs', *Endocrinology, 53*, 134-62

Canfield, W., Nesheim, M., Kisiel, W. and Mann, K.G. (1978) 'Proteolytic inactivation of bovine factor Va by bovine activated protein C', *Circulation, 58*, II-210

Carlson, T.H., Fradl, D.C., Leonard, B.D., Wentland, S.H. and Reeve, E.B. (1977) 'Fibrinogen synthesis stimulation by prostaglandin E_1 and some other vasodilators', *American Journal of Physiology, 233*, Hl-9

Carrell, R.W., Jeppsson, J.-O., Laurell, C.-B., Brennan, S.O., Owen, M.C., Vaughan, L. and Boswell, D.R. (1982) 'Structure and variation of human $α_1$-antitrypsin', *Nature, 298*, 329-34

Castillo, R., Maragall, S., Rodés, J., Clemente, C., Profitós, J. and Ordinas, A. (1977) 'Increased factor VIII complex and defective ristocetin-induced platelet aggregation in liver disease', *Thrombosis Research, 11*, 899-906

Chan, J.Y.C. (1976) 'Purification of factor XII (Hageman factor) from human plasma', *Thrombosis Research, 8*, 337-49

Chan, J.Y.C., Burrowes, C.E., Habal, F.M. and Movat, H.Z. (1977) 'The inhibition of activated factor XII (Hageman factor) by antithrombin III: the effect of other plasma proteinase inhibitors', *Biochemical and Biophysical Research Communications, 74*, 150-8

Chan, J.Y.C., Movat, H.Z. (1976) 'Purification of factor XII (Hageman factor) from human plasma', *Thrombosis Research, 8*, 337-49

Chan, V. and Chan, T.K. (1979) 'Antithrombin III in fresh and cultured human endothelial cells: a natural anticoagulant from the vascular endothelium', *Thrombosis Research, 15*, 209-13

Chan, V., Chan, T.K., Wong, V., Tso, S.C. and Todd, D. (1979) 'The determination of antithrombin III by radioimmunoassay and its clinical application', *British Journal of Haematology, 41*, 563-72

Chargaff, E., Bendich, A. and Cohen, S.S. (1944) 'The thromboplastic protein: structure, properties, disintegration', *Journal of Biological Chemistry, 156*, 161-78

Chen, Y., Briese, F.W. and Reeve, E.B. (1974) 'Relation of fibrinogen secretion (synthesis) to ACTH dose', *American Journal of Physiology, 227*, 927-61

Chesney, C.M., Pifer, D. and Colman, R.W. (1981) 'Subcellular localisation and secretion of factor V from human platelets', *Proceedings of the National Academy of Sciences, 78*, 5180-4

Chuang, T.F., Sargeant, R.B. and Hougie, C. (1974) 'The effect of calcium ions on the properties of factor IX and its activated form', *British Journal of Haematology, 27*, 281-7

Claeys, H. and Collen, D. (1978) 'Purification and characterization of bovine coagulation factor XII (Hageman factor)', *European Journal of Biochemistry, 87*, 69-74

Cofrancesco, E., Radaelli, F., Pogliani, E., Amici, N., Torri, G.G. and Casu, B. (1979) 'Correlation of sulfate content and degree of carboxylation of heparin and related glycosaminoglycans with anticomplement activity. Relationships to the anticoagulant and platelet-aggregating activities', *Thrombosis Research, 14*, 179-87

Cohen, I., Blankenberg, T.A., Borden, D,. and Veis, A. (1979) 'Factor XIIIa-catalysed cross-linking of platelet and muscle actin. Regulation by adenosine triphosphate', *Thrombosis and Haemostasis, 42*, 395

Collen, D., Schetz, J., de Cock, F. and Holmer, E. (1977) 'Metabolism of antithrombin III (heparin cofactor) in man: effects of venous thrombosis and of heparin administration', *European Journal of Clinical Investigation, 7*, 27-35

Colman, R.W., Bagdasarian, A., Talamo, R.C., Scott, C.F., Seavey, M., Guimaràes, J.A., Pierce, J.V. and Kaplan, A.P. (1975) 'Williams trait. Human kininogen deficiency with diminished levels of plasminogen proactivator and prekallikrein associated with abnormalities of the Hageman factor-dependent pathways', *Journal of Clinical Investigation, 56*, 1650-62

Comp, P.C. and Esmon, C.T. (1979) 'Activated protein C inhibits platelet prothrombin-converting activity', *Blood, 54*, 1272-81

Comp, P.C. and Esmon, C.T. (1981) 'Generation of fibrinolytic activity by the infusion of activated protein C into dogs', *Journal of Clinical Investigation, 68*, 1221-8

Comp, P.C., Jacocks, R.M., Ferrell, G.L. and Esmon, C.T. (1982) 'Activation of protein C *in vivo*', *Journal of Clinical Investigation, 70*, 127-34

Cottrell, B.A., Strong, D.D., Watt, K.W.K. and Doolittle, R.F. (1979) 'Amino acid sequence studies on the α chain of human fibrinogen. Exact location of cross-linking acceptor sites', *Biochemistry, 18*, 5405-10

Counts, R.B., Paskell, S.L. and Elgee, S.K. (1978) 'Disulfide bonds and the quaternary structure of factor VIII/von Willebrand factor', *Journal of Clinical Investigation, 62*, 702-9

Crawford, I.P. (1973) 'Purification and properties of normal human α_1-antitrypsin', *Archives of Biochemistry and Biophysics, 156*, 215-22

Credo, R.B., Curtis, C.G. and Lorand, L. (1978) 'Ca^{2+}-related regulatory function of fibrinogen', *Proceedings of the National Academy of Sciences, 75*, 4234-7

Credo, R.B., Curtis, C.G. and Lorand, L. (1981) 'α-chain domain of fibrinogen controls generation of fibrinoligase (coagulation factor XIIIa). Calcium ion regulatory aspects', *Biochemistry, 20*, 3770-8

Dahlbäck, B. (1980) 'Human coagulation factor V purification and thrombin-catalyzed activation', *Journal of Clinical Investigation, 66*, 583-91

Dahlbäck, B. and Stenflo, J. (1978) 'Binding of bovine coagulation factor Xa to platelets', *Biochemistry, 17*, 4938-45

Dahlbäck, B. and Stenflo, J. (1980) 'Inhibitory effect of activated protein C on activation of prothrombin by platelet-bound factor Xa', *European Journal of Biochemistry, 107*, 331-5

Dam, H. (1935) 'The antihaemorrhagic vitamin of the chick', *Journal of Biochemistry, 29*, 1273-85

Damus, P.S., Hicks, M. and Rosenberg, R.D. (1973) 'Anticoagulant action of heparin', *Nature, 246*, 355-7

Danielsson, A. and Björk, I. (1978) 'The binding of low-affinity and high-affinity heparin to antithrombin', *European Journal of Biochemistry, 90,* 7-12

Danishefsky, I., Zweben, A. and Slomiany, B.L. (1978) 'Human antithrombin III. Carbohydrate components and associated glycolipid', *Journal of Biological Chemistry, 253,* 32-7

Debanne, M.T., Bell, R. and Dolovich, J. (1975) 'Uptake of proteinase-α-macroglobulin complexes by macrophages', *Biochimica et Biophysica Acta, 411,* 295-304

Debanne, M.T., Bell, R. and Dolovich, J. (1976) 'Characteristics of the macrophage uptake of proteinase-α-macroglobulin complexes', *Biochimica et Biophysica Acta, 428,* 466-75

Deuel, T.F., Senior, R.M., Chang, D., Griffin, G.L., Heinrikson, R.L. and Kaiser, E.T. (1981) 'Platelet factor 4 is chemotactic for neutrophils and monocytes', *Proceedings of the National Academy of Sciences, 78,* 4584-7

van Dieijen, G., Tans, G., Rosing, J. and Hemker, H.C. (1981) 'The role of phospholipid and factor VIIIa in the activation of bovine factor X', *Journal of Biological Chemistry, 256,* 3433-42

Di Scipio, R.G. and Davie, E.W. (1979) 'Characterization of protein S, a γ-carboxyglutamic acid containing protein from bovine and human plasma', *Biochemistry, 18,* 899-904

Di Scipio, R.G., Hermodson, M.A., Yates, S.G. and Davie, E.W. (1977) 'A comparison of human prothrombin, factor IX (Christmas factor), factor X (Stuart factor), and protein S', *Biochemistry, 16,* 698-706

Di Scipio, R.G., Kurachi, K. and Davie, E.W. (1978) 'Activation of human factor IX (Christmas factor)', *Journal of Clinical Investigation, 61,* 1528-38

Dodds, W.J. (1969) 'Hepatic influence on splenic synthesis and release of coagulation activities', *Science, 166,* 882-3

Dodds, W.J. and Hoyer, L.W. (1974) 'Coagulation activities in perfused organs: regulation by addition of animal plasmas', *British Journal of Haematology, 26,* 497-509

Donaldson, V.H., Kleniewski, J., Saito, H. and Sayed, J.K. (1977) 'Prekallikrein deficiency in a kindred with kininogen deficiency and Fitzgerald trait clotting defect. Evidence that high molecular weight kininogen and prekallikrein exist as a complex in normal human plasma', *Journal of Clinical Investigation, 60,* 571-83

Donovan, J.W. and Mihalyi, E. (1974) 'Conformation of fibrinogen: calorimetric evidence for a three-nodular structure', *Proceedings of the National Academy of Sciences, 71,* 4125-8

Doolittle, R.F. (1981) 'Fibrinogen and fibrin', in A.L. Bloom and D.P. Thomas (eds.) *Haemostasis and Thrombosis,* Churchill Livingstone, Edinburgh, pp. 163-91

Doolittle, R.F., Cassman, K.G., Chen, R., Sharp, J.J. and Wooding, G.L. (1972) 'Correlation of the mode of fibrin polymerization with the pattern of cross-linking', *Annals of the New York Academy of Sciences, 202,* 114-26

Doolittle, R.F., Cassman, K.G., Cottrell, B.A., Friezner, S.J., Hucko, J.T. and Takagi, T. (1977) 'Amino acid sequence studies on the α-chain of human fibrinogen. Characterization of eleven cyanogen bromide fragments', *Biochemistry, 16,* 1703-9

Doolittle, R.F., Goldbaum, D.M. and Doolittle, L.R. (1978) 'Designation of sequences involved in the "coiled coil" interdomainal connector in fibrinogen: construction of an atomic scale model', *Journal of Molecular Biology, 120,* 311-25

Doolittle, R.F., Watt, K.W.K., Cottrell, B.A., Strong, D.D. and Riley, M. (1979) 'The amino acid sequence of the α-chain of human fibrinogen', *Nature, 280,* 464-8

Doucet-de Bruine, M.H.M., Sixma, J.J., Over, J. and Beeser-Visser, N.H. (1978) 'Heterogeneity of human factor VIII. II. Characterization of forms of factor VIII binding to platelets in the presence of ristocetin', *Journal of Laboratory and Clinical Medicine, 92,* 96-107

Dunn, J.T., Silverberg, M. and Kaplan, A.P. (1982) 'The cleavage and formation of acvtivated human Hageman factor by autodigestion and by kallikrein', *Journal of Biological Chemistry, 257,* 1779-84

Egeberg, O. (1965) 'Inherited antithrombin deficiency causing thrombophilia', *Thrombosis et Diathesis Haemorrhagica, 13,* 516-30

Einarsson, R. and Andersson, L.-O. (1977) 'Binding of heparin to human antithrombin III as studied by measurements of tryptophan fluorescence', *Biochimica et Biophysica Acta, 490,* 104-11

Enfield, D.L., Ericsson, L.H., Walsh, K.A., Neurath, H. and Titani, K. (1975) 'Bovine factor X_1 (Stuart factor). Primary structure of the light chain', *Proceedings of the National Academy of Sciences, 72,* 16-9

Esmon, C.T. (1979) 'The subunit structure of thrombin-activated factor V. Isolation of activated factor V, separation of subunits, and reconstitution of biological activity', *Journal of Biological Chemistry, 254,* 964-73

Esmon, C.T. and Jackson, C.M. (1974) 'The conversion of prothrombin to thrombin IV. The function of the fragment 2 region during activation in the presence of factor V', *Journal of Biological Chemistry, 249,* 7791-7

Esmon, C.T., Stenflo, J., Suttie, J.W. and Jackson, C.M. (1976) 'A new vitamin K-dependent protein: A phospholipid-binding zymogen of a serine esterase', *Journal of Biological Chemistry, 251,* 3052-6

Esmon, N.L., Owen, W.G. and Esmon, C.T. (1982) 'Isolation of a membrane-bound cofactor for thrombin-catalyzed activation of protein C', *Journal of Biological Chemistry, 257,* 859-64

Fair, B.D., Saito, H., Ratnoff, O.D. and Rippon, W.B. (1977) 'Detection by fluorescence of structural changes accompanying the activation of Hageman factor (factor XII)', *Proceedings of the Society for Experimental Biology and Medicine, 155,* 199-202

Fass, D.N., Knutson, G.J. and Bowie, E.J.W. (1978) 'Porcine Willebrand factor: a population of multimer', *Journal of Laboratory and Clinical Medicine, 91,* 307-20

von Felton, A. and Straub, P.W. (1969) 'Coagulation studies of cord blood, with special reference to "fetal fibrinogen"', *Thrombosis et Diathesis Haemorrhagica, 22,* 273-80

Fenton, J.W., Fasco, M.J., Stackrow, A.B., Aronson, D.L., Young, A.M. and Finlayson, J.S. (1977) 'Human thrombins. Production, evaluation and properties of α-thrombin', *Journal of Biological Chemistry, 252,* 3587-98

Ferry, J.D. (1952) 'The mechanism of polymerization of fibrin', *Proceedings of the National Academy of Sciences, 38,* 566-9

Fish, W. and Björk, I. (1979a) 'Production by thrombin of a proteolytically modified form of antithrombin and release of the same form from the antithrombin-thrombin complex', *Thrombosis and Haemostasis, 42,* 129

Fish, W.W. and Björk, I. (1979b) 'Release of a two-chain form of antithrombin from the antithrombin-thrombin complex', *European Journal of Biochemistry, 101,* 31-8

Forbes, C.D., Pensky, J. and Ratnoff, O.D. (1970) 'Inhibition of activated Hageman factor and activated plasma thromboplastin antecedent by purified Cl inactivator', *Journal of Laboratory and Clinical Medicine, 76,* 809-15

Fowler, W.E. and Erickson, H.P. (1979) 'The trinodular structure of fibrinogen: confirmation by both shadowing and negative stain electron microscopy', *Journal of Molecular Biology, 134,* 241-9

Fowler, W.E., Fretto, L.J., Erickson, H.P. and McKee, P.A. (1980) 'Electron microscopy of plasmic fragments of human fibrinogen as related to trinodular structure of the intact molecule', *Journal of Clinical Investigation, 66*, 50-6

Franks, J.J., Kirsch, R.E., Frith, L.O., Purves, I.R., Franks, W.T., Franks, J.A., Mason, P. and Saunders, S.J. (1981) 'Effect of fibrinogenolytic products D and E on fibrinogen and albumin synthesis in the rat', *Journal of Clinical Investigation, 67*, 575-80

Franzèn, L.-E., Svensson, S. and Larm, O. (1980) 'Structural studies on the carbohydrate portion of human antithrombin III', *Journal of Biological Chemistry, 255*, 5090-3

Fretto, L.J., Ferguson, E.W., Steinman, H.M. and McKee, P.A. (1978) 'Localization of the α-chain cross-link acceptor sites of human fibrin', *Journal of Biological Chemistry, 253*, 2184-95

Fryklund, L., Borg, H. and Andersson, L.-O. (1976) 'Amino-terminal sequence of human factor IX: presence of γ-carboxyl glutamic acid residues', *FEBS Letters, 65*, 187-9

Fujikawa, K., Legaz, M.E. and Davie, E.W. (1972) 'Bovine factors X_1 and X_2 (Stuart factor). Isolation and characterization', *Biochemistry, 11*, 4882-91

Fujikawa, K., Thompson, A.R., Legaz, M.E., Meyer, R.G. and Davie, E.W. (1973) 'Isolation and characterization of bovine factor IX (Christmas factor)', *Biochemistry, 12*, 4938-45

Fujikawa, K., Coan, M.H., Legaz, M.E. and Davie, E.W. (1974a) 'The mechanism of activation of bovine factor X (Stuart factor) by intrinsic and extrinsic pathways', *Biochemistry, 13*, 5290-9

Fujikawa, K., Legaz, M.E., Kato, H. and Davie, E.W. (1974b) 'The mechanism of activation of bovine factor IX (Christmas factor) by bovine factor XIa (activated plasma thromboplastin antecedent)' *Biochemistry, 13*, 4508-16

Fujikawa, K., Walsh, K.A. and Davie, E.W. (1977a) 'Isolation and characterization of bovine factor XII (Hageman factor)', *Biochemistry, 16*, 2270-8

Fujikawa, K., Kurachi, K. and Davie, E.W. (1977b) 'Characterization of bovine factor XIIa (activated Hageman factor)', *Biochemistry, 16*, 4182-8

Fujikawa, K., Heimark, R.L., Kurachi, K. and Davie, E.W. (1980) 'Activation of bovine factor XII (Hageman factor) by plasma kallikrein', *Biochemistry, 19*, 1322-30

Fuller, G.M. and Ritchie, D.G. (1981) 'A regulatory pathway for fibrinogen biosynthesis involving an indirect feedback loop', *Annals of the New York Academy of Sciences, 389*, 308-22

Gaffney, P.J. (1972) 'Localisation of carbohydrate in the subunits of human fibrinogen and its plasmin-induced fragments', *Biochimica et Biophysica Acta, 263*, 453-8

Galanakis, D.R. and Mosesson, M.W. (1977) 'Comparative studies on fetal fibrinogen from full term and premature infants', *Thrombosis and Haemostasis, 38*, 102

Galanakis, D.R. and Mosesson, M.W, (1979) 'Correction of the delayed fibrin aggregation of fetal fibrinogen by partial removal of sialic acid', *Thrombosis and Haemostasis, 42*, 79

Gallimore, M.J., Fareid, E. and Stormorken, H. (1978) 'The purification of human plasma kallikrein with weak plasminogen activator activity', *Thrombosis Research, 12*, 409-20

Gallimore, M.J., Amundsen, E., Larsbraaten, M., Lyngaas, K. and Fareid, E. (1979) 'Studies on plasma inhibitors of plasma kallikrein using chromogenic peptide substrate assays', *Thrombosis Research, 16*, 695-703

Gallin, J.I. and Kaplan, A.P. (1974) 'Mononuclear cell chemotactic activity of kallikrein and plasminogen activator and its inhibition by CĪ inhibitor and

α_2-macroglobulin', *Journal of Immunology, 113*, 1928-34

Ganrot, P.O. and Niléhn, J.E. (1968) 'Plasma prothrombin during treatment with Dicoumarol II. Demonstration of an abnormal prothrombin fraction', *Scandinavian Journal of Clinical and Laboratory Investigation, 22*, 27-8

Gans, H. and Tan, B.H. (1967) 'α_1-antitrypsin, an inhibitor for thrombin and plasmin', *Clinica Chimica Acta, 17*, 111-7

Gedde-Dahl, T.W., Jeremic, M. and Weisert, O. (1975) 'Factor V (proaccelerin) concentration in 1016 blood donors. The effects of age, sex, and ABO blood groups', *Scandinavian Journal of Clinical and Laboratory Investigation, 35*, 25-30

Giddings, J.C., Shearn, S.A.M. and Bloom, A.L. (1975) 'The immunological localization of factor V in human tissue', *British Journal of Haematology, 29*, 57-65

Gigli, I., Mason, J.W., Colman, R.W. and Austen, K.F. (1970) 'Interaction of plasma kallikrein with Cl-inactivator', *Journal of Immunology, 104*, 574-81

Ginsberg, M.H., Jaques, B., Cochrane, C.G. and Griffin, J.H. (1980) 'Urate crystal-dependent cleavage of Hageman factor in human plasma and in synovial fluid', *Journal of Laboratory and Clinical Medicine, 95*, 497-506

Gjønnaess, H. (1972) 'Cold promoted activation of factor VIII. I. Evidence for the existence of an activator', *Thrombosis et Diathesis Haemorrhagica, 28*, 155-68

Godal, H.C., Rygh, M. and Laake, K. (1974) 'Progressive inactivation of purified factor VII by heparin and antithrombin III', *Thrombosis Research, 5*, 773-5

Green, A.J. and Ratnoff, O.D. (1974) 'Elevated antihemophilic factor (AHF, factor VIII) procoagulant activity and AHF-like antigen in alcoholic cirrhosis of the liver', *Journal of Laboratory and Clinical Medicine, 83*, 189-97

Greenberg, C.S. and Shulman, M.A, (1982) 'The zymogen forms of blood coagulation factor XII binds specifically to fibrinogen', *Journal of Biological Chemistry, 257*, 6096-101

Griffin, J.H. (1977) 'New hypothesis for the molecular mechanism of surface-dependent activation of Hageman factor (factor XII)', *Thrombosis and Haemostasis, 38*, 50

Griffin, J.H. and Cochrane, C.G. (1976) 'Mechanisms for the involvement of high molecular weight kininogen in surface-dependent reactions of Hageman factor', *Proceedings of the National Academy of Sciences, 73*, 2554-8

Guillin, M.-C. and Menache, D. (1973) 'Fetal fibrinogen and fibrinogen Paris I: comparative fibrin monomers aggregation studies', *Thrombosis Research, 3*, 117-35

Hall, C. and Slayter, H. (1959) 'The fibrinogen molecule: its size, shape, and mode of polymerization', *Journal of Biophysical and Biochemical Cytology, 5*, 11-5

Han, Y.N., Komiya, M., Iwanaga, S. and Suzukui, Y. (1975a) 'Studies on the primary structure of bovine high-molecular weight kininogen. Amino acid sequence of a fragment ("Histidine-rich peptide") released by plasma kallikrein', *Journal of Biochemistry, 77*, 55-68

Han, Y.N., Komiya, M., Kato, H., Iwanaga, S. and Suzuki, T. (1975b) 'Primary structure of bovine high molecular weight kininogen: chemical compositions of kinin-free kininogen and peptide fragments released by plasma kallikrein', *FEBS Letters, 57*, 254-8

Han, Y.N., Kato, H., Iwanaga, S. and Suzuki, T. (1976) 'Bovine plasma high molecular weight kininogen: the amino acid sequence of fragment 1 (glycopeptide) released by the action of plasma kallikrein and its location in the precursor protein', *FEBS Letters, 63*, 197-200

Harmon, J., Jamieson, G.A. and Rock, G. (1981) 'Determination of molecular size of VIII:C in whole plasma by electron irradiation', *Thrombosis and Haemostasis, 46*, 255

Harpel, P.C,. (1970) 'Human plasma alpha$_2$-macroglobulin. An inhibitor of plasma kallikrein', *Journal of Experimental Medicine, 132,* 329-52

Harpel, P.C. (1972) 'Studies on the interaction between collagen and a plasma kallikrein-like activity. Evidence for a surface-active enzyme system', *Journal of Clinical Investigation, 51,* 1813-22

Harpel, P.C. (1973) 'Studies on human plasma α_2-macroglobulin-enzyme interactions', *Journal of Experimental Medicine, 138,* 508-21

Hathaway, W.E., Belhasen, L.P. and Hathaway, H.S. (1965) 'Evidence for a new thromboplastin factor: Case report, coagulation studies and physicochemical studies', *Blood, 26,* 521-32

Hatton, M.W.C., Berry, L.R. and Regoeczi, E. (1978) 'Inhibition of thrombin by antithrombin III in the presence of certain glycosaminoglycans found in the mammalian aorta', *Thrombosis Research, 13,* 655-70

Heber, H., Geiger, R. and Heimburger, N. (1978) 'Human plasma kallikrein: purification, enzyme characterization and quantitative determination in plasma', *Hoppe-Seyler's Zeitschrift fur Physiologische Chemie, 359,* 659-69

Heck, L.W. and Kaplan, A.P. (1974) 'Substrates of Hageman factor. I. Isolation and characterization of human factor XI (PTA) and inhibition of the activated enzyme by α_1-antitrypsin', *Journal of Experimental Medicine, 140,* 1615-30

Hedner, U. (1979) 'Inhibition of Hageman factor by a fibrinolytic inhibitor', *Thrombosis and Haemostasis, 42,* 262

Hedner, U. and Martinsson, G. (1978) 'Inhibition of activated Hageman factor (factor XIIa) by an inhibitor of the plasminogen activation (PA inhibitor)', *Thrombosis Research, 12,* 1015-23

Heimark, R.L. and Davie, E.W. (1979) 'Isolation and characterization of bovine plasma prekallikrein (Fletcher factor)', *Biochemistry, 18,* 5743-50

Heimark, R.L., Kurachi, K., Fujikawa, K. and Davie, E.W. (1980) 'Surface activation of blood coagulation, fibrinolysis and kinin formation', *Nature, 286,* 456-60

Heimburger, N., Haupt, H. and Schwick, H.G. (1971) 'Proteinase inhibitors of human plasma', in H. Fritz and H. Tschesche (eds.) *Proceedings of the International Research Conference on Proteinase Inhibitors,* Walter de Gruyter, Berlin and New York, pp. 1-22

Hemker, H.C. and Muller, A.D. (1968) 'Kinetic aspects of the interaction of blood-clotting enzymes. VI. Localization of the site of blood-coagulation inhibition by the protein induced by vitamin K absence (PIVKA)', *Thrombosis et Diathesis Haemorrhagica, 20,* 78-87

Hemker, H.C., Veltkamp, J.J., Hensen, A. and Loeliger, E.A. (1963) 'Nature of prothrombin biosynthesis: preprothrombinaemia in vitamin K-deficiency', *Nature, 200,* 589-90

Henschen, A. and Lottspeich, F. (1977) 'Amino acid sequence of human fibrin. Preliminary note on the completion of the β chain sequence', *Hoppe-Seyler's Zeitschrift fur Physiologische Chemie, 358,* 1643-6

Hessel, B., Makino, M., Iwanaga, S. and Blombäck, B. (1979) 'Primary structure of human fibrinogen and fibrin. Structural studies on NH$_2$-terminal part of Bβ chain', *European Journal of Biochemistry, 98,* 521-34

Hjort, P., Rapaport, S.I. and Owren, P.A. (1955) 'Evidence that platelet accelerator (platelet factor I) is adsorbed plasma proaccelerin', *Blood, 10,* 1139-50

Højrup, P., Roepstorff, P. and Petersen, T.E. (1982) 'Amino-acid sequence of the vitamin-K-dependent part of protein Z', *European Journal of Biochemistry, 126,* 343-8

Holmer, E., Söderström, G. and Andersson, L.-O. (1979) 'Studies on the mechanism of the rate-enhancing effect of heparin on the thrombin-antithrombin III reaction', *European Journal of Biochemistry, 93*, 1-5

Höök, C., Björk, I., Hopwood, J. and Lindahl, U. (1976) 'Anticoagulant activity of heparin: separation of high-activity and low-activity heparin species by affinity chromatography on immobilised antithrombin', *FEBS Letters, 66*, 90-3

Hopwood, J., Höök, M., Linker, A. and Lindahl, U. (1976) 'Anticoagulant activity of heparin: isolation of antithrombin-binding sites', *FEBS Letters, 69*, 51-4

Horowitz, H.I. and Fujimoto, M.M. (1965) 'Association of factors XI and XII with blood platelets', *Proceedings of the Society for Experimental Biology and Medicine, 119*, 487-92

Hougie, C., Denson, K.W. and Biggs, R. (1967) 'A study of the reaction product of factor VIII and factor IX by gel filtration', *Thrombosis et Diathesis Haemorrhagica, 18*, 211-22

Howard, J.B. and Nelsestuen, G.L. (1975) 'Isolation and characterization of vitamin K-dependent region of bovine clotting factor X', *Proceedings of the National Academy of Sciences, 72*, 1281-5

Hoyer, L.W. and Trabold, N.C. (1981) 'The effect of thrombin on human factor VIII. Cleavage of the factor VIII procoagulant protein during activation', *Journal of Laboratory and Clinical Medicine, 97*, 50-64

Hultin, M.B. and Nemerson, Y. (1978) 'Activation of factor X by factors IXa and VIII; a specific assay for factor IXa in the presence of thrombin-activated factor VIII', *Blood, 52*, 928-40

Hurst, R.E., Menter, J.M., West, S.S., Settine, J.M. and Coyne, E.H. (1979) 'Structural basis for the anticoagulant activity of heparin. I. Relationship to the number of charged groups', *Biochemistry, 18*, 4283-7

Hvatum, M. and Prydz, H. (1969) 'Studies on tissue thromboplastin — its splitting into two separable parts', *Thrombosis et Diathesis Haemorrhagica, 21*, 217-22

Ikari, N., Sugo, T., Fujii, S. and Iwanaga, S. (1981) 'The role of bovine high-molecular weight (HMW) kininogen in contact-mediated activation of bovine factor XII: interaction of HMW kininogen with kaolin and plasma prekallikrein', *Journal of Biochemistry, 89*, 1699-709

Ittyerah, T.R., Weidner, N., Wochner, R.D. and Sherman, L.A. (1979) 'Effect of fibrin degradation products and thrombin on fibrinogen synthesis', *British Journal of Haematology, 43*, 661-8

Iwamoto, M. and Abiko, Y. (1970) 'Plasminogen-plasmin system. VI. Preparation of α_2-macroglobulin antiplasmin from human plasma', *Biochimica et Biophysica Acta, 214*, 402-10

Jackson, C.M. (1972) 'Characterisation of two glycoprotein variants of bovine factor X and demonstration that the factor X zymogen contains two polypeptide chains', *Biochemistry, 11*, 4873-81

Jackson, C.M. and Hanahan, D.J. (1968) 'Studies on bovine factor X. II. Observations on some alterations in zone electrophoretic and chromatographic behavior occurring during purification', *Biochemistry, 7*, 4506-17

Jaffe, E.A. and Nachman, R.L. (1975) 'Subunit structure of factor VIII antigen synthesized by cultured human endothelial cells', *Journal of Clinical Investigation, 56*, 698-702

Jaffe, E.A., Hoyer, L.W. and Nachman, R.L. (1973) 'Synthesis of antihemophilic factor antigen by cultured human endothelial cells', *Journal of Clinical Investigation, 52*, 2757-64

Jaffe, E.A., Hoyer, L.W. and Nachman, R.L. (1974) 'Synthesis of von Willebrand factor by cultured human endothelial cells', *Proceedings of the National Academy of Sciences, 71*, 1906-9

Jeejeebhoy, K.N., Bruce-Robertson, A., Sodtke, U. and Foley, M. (1970) 'The effect of growth hormone on fibrinogen synthesis', *Biochemical Journal, 119*, 243-9

Jeejeebhoy, K.N., Ho, J., Greenberg, G.R., Phillips, M.J., Bruce-Robertson, A. and Sodtke, U. (1975) 'Albumin, fibrinogen and transferrin synthesis in isolated rat hepatocyte suspensions', *Biochemical Journal, 146*, 141-55

Jesty, J. (1978) 'The inhibition of activated bovine coagulation factors X and VII by antithrombin III', *Archives of Biochemistry and Biophysics, 185*, 165-73

Jesty, J. (1979) 'Dissociation of complexes and their derivatives formed during inhibition of bovine thrombin and activated factor X by antithrombin III', *Journal of Biological Chemistry, 254*, 1044-9

Jesty, J. and Nemerson, Y. (1974) 'Purification of factor VII from bovine plasma. Reaction with tissue factor and activation of factor X', *Journal of Biological Chemistry, 249*, 509-15

Jesty, J. and Silverberg, S.A. (1979) 'Kinetics of the tissue factor-dependent activation of coagulation factors IX and X in a bovine plasma system', *Journal of Biological Chemistry, 254*, 12337-45

Jesty, J., Spencer, A.K. and Nemerson, Y. (1974) 'The mechanism of activation of factor X. Kinetic control of alternative pathways leading to the formation of activated factor X', *Journal of Biological Chemistry, 249*, 5614-22

Jesty, J., Spencer, A.K., Nakashima, Y., Nemerson, Y and Konigsberg, W. (1975) 'The activation of coagulation factor X. Identity of cleavage sites in the alternative activation pathways and characterization of the COOH-terminal peptide', *Journal of Biological Chemistry, 250*, 4497-504

Johnson, E.A. and Mulloy, B. (1976) 'The molecular weight range of mucosal heparins', *Carbohydrate Research, 1951*, 119-127

Jones, J.M., Creeth, J.M. and Kekwick, R.A. (1972) 'Thiol reduction of human α_2-macroglobulin. The subunit structure', *Biochemical Journal, 127*, 187-97

Jordan, R., Beeler, D. and Rosenberg, R. (1979) 'Fractionation of low molecular weight heparin species and their interaction with antithrombin', *Journal of Biological Chemistry, 254*, 2902-13

Jordan, R.E., Favreau, L.V., Braswell, E.H. and Rosenberg, R.D. (1982) 'Heparin with two binding sites for antithrombin or platelet factor 4', *Journal of Biological Chemistry, 257*, 400-6

Josso, F. and Beguin, S. (1981) 'Changes in the antithrombin III activity at the interface plasma-phospholipids', *Thrombosis and Haemostasis, 46*, 285

Kalousek, F., Konigsberg, W. and Nemerson, Y. (1975) 'Activation of factor IX by activated factor X: a link between the extrinsic and intrinsic coagulation systems', *FEBS Letters, 50*, 382-5

Kampschmidt, R.F. and Upchurch, H.F. (1974) 'Effect of leucocyte endogenous mediator on plasma fibrinogen and haptoglobin', *Proceedings of the Society for Experimental Biology and Medicine, 146*, 904-7

Kampschmidt, R.F., Upchurch, H.F. and Pulliam, L.A. (1981) 'Characterization of a leukocyte-derived endogenous mediator responsible for increased plasma fibrinogen', *Annals of the New York Academy of Sciences, 389*, 338-51

Kane, W.H., Lindhout, M.J., Jackson, C.M. and Majerus, P.W. (1980) 'Factor Va-dependent binding of factor Xa to human platelets', *Journal of Biological Chemistry, 255*, 1170-4

Kato, H., Sugo, T., Ikari, N., Hashimoto, N., Iwanaga, S. and Fujii, S. (1979) 'Role of bovine HMW in kininogen in contact-mediated activation of factor XII: Demonstration of a nicked form, "active kininogen", with maximal co-factor activity by limited proteolysis', *Thrombosis and Haemostasis, 42*, 262

Kerbiriou, D.M. and Griffin, J.H. (1979) 'Human high molecular weight kininogen. Studies of structure-function relationships and of proteolysis of the molecule occurring during contact activation of plasma', *Journal of Biological Chemistry, 254*, 12020-7

Kerbiriou, D.H., Bouma, B.N. and Griffin, J.H. (1980) 'Immunochemical studies of human high molecular weight kininogen and of its complexes with plasma prekallikrein or kallikrein', *Journal of Biological Chemistry, 255*, 3952-8

Kessler, C.M. and Bell, W.R. (1979a) 'Regulation of fibrinogen biosynthesis: effect of fibrin degradation products, low-molecular-weight peptides of fibrinogenolysis, and fibrinopeptides A and B', *Journal of Laboratory and Clinical Medicine, 93*, 758-67

Kessler, C.M. and Bell, W.R. (1979b) 'The effect of homologous thrombin and fibrinogen degradation products on fibrinogen synthesis in rabbits', *Journal of Laboratory and Clinical Medicine, 93*, 768-82

Kiesselbach, T.H. and Wagner, R.H. (1972) 'Demonstration of factor XIII in human megakaryocytes by a fluorescent antibody technique', *Annals of the New York Academy of Sciences, 202*, 318-28

Kisiel, W. (1979) 'Human plasma protein C. Isolation, characterization, and mechanism of activation by α-thrombin', *Journal of Clinical Investigation, 64*, 761-9

Kisiel, W. and Davie, E.W. (1975) 'Isolation and characterization of bovine factor VII', *Biochemistry, 14*, 4928-34

Kisiel, W. and McMullen, B.A. (1981) 'Isolation and characterization of human factor VIIa', *Thrombosis Research, 22*, 375-80

Kisiel, W., Ericsson, L.H. and Davie, E.W. (1976) 'Proteolytic activation of protein C from bovine plasma', *Biochemistry, 15*, 4893-900

Kisiel, W., Fujikawa, K. and Davie, E.W. (1977a) 'Activation of bovine factor VII (proconvertin) by factor XIIa (activated Hageman factor)', *Biochemistry, 16*, 4189-94

Kisiel, W., Canfield, W.M., Ericsson, L.H. and Davie, E.W. (1977b) 'Anticoagulant properties of bovine plasma protein C following activation by thrombin', *Biochemistry, 16*, 5824-31

Koide, T. (1979) 'Isolation and characterization of antithrombin III from human, porcine and rabbit plasma, and rat serum', *Journal of Biochemistry, 86*, 1841-50

Koide, T., Kato, H. and Davie, E.W. (1977) 'Isolation and characterization of bovine factor XI (plasma thromboplastin antecedent)', *Biochemistry, 16*, 2279-86

Koide, T., Odani, S. and Ono, T. (1982) 'The *N*-terminal sequence of human plasma histidine-rich glycoprotein homologous to antithrombin with high affinity for heparin', *FEBS Letters, 141*, 222-4

Komiya, M., Kato, H. and Suzuki, T. (1974) 'Bovine plasma kininogens I. Further purification of high molecular weight kininogen and its physicochemical properties', *Journal of Biochemistry, 76*, 811-22

Kosow, D.P. and Orthner, C.L. (1979) 'Kinetics of the activation of human prothrombin by human coagulation factor Xa. Initial rate studies in the presence of Ca^{2+} and phospholipid', *Journal of Biological Chemistry, 254*, 9448-52

Koutts, J., Walsh, P.N., Plow, E.F., Fenton, J.W., Bouma, B.N. and Zimmerman, T.S. (1978) 'Active release of human platelet factor VIII-related antigen by adenosine diphosphate, collagen, and thrombin', *Journal of Clinical Investigation, 62*, 1255-63

Kraemer, P.M. (1977) 'Heparin releases heparan sulphate from the cell surface', *Biochemical and Biophysical Research Communications, 78*, 1334-40

Kropatkin, M.L. and Izak, G. (1968) 'Studies on the hypercoagulable state', *Thrombosis et Diathesis Haemorrhagica, 19*, 547-55

Kudryk, B.J., Collen, D., Woods, K.R. and Blombäck, B. (1974) 'Evidence for localization of polymerization sites in fibrinogen', *Journal of Biological Chemistry, 249*, 3322-5

126

Kueppers, F. and Fallat, R.J. (1969) 'α_1-antitrypsin deficiency: a defect in protein synthesis', *Clinica Chimica Acta, 24*, 401-3

Kurachi, K. and Davie, E.W, (1977) 'Activation of human factor XI (plasma thromboplastin antecedent) by factor XIIa (activated Hageman factor)', *Biochemistry, 16*, 5831-9

Kurachi, K., Schmer, G., Herodson, M.A., Teller, D.C. and Davie, E.W. (1976a) 'Characterization of human, bovine, and horse antithrombin III', *Biochemistry, 15*, 368-73

Kurachi, K., Fujikawa, K., Schmer, G. and Davie, E.W. (1976b) 'Inhibition of bovine factor IXa and factor Xaβ by antithrombin III', *Biochemistry, 15*, 373-7

Kurachi, K., Fujikawa, K. and Davie, E.W. (1980) 'Mechanism of activation of bovine factor XI by factor XII and factor XIIa', *Biochemistry, 19*, 1330-8

Laake, K. and Ellingsen, R. (1974) 'Purification and some characteristics of factor VII in human citrated plasma, glass-activated serum, and cold-activated plasma', *Thrombosis Research, 5*, 539-56

Laake, K. and Østerud, B. (1974) 'Activation of purified plasma VII by human plasmin, plasma kallikrein, and activated components of the human intrinsic blood coagulation system', *Thrombosis Research, 5*, 759-72

Laki, K. and Gladner, J.A. (1964) 'Chemistry and physiology of the fibrinogen-fibrin transition', *Physiological Reviews, 44*, 127-60

Lanchantin, G.F., Plesset, M.L., Friedmann, J.A. and Hart, D.W. (1966) 'Dissociation of esterolytic and clotting activities of thrombin by trypsin-binding macroglobulin', *Proceedings of the Society of Experimental Biology and Medicine, 121*, 444-9

Lane, J.L., Bird, P. and Rizza, C.R. (1975) 'A new assay for the measurement of total progressive antithrombin', *British Journal of Haematology, 30*, 103-115

Larrieu, M.J., Soulier, J.P. and Minkowski, A. (1952) 'Le sang du cordon ombilical: Etude complète de sa coagulabilité, comparison avec le sang maternal', *Etudes Néonatales, 1*, 39-60

Larrieu, M.J., Rigollot, C. and Marder, V.J. (1972) 'Comparative effects of fibrinogen degradation fragments D and E on coagulation', *British Journal of Haematology, 22*, 719-33

Larrieu, M.J., Dray, L. and Ardaillou, N. (1975) 'Biological effects of fibrinogen-fibrin degradation products', *Thrombosis et Diatheses Haemorrhagica, 34*, 686-92

Latallo, Z.S., Budzynski, A.Z., Lipinski, B. and Kowalski, E. (1964) 'Inhibition of thrombin and of fibrin polymerization, two activities derived from plasmin-digested fibrinogen', *Nature, 203*, 1184-5

Lau, H.K. and Rosenberg, R.D. (1980) 'The isolation and characterization of a specific antibody population directed against the thrombin-antithrombin complex', *Journal of Biological Chemistry, 255*, 5885-93

Laudano, A.P. and Doolittle, R.F. (1978) 'Synthetic peptide derivatives that bind to fibrinogen and prevent the polymerization of fibrin monomers', *Proceedings of the National Academy of Sciences, 75*, 3085-9

Laudano, A.P. and Doolittle, R.F. (1980) 'Studies on synthetic peptides that bind to fibrinogen and prevent fibrin polymerization. Structural requirements, number of binding sites, and species differences', *Biochemistry, 19*, 1013-9

Laurent, T.C. and Blombäck, B. (1958) 'On the significance of the release of two different peptides from fibrinogen during clotting', *Acta Chemica Scandinavica, 12*, 137-46

Laurent, T.C., Tengblad, A., Thunberg, L., Höök, M. and Lindahl, U. (1978) 'On the molecular weight dependence of the anticoagulant activity of heparin', *Biochemical Journal, 175*, 691-701

Learned, L.A., Bloom, J.W. and Hunter, M.J. (1976) 'The antithrombin activity of α-1-protease inhibitor: the antitrypsin activity of antithrombin III', *Thrombosis Research, 8*, 99-109

Lee, A.K.Y., Chan, V. and Chan, T.K. (1979) 'The identification and localization of antithrombin III in human tissues', *Thrombosis Research, 14*, 209-17

Lee, S.Y. and Chung, S.I. (1976) 'Biosynthesis and degradation of plasma protransglutaminase (factor XIII)', *Federation Proceedings, 35*, 1486

Legaz, M.E., Schmer, G., Counts, R.B. and Davie, E.W. (1973) 'Isolation and characterization of human factor VIII (antihemophilic factor)', *Journal of Biological Chemistry, 248*, 3946-55

Liang, T.J. and Grieninger, G. (1981) 'Direct effect of insulin on the synthesis of specific plasma proteins: Biphasic response of hepatocytes cultured in serum- and hormone-free medium', *Proceedings of the National Academy of Sciences, 78*, 6972-6

Lindahl, U., Bäckström, G., Höök, M., Thunberg, L., Fransson, L.-A. and Linker, A. (1979) 'Structure of the antithrombin-binding site in heparin', *Proceedings of the National Academy of Sciences, 76*, 3198-202

Link, K.P. (1959) 'The discovery of dicumarol and its sequels', *Circulation, 19*, 97-107

Lipscomb, M.S. and Walsh, P.N. (1979) 'Human platelets and factor XI. Localization in platelet membranes of factor XI-like activity and its functional distinction from plasma factor XI', *Journal of Clinical Investigation, 63*, 1006-14

Liu, D.T.H. and McCoy, L.E. (1975) 'Tissue extract thromboplastin: Quantitation, fractionation and characterization of protein components', *Thrombosis Research, 7*, 199-211

Lollar, P., Hoak, T.C. and Owen, W.G. (1980) 'Binding of thrombin to cultured human endothelial cells. Nonequilibrium aspects', *Journal of Biological Chemistry, 255*, 10279-83

Loskutoff, D.J. (1979) 'Effect of thrombin on the fibrinolytic activity of cultured bovine endothelial cells', *Journal of Clinical Investigation, 64*, 329-32

Losowsky, M.S. and Walls, W.D. (1979) 'Mechanisms of acquired defects of the stabilization of fibrin', *British Journal of Haematology, 20*, 377-83

Lottspeich, F. and Henschen, A. (1977a) 'Amino acid sequence of human fibrin. Preliminary note on the completion of the γ-chain sequence', *Hoppe-Seyler's Zeitschrift fur Physiologische Chemie, 358*, 935-8

Lottspeich, F. and Henschen, A. (1977b) 'Amino acid sequence of human fibrin. Preliminary note on three internal peptides obtained by cyanogen bromide cleavage of the β-chain', *Hoppe-Seyler's Zeitschrift fur Physiologische Chemie, 358*, 1639-42

Lundblad, R.L., Noyes, C.M., Mann, K.G. and Kingdon, H.S. (1979) 'The covalent differences between bovine α- and β-thrombin', *Journal of Biological Chemistry, 254*, 8524-8

Lutcher, C.L. (1976) 'Reid trait: A new expression of high molecular weight kininogen (HMW kininogen) deficiency', *Clinical Research, 24*, 47

MacGregor, I.R., Lane, D.A. and Kakkar, V.V. (1979) 'Evidence for a plasma inhibitor of the heparin accelerated inhibition of factor Xa by antithrombin III', *Biochimica et Biophysica Acta, 586*, 584-93

Machovich, R. (1975) 'Mechanism of action of heparin through thrombin on blood coagulation', *Biochimica et Biophysica Acta, 412*, 13-7

Machovich, R., Blasko, G. and Palos, L.A. (1975) 'Action of heparin on thrombin-antithrombin reaction', *Biochimica et Biophysica Acta, 379*, 193-200

Mackie, M., Bennett, B., Ogston, D. and Douglas, A.S. (1978) 'Familial thrombosis: inherited deficiency of antithrombin III', *British Medical Journal, 1*, 136-8

Magnusson, S., Sottrup-Jensen, L. and Petersen, T.E. (1974) 'Primary structure of the vitamin K-dependent part of prothrombin', *FEBS Letters, 44*, 189-93

Magnusson, S., Petersen, T.E., Sottrup-Jensen, L. and Claeys, H. (1975) 'Complete primary structure of prothrombin: Isolation, structure and reactivity of ten carboxylated glutamic acid residues and regulation of prothrombin activation by thrombin', in *Proteases and Biological Control*, Vol. 2, Cold Spring Harbor, pp. 123-49

Mandle, R. and Kaplan, A.P. (1977) 'Hageman factor substrates. Human plasma prekallikrein: mechanism of activation by Hageman factor and participation in Hageman factor-dependent fibrinolysis', *Journal of Biological Chemistry, 252*, 6097-104

Mandle, R., Colman, R.W. and Kaplan, A.P. (1976) 'Identification of prekallikrein and HMW kininogen as circulating complex in human plasma', *Proceedings of the National Academy of Sciences, 73*, 4179-83

Marciniak, E. (1970) 'Coagulation inhibitor elicited by thrombin', *Science, 170*, 452-3

Marciniak, E. (1972) 'Inhibitor of human blood coagulation elicited by thrombin', *Journal of Laboratory and Clinical Medicine, 79*, 924-34

Marciniak, E. (1973) 'Factor-Xa inactivation by antithrombin III: evidence for biological stabilization of factor Xa by factor V-phospholipid complex', *British Journal of Haematology, 24*, 391-400

Marciniak, E. (1974) 'Anticoagulant properties of low density lipoproteins', *Federation Proceedings, 33*, 217

Marciniak, E. and Tsukamura, S. (1972) 'Two progressive inhibitors of factor Xa in human blood', *British Journal of Haematology, 22*, 341-51

Marciniak, E., Farley, C.H. and DeSimone, P.A. (1974) 'Familial thrombosis due to antithrombin III deficiency', *Blood, 43*, 219-31

Marder, V.J. and Shulman, N.R. (1969) 'High molecular weight derivatives of human fibrinogen produced by plasmin. II. Mechanisms of their anticoagulant activity', *Journal of Biological Chemistry, 244*, 2120-4

Margolis, J. (1958) 'Activation of plasma by contact with glass: evidence for a common reaction which releases plasma kinin and initiates coagulation', *Journal of Physiology, 144*, 1-22

Marlar, R.A. and Griffin, J.H. (1980) 'Deficiency of protein C inhibitor in combined factor V/VIII deficiency disease', *Journal of Clinical Investigation, 66*, 1186-9

Matheson, N.R. and Travis, J. (1976) 'Inactivation of human thrombin in the presence of human α_1-proteinase inhibitor', *Biochemical Journal, 159*, 495-502

Matsuo, T., Ohki, Y., Kondo, S. and Matsuo, O. (1979) 'Familial antithrombin III deficiency in a Japanese family', *Thrombosis Research, 16*, 815-23

Mattock, P. and Esnouf, M.P. (1973) 'A form of bovine factor X with a single polypeptide chain', *Nature New Biology, 242*, 90-2

McConnell, D.J. (1972) 'Inhibitors of kallikrein in human plasma', *Journal of Clinical Investigation, 51*, 1611-23

McDonagh, J., Messel, H., McDonagh, R.P., Murano, G. and Blombäck, B. (1972) 'Molecular weight analysis of fibrinogen and fibrin chains by an improved sodium dodecyl sulphate gel electrophoresis method', *Biochimica et Biophysica Acta, 257*, 135-42

McMillin, C.R., Saito, H., Ratnoff, O.D. and Walton, A.G. (1974) 'The secondary structure of human Hageman factor (factor XII) and its aeteration by activating agents', *Journal of Clinical Investigation, 54*, 1312-22

Meier, H.L., Pierce, J.V., Colman, R.W. and Kaplan, A.P. (1977) 'Activation and function of human Hageman factor. The role of high molecular weight · kininogen and prekallikrein', *Journal of Clinical Investigation, 60*, 18-31

Miletich, J.P., Jackson, C.M. and Majerus, P.W. (1977) 'Interaction of coagulation factor Xa with human platelets', *Proceedings of the National Academy of Sciences, 74*, 4033-6

Miletich, J.P., Jackson, C.M. and Majerus, P.W. (1978a) 'Properties of the factor Xa binding site on human platelets', *Journal of Biological Chemistry, 253*, 6908-16

Miletich, J.P., Majerus, D.W. and Majerus, P.W. (1978b) 'Patients with congenital factor V deficiency have decreased factor Xa binding sites on their platelets', *Journal of Clinical Investigation, 62*, 824-31

Miller, L.L. and Bale, W.F. (1954) 'Synthesis of all plasma protein fractions except gamma globulins by the liver', *Journal of Experimental Medicine, 99*, 125-32

Miller, L.L., Bly, C.G., Watson, M.L. and Bale, W.F. (1951) 'Dominant role of liver in plasma protein synthesis; direct study of isolated perfused rat liver with the aid of lysine-ε-C^{14}', *Journal of Experimental Medicine, 94*, 431-53

Miller-Andersson, M., Andersson, L.-O. and Borg, H. (1973) 'Isolation and characterization of a new immediate antithrombin from human plasma', *Abstracts IVth International Congress on Thrombosis and Haemostasis*, Vienna, p. 216

Miller-Andersson, M., Borg, H. and Andersson, L.-O. (1974) 'Purification of antithrombin III by affinity chromatography', *Thrombosis Research, 5*, 439-52

Morrison-Silverberg, S.A. and Jesty, J. (1981) 'The role of activated factor X in the control of blood coagulation', *Journal of Biological Chemistry, 256*, 1625-30

Mosesson, M.W., Escaig, J. and Feldmann, G. (1979) 'Electron microscopy of metal-shadowed fibrinogen molecules deposited on carbon films at different concentrations', *Thrombosis and Haemostasis, 42*, 88

Moskowitz, R.W., Schwartz, H.J., Michel, B., Ratnoff, O.D. and Astrup, T. (1970) 'Generation of kinin-like agents by chondroitin sulphate, chitin sulphate, and human articular cartilage: Possible pathophysiologic implications', *Journal of Laboratory and Clinical Medicine, 76*, 790-8

Mui, P.T.K. and Ganguly, P. (1977) 'Cross-linking of actin and fibrin by fibrin-stabilizing factor', *American Journal of Physiology, 233*, H346-9

Murano, G., Seegers, W.H. and Zolton, R.P. (1974) 'Autoprothrombin II-A: A competitive inhibitor of autoprothombin C (Factor Xa). A review with additions', *Thrombosis et Diathesis Haemorrhagica*, Suppl. 57, 305-14

Nachman, R.L. and Jaffe, E.A. (1975) 'Subcellular platelet factor VIII antigen and von Willebrand factor', *Journal of Experimental Medicine, 141*, 1101-13

Nachman, R.L., Levine, R. and Jaffe, E.A. (1977) 'Synthesis of factor-VIII antigen by cultured guinea pig megakaryocytes', *Journal of Clinical Investigation, 60*, 914-21

Nakayasu, T. and Nagasawa, S. (1979) 'Studies on human kininogens I. Isolation, characterization, and cleavage by plasma kallikrein of high molecular weight (HMW)-kininogen', *Journal of Biochemistry, 85*, 249-58

Nemerson, Y. (1968) 'The phospholipid requirement of tissue factor in blood coagulation', *Journal of Clinical Investigation, 47*, 72-80

Nemerson, Y. and Pitlick, F.A. (1970) 'Purification and characterization of the protein component of tissue factor', *Biochemistry, 9*, 5100-5

Nesheim, M.E. and Mann, K.G. (1979) 'Thrombin-catalyzed activation of single chain bovine factor V', *Journal of Biological Chemistry, 254*, 1326-34

Nesheim, M.E., Myrmel, K.H., Hibbard, L. and Mann, K.G. (1979) 'Isolation and characterization of single chain bovine factor V', *Journal of Biological Chemistry, 254*, 508-17

Nesheim, M.E., Kettner, C., Shaw, E. and Mann, K.G. (1981) 'Cofactor dependence of factor Xa incorporation into the prothrombinase complex', *Journal of Biological Chemistry, 256*, 6537-40

Nesheim, M.E., Canfield, W.M., Kisiel, W. and Mann, K.G. (1982) 'Studies of the capacity of factor Xa to protect factor Va from inactivation by activated protein C', *Journal of Biological Chemistry, 257*, 1443-7

Niewiarowski, S. and Kowalski, E. (1958) 'Un nouvel anticoagulant dérivé du fibrinogène', *Revue d'Hématologie, 13*, 320-7

Niewiarowski, S. and Latallo, Z. (1957) 'Influence of plasmin on the first coagulation stage', *Bulletin de l'Académie Polonaise des Sciences, 5*, 219-26

Niewiarowski, S. and Prou-Wartelle, O. (1959) 'Rôle du facteur contact (facteur Hageman) dans la fibrinolyse', *Thrombosis et Diathesis Haemorrhagica, 3*, 593-606

Nordenman, B., Nyström, C. and Björk, I. (1977) 'The size and shape of human and bovine antithrombin III', *European Journal of Biochemistry, 78*, 195-203

Norman, J.C., Lambilliotte, J.-P., Kojima, Y. and Sise, H.S. (1967) 'Antihemophilic factor release by perfused liver and spleen: relationship to hemophilia', *Science, 158*, 1060-1

Nossel, H.L. (1966) 'Activation of factors XII (Hageman) and XI (PTA) by skin contact', *Proceedings of the Society for Experimental Biology and Medicine, 122*, 16-7

Nossel, H.L. and Niemetz, J. (1965) 'A normal inhibitor of the blood coagulation contact reaction product', *Blood, 25*, 712-23

Ødegard, O.R., Abildgaard, U., Lie, M. and Miller-Andersson, M. (1977) 'Inactivation of bovine and human thrombin and factor Xa by antithrombin III studied with amidolytic methods', *Thrombosis Research, 11*, 2-5-16

Ogston, D., Ogston, C.M. and Ratnoff, O.D. (1969) 'Studies on the clot-promoting effect of skin', *Journal of Laboratory and Clinical Medicine, 73*, 70-7

Oh-ishi, S., Katori, M., Han, Y.N., Iwanaga, S., Kato, H. and Suzuki, T. (1977) 'Possible physiological role of new peptide fragments released from bovine high molecular weight kininogen by plasma kallikrein', *Biochemical Pharmacology, 26*, 115-20

Olexa, S.A., Budzynski, A.Z. (1981) 'Localisation of a fibrin polymerisation site', *Journal of Biological Chemistry, 256*, 3544-9

Olson, J.P., Miller, L.L. and Troup, S.B. (1966) 'Synthesis of clotting factors by the isolated perfused rat liver', *Journal of Clinical Investigation, 45*, 690-701

Østerud, B. and Rapaport, S.I. (1977) 'Activation of factor IX by the reaction product of tissue factor and factor VII: additional pathway for initiating blood coagulation', *Proceedings of the National Academy of Sciences, 74*, 5260-4

Østerud, B., Rapaport, S.I., Schiffman, S. and Chong, M.M.Y. (1971) 'Formation of intrinsic factor X-activator activity, with special reference to the role of thrombin', *British Journal of Haematology, 21*, 643-60

Østerud, B., Berre, A., Otnaess, A.-B., Bjørklid, E. and Prydz, H. (1972) 'Activation of the coagulation factor VII by tissue thromboplastin and calcium', *Biochemistry, 11*, 2853-7

Østerud, B., Miller-Andersson, M., Abildgaard, U. and Prydz, H. (1976) 'The effect of antithrombin III on the activity of the coagulation factors VII, IX and X', *Thrombosis and Haemostasis, 35*, 295-304

Østerud, B., Rapaport, S.I. and Lavine, K.K. (1977) 'Factor V activity of platelets: evidence for an activated factor V molecule and for a platelet activator', *Blood, 49*, 819-34

Østerud, B., Bouma, B.N. and Griffin, J.H. (1978) 'Human blood coagulation factor IX. Purification, properties, and mechanism of activation by activated factor XI', *Journal of Biological Chemistry, 253*, 5946-51

Østerud, B., Harper, E., Rapaport, S.I. and Lavine, K. (1979) 'Evidence against collagen activation of platelet associated factor XI as a mechanism for initiating intrinsic clotting', *Scandinavian Journal of Haematology, 22*, 205-13

Otis, P.T. and Rapaport, S.I. (1973) 'Failure of fibrinogen degradation products to increase plasma fibrinogen in rabbits', *Proceedings of the Society for Experimental Biology and Medicine, 144*, 124-9

Owen, C.A. and Bowie, E.J.W. (1977) 'Generation of coagulation factors V, XI, and XII by the isolated rat liver', *Haemostasis, 6*, 205-12

Owen, C.A. and Bowie, E.J.W. (1981) 'Generation of plasmatic coagulation factors by the isolated rat liver perfused with completely synthetic blood substitute', *Thrombosis Research, 22*, 259-66

Owen, W.G. and Esmon, C.T. (1981) 'Functional properties of an endothelial cell cofactor for thrombin-catalyzed activation of protein C', *Journal of Biological Chemistry, 256*, 5532-5

Owen, W.G., Esmon, C.T. and Jackson, C.M. (1974) 'The conversion of prothrombin to thrombin I. Characterization of the reaction products formed during the activation of bovine prothrombin', *Journal of Biological Chemistry, 249*, 594-605

Pearson, J.D. and Gordon, J.L. (1979) 'Vascular endothelial and smooth muscle cells in culture selectively release adenine nucleotides', *Nature, 281*, 384-6

Pepper, D.S., Gaffney, P.J. and Blume, H.D. (1974) 'The distribution of carbohydrate in the plasmin resistant core fragments (D, E) of human fibrinogen', *Biochimica et Biophysica Acta, 365*, 203-7

Petersen, T.E., Dudek-Wojciechowska, G., Sottrup-Jensen, L. and Magnusson, S. (1979) 'Primary structure of antithrombin III (heparin cofactor). Partial homology between α_1-antitrypsin and antithrombin III', in D. Collen, B. Wiman and M. Verstraete (eds.) *The Physiological Inhibitors of Blood Coagulation*, Elsevier/North-Holland, Biomedical Press, pp. 43-54

Petersen, T.E., Thogersen, H.C., Sottrup-Jensen, L., Magnusson, S. and Jörnvall, H. (1980) 'Isolation and *N*-terminal amino acid sequence of protein Z, a γ-carboxyglutamic acid containing protein from bovine plasma', *FEBS Letters, 114*, 278-82

Philip, G., Moran, J. and Colman, R.W. (1970) 'Dissociation and association of the oligomeric forms of factor V', *Biochemistry, 9*, 2212-8

Pickart, L.R. and Pilgeram, L.O. (1967) 'The role of thrombin in fibrinogen biosynthesis', *Thrombosis et Diathesis Haemorrhagica, 17*, 358-64

Pickart, L.R. and Thaler, M.M. (1976) 'Free fatty acids and albumin as mediators of thrombin-stimulated fibrinogen synthesis', *American Journal of Physiology, 230*, 996-1002

Piepkorn, M.W., Lagunoff, D. and Schmer, G. (1978) 'Heparin binding to antithrombin III: variation in binding sites and affinity', *Biochemical and Biophysical Research Communications, 85*, 851-6

Pifer, D.D., Colman, R.W. and Chesney, C.M. (1977) 'Subcellular localization and secretion of factor V by human platelets', *Thrombosis and Haemostasis, 38*, 126

Pisano, J.J., Finlayson, J.S., Peyton, M.P. and Nagai, Y. (1971) 'ε (γ-glutamyl) lysine in fibrin: lack of crosslink formation in factor XIII deficiency', *Proceedings of the National Academy of Sciences, 68*, 770-2

Pool, J.G. and Robinson, J. (1959) '*In vitro* synthesis of coagulation factors by rat liver slices', *American Journal of Physiology, 196*, 423-8

Pouit, L., Marcille, G., Suscillon, M. and Hollard, D. (1972) 'Etude en microscopie electronique de différentes étapes de la fibrinoformation',

Thrombosis et Diathesis Haemorrhagica, 27, 559-72

Proud, D., Pierce, J.V. and Pisano, J.J. (1980) 'Radioimmunoassay of human high molecular weight kininogen in normal and deficient plasmas', *Journal of Laboratory and Clinical Medicine, 95*, 563-74

Prowse, C.V., Mattock, P., Esnouf, M.P. and Russell, A.M. (1976) 'A variant of prothrombin induced in cattle by prolonged administration of warfarin', *Biochimica et Biophysica Acta, 434*, 265-79

Radcliffe, R.D. and Barton, P.G. (1973) 'Comparisons of the molecular forms of activated bovine factor X. Products of activation with Russell's viper venom, insoluble trypsin, sodium citrate, tissue factor, and the intrinsic system', *Journal of Biological Chemistry, 248*, 6788-95

Radcliffe, R. and Nemerson, Y. (1975) 'Activation and characterization of a single chain form of factor VII', *Journal of Biological Chemistry, 250*, 388-95

Radcliffe, R. and Nemerson, Y. (1976) 'Mechanism of activation of bovine factor VII. Products of cleavage by factor Xa', *Journal of Biological Chemistry, 251*, 4797-802

Radcliffe, R., Bagdasarian, A., Colman, R. and Nemerson, Y. (1977) 'Activation of bovine factor VII by Hageman factor fragments', *Blood, 50*, 611-7

Ratnoff, O.D. (1968) 'Activation of Hageman factor by L-homocystine', *Science, 162*, 1007-9

Ratnoff, O.D. and Colopy, J.E. (1955) 'A familial hemorrhagic trait associated with a deficiency of a clot-promoting fraction of plasma', *Journal of Clinical Investigation, 34*, 602-13

Ratnoff, O.D. and Crum, J.D. (1964) 'Activation of Hageman factor by solutions of ellagic acid', *Journal of Laboatory and Clinical Medicine, 63*, 359-77

Ratnoff, O.D. and Rosenblum, J.M. (1958) 'Role of Hageman factor in the initiation of clotting by glass: evidence that glass frees Hageman factor from inhibition', *American Journal of Medicine, 25*, 160-8

Ratnoff, O.D. and Saito, H. (1979a) 'Interactions among Hageman factor, plasma kallikrein, high molecular weight kininogen, and plasma thromboplastin antecedent', *Proceedings of the National Academy of Sciences, 76*, 958-61

Ratnoff, O.D. and Saito, H. (1979b) 'Amidolytic properties of single-chain activated Hageman factor', *Proceedings of the National Academy of Sciences, 76*, 1461-3

Ratnoff, O.D., Pensky, J., Ogston, D. and Naff, G.B. (1969) 'The inhibition of plasmin, plasma kallikrein, plasma permeability factor and the C'1r subcomponent of complement by serum C'1 esterase inhibitor', *Journal of Experimental Medicine, 129*, 315-31

Ratnoff, O.D., Pensky, J., Donaldson, V.H. and Amir, J. (1972) 'The inhibitory properties of plasma against activated plasma thromboplastin antecedent (factor XIa) in hereditary angioneurotic edema', *Journal of Laboratory and Clinical Medicine, 80*, 803-9

Regoeczi, E. (1970) 'Fibrinogen catabolism: kinetics of catabolism following sudden elevation of the pool with exogenous fibrinogen', *Clinical Science, 38*, 111-21

Revak, S.D. and Cochrane, C.G. (1976) 'The relationship of structure and function in human Hageman factor. The association of enzymatic and binding activities with separate regions of the molecule', *Journal of Clinical Investigation, 57*, 852-60

Revak, S.D., Cochrane, C.G., Johnston, A.R. and Hugli, T.E. (1974) 'Structural changes accompanying enzymatic activation of human Hageman factor', *Journal of Clinical Investigation, 54*, 619-27

Revak, S.D., Cochrane, C.G. and Griffin, J.H. (1977) 'The binding and cleavage characteristics of human Hageman factor during contact activation. Comparison of normal plasma with plasma deficient in factor XI, prekallikrein, or high molecular weight kininogen', *Journal of Clinical Investigation, 59,* 1167-75

Revak, S.D., Cochrane, C.G., Bouma, B.N. and Griffin, J.H. (1978) 'Surface and fluid phase activities of two forms of activated Hageman factor produced during contact activation of plasma', *Journal of Experimental Medicine, 147,* 719-29

Rider, D.M., McDonagh, R.P. and McDonagh, J. (1978) 'A possible contributory role of the platelet in the formation of plasma factor XIII', *British Journal of Haematology, 39,* 579-88

Rimon, A., Shamash, Y. and Shapiro, B. (1966) 'The plasmin inhibitor of human plasma, IV. Its action on plasmin, trypsin, chymotrypsin and thrombin', *Journal of Biological Chemistry, 241,* 5102-7

Rinderknecht, H. and Geokas, M.C. (1973) 'On the physiological role of α_2-macroglobulin', *Biochimica et Biophysica Acta, 295,* 233-44

Rosenberg, R.D. and Damus, P.S. (1973) 'The purification and mechanism of action of human antithrombin-heparin cofactor', *Journal of Biological Chemistry, 248,* 6490-505

Rosenberg, R.D. and Lam, L. (1979) 'Correlation betwee structure and function of heparin', *Proceedings of the National Academy of Sciences, 76,* 1218-22

Rosenberg, J.S., McKenna, P. and Rosenberg, R.D. (1975) 'Inhibition of human factor IXa by human antithrombin-heparin cofactor', *Journal of Biological Chemistry, 250,* 8883-8

Rosing, J., Tans, G., Covers-Riemslag, J.W.P., Zwaal, R.F.A. and Hemker, H.C. (1980) 'The role of phospholipids and factor Va in the prothrombinase complex', *Journal of Biological Chemistry, 255, 274-83*

Ruggeri, Z.M. and Zimmerman, T.S. (1980) 'Variant von Willebrand's disease: Characterization of two subtypes by analysis of multimeric composition of factor VIII/von Willebrand factor in plasma and platelets', *Journal of Clinical Investigation, 65,* 1318-25

Rupp, R.G. and Fuller, G.M. (1979) 'The effects of leukocytic and serum factors on fibrinogen biosynthesis in cultured hepatocytes', *Experimental Cell Research, 118,* 23-30

Saito, H. (1977) 'Purification of high molecular weight kininogen and the role of this agent in blood coagulation', *Journal of Clinical Investigation, 60,* 584-94

Saito, H. and Goldsmith, G.H. (1977) 'Plasma thromboplastin antecedent (PTA, factor XI): a specific and sensitive radioimmunoassay', *Blood, 50,* 377-85

Saito, H., Ratnoff, O.D., Donaldson, V.H., Haney, G. and Pensky, J. (1974) 'Inhibition of the adsorption of Hageman factor (Factor XII) to glass by normal human plasma', *Journal of Laboratory and Clinical Medicine, 84,* 62-73

Saito, H., Ratnoff, O.D., Waldmann, R. and Abraham, J.P. (1975) 'Fitzgerald trait. Deficiency of a hitherto unrecognized agent, Fitzgerald factor, participating in surface-mediated reaction of clotting, fibrinolysis, generation of kinins, and the property of diluted plasma enhancing vascular permeability (PF/Dil)', *Journal of Clinical Investigation, 55,* 1082-9

Saito, H., Ratnoff, O.D. and Pensky, J. (1976) 'Radio-immunoassay of human Hageman factor (factor XII)', *Journal of Laboratory and Clinical Medicine, 88,* 506-14

Saito, H., Poon, M.G., Vicic, W., Goldsmith, G.H. and Menitove, J.E. (1978) 'Human plasma prekallikrein (Fletcher factor) clotting activity and antigen in health and disease', *Journal of Laboratory and Clinical Medicine, 92,* 84-94

Saito, H., Goldsmith, G.H., Moroi, M. and Aoki, N. (1979) 'Inhibitory spectrum of α_2-plasmin inhibitor', *Proceedings of the National Academy of Sciences, 76,* 2013-7

Sakariassen, K.S., Bolhuis, P.A. and Sixma, J.J. (1979) 'Human blood platelet adhesion to artery subendothelium is mediated by factor VIII-von Willebrand factor bound to the subendothelium', *Nature, 279*, 638-8

Saraswathi, S., Rawala, R. and Colman, R.E. (1978) 'Subunit structure of bovine factor V. Influence of proteolysis during blood collection', *Journal of Biological Chemistry, 253*, 1024-9

Schapira, M., Scott, C.F. and Colman, R.W. (1981a) 'Relative importance of plasma protease inhibitors in the inactivation of kallikrein in human plasma', *Thrombosis and Haemostasis, 46*, 259

Schapira, M., Scott, C.F. and Colman, R.W. (1981b) 'Protection of human plasma kallikrein from inactivation by Cl inhibitor and other protease inhibitors. The role of high molecular weight kininogen', *Biochemistry, 20*, 2738-43

Schapira, M., Scott, C.F., James, A., Silver, L.D., Kueppers, F., James, H.L. and Colman, R.W. (1982) 'High molecular weight kininogen or its light chain protects human plasma kallikrein from inactivation by plasma protease inhibitors', *Biochemistry, 21*, 567-72

Schiffman, S. and Lee, P. (1974) 'Preparation, characterization, and activation of a highly purified factor XI: evidence that a hitherto unrecognized plasma activity participates in the interaction of factors XI and XII', *British Journal of Haematology, 27*, 101-14

Schmaier, A.H., Kuchibhotla, J. and Colman, R.W. (1981) 'High molecular weight kininogen — a secreted platelet coagulant protein', *Thrombosis and Haemostasis, 46*, 96

Schreiber, A.D., Kaplan, A.P. and Austen, K.F. (1973) 'Inhibition by Cl INA of Hageman factor fragment activation of coagulation, fibrinolysis and kinin generation', *Journal of Clinical Investigation, 52*, 1402-9

Schwartz, M.L., Pizzo, S.V., Hill, R.L. and McKee, P.A. (1973) 'Human factor XIII from plasma and platelets. Molecular weights, subunit structures, proteolytic activation, and cross-linking of fibrinogen and fibrin', *Journal of Biological Chemistry, 248*, 1395-407

Scott, C.F. and Colman, R.W. (1980) 'Function and immunochemistry of prekallikrein-high molecular weight kininogen complex in plasma', *Journal of Clinical Investigation, 65*, 413-21

Scott, C.F., Schapira, M. and Colman, R.W. (1982a) 'Effect of heparin on the inactivation rate of human factor XIa by antithrombin III', *Blood, 60*, 940-7

Scott, C.F., Schapira, M., James H.L., Cohen, A.B. and Colman, R.W. (1982b) 'The inactivation of factor XIa by plasma protease inhibitors. Predominant role of alpha$_1$-protease inhibitor and protective effect of high molecular weight kininogen', *Journal of Clinical Investigation, 69*, 844-52

Scully, M.F., Ellis, V., MacGregor, I.R. and Kakkar, V.V. (1981) 'Inhibition of human factor Xa by various plasma protease inhibitors', *Thrombosis and Haemostasis, 46*, 90

Seegers, W.H., Johnson, J.F. and Fell, C. (1954) 'Antithrombin reaction related to prothrombin activation', *American Journal of Physiology, 176*, 97-103

Seegers, W.H., Novoa, H.E. and Henry, R. (1976) 'Relationship of "new" vitamin K-dependent protein C and "old" autoprothrombin IIa', *Thrombosis Research, 8*, 543-53

Seligsohn, U. and Klein, B. (1975) 'The effect of leucocyte extract on fibrinogen synthesis in rabbits', *Thrombosis Research, 7*, 17-24

Seligsohn, U., Rapaport, S.I. and Kuefler, P.R. (1973) 'Extra-adrenal effect of ACTH on fibrinogen synthesis', *American Journal of Physiology, 224*, 1172-9

Seligsohn, U., Østerud, B., Brown, S.F., Griffin, J.H. and Rapaport, S.I. (1979) 'Activation of human factor VII in plasma and in purified systems. Roles of activated factor IX, kallikrein, and activated factor XII', *Journal of Clinical Investigation, 64,* 1056-65

Shamash, Y. and Rimon, A. (1966) 'The plasmin inhibitors of human plasma. III. Purification and partial characterization', *Biochimica et Biophysica Acta, 121,* 35-41

Shapiro, G.A., Andersen, J.C., Pizzo, S.V. and McKee, P.A. (1973) 'The subunit structure of normal and hemophilic factor VIII', *Journal of Clinical Investigation, 52,* 2198-210

Shapiro, S.S. and Anderson, D.B. (1977) 'Thrombin inhibition in normal plasma', in R.L. Lundblad (ed.) *Chemistry and Biology of Thrombin,* Ann Arbor Science, Michigan, pp. 361-74

Shaw, E., Giddings, J.C., Peake, I.R. and Bloom, A.L. (1979) 'Synthesis of procoagulant factor VIII, factor VIII related antigen and other coagulation factors by the isolated perfused rat liver', *British Journal of Haematology, 41,* 585-96

Silverberg, M., Dunn, J.T., Garen, L. and Kaplan, A.P. (1980) 'Autoactivation of human Hageman factor', *Journal of Biological Chemistry, 255,* 7281-6

Smith, G.F. (1977) 'The heparin-thrombin complex in the mechanism of thrombin inactivation by heparin', *Biochemical and Biophysical Research Communications, 77,* 111-7

Smith, G.F. and Craft, T.J. (1976) 'Heparin reacts stoichiometrically with thrombin during thrombin inhibition in human plasma', *Biochemical and Biophysical Research Communications, 71,* 738-45

Smith, G.F. and Sundboom, J.L. (1981a) 'Heparin and protease inhibition. I. Heparin complexes with thrombin, plasmin, and trypsin', *Thrombosis Research, 22,* 103-14

Smith, G.F. and Sundboom, J.L. (1981b) 'Heparin and protease inhibition. II. The role of heparin in the AT III inactivation of thrombin, plasmin and trypsin', *Thrombosis Research, 22,* 115-33

Sodetz, J.M., Pizzo, S.V. and McKee, P.A. (1977) 'Relationship of sialic acid to function and *in vivo* survival of human factor VIII/von Willebrand factor protein', *Journal of Biological Chemistry, 252,* 5538-46

Sodetz, J.M., Paulson, J.C., Pizzo, S.V. and McKee, P.A. (1978) 'Carbohydrate on human factor VIII/von Willebrand factor. Impairment of function by removal of specific galactose residues', *Journal of Biological Chemistry, 253,* 7202-6

Sottrup-Jensen, L., Hansen, H.F., Mortensen, S.B., Petersen, T.E., Magnusson, S. and Jörnvall, H. (1981) 'Mechanism of proteinase-binding to α_2-macroglobulin', *Thrombosis and Haemostasis, 46,* 87

Stead, N., Kaplan, A.P. and Rosenberg, R.D. (1976) 'Inhibition of activated factor XII by antithrombin-heparin cofactor', *Journal of Biological Chemistry, 251,* 6481-8

Steinbuch, M., Blatrix, C. and Josso, F. (1967) 'α_2-macroglobulin as progressive antithrombin', *Nature, 216,* 500-1

Stenflo, J. (1970) 'Dicoumarol-induced prothrombin in bovine plasma', *Acta Chemica Scandinavica, 24,* 3762-3

Stenflo, J. (1976) 'A new vitamin K-dependent protein. Purification from bovine plasma and preliminary characterization', *Journal of Biological Chemistry, 251,* 355-63

Stenflo, J., Fernlund, P., Egan, W. and Roepstorff, P. (1974) 'Vitamin K dependent modifications of glutamic acid residues in prothrombin', *Proceedings of the National Academy of Sciences, 71,* 2730-3

Strong, D.D., Watt, K.W.K., Cottrell, B.A. and Doolittle, R.F. (1979) 'Amino acid sequence studies on the α chain of human fibrinogen. Complete sequence of the largest cyanogen bromide fragment', *Biochemistry, 18*, 5399-404

Sugo, T., Ikari, N., Kato, H., Iwanaga, S. and Fujii, S. (1980) 'Functional sites of bovine high molecular weight kininogen as a cofactor in kaolin-mediated activation of factor XII (Hageman factor)', *Biochemistry, 19*, 3215-20

Sugo, T., Kato, H., Iwanaga, S. and Fujii, S. (1981) 'The accelerating effect of bovine plasma HMW kininogen on the surface-mediated activation of factor XII: generation of a derivative form (active kininogen) with maximal cofactor activity by limited proteolysis', *Thrombosis Research, 24*, 329-37

Suomela, H., Blombäck, M. and Blombäck, B. (1977) 'The activation of factor X evaluated by using synthetic substrates', *Thrombosis Research, 10*, 267-81

Suttie, J.W. and Jackson, C.M. (1977) 'Prothrombin structure, activation and biosynthesis', *Physiological Reviews, 57*, 1-70

Takagi, T. and Doolittle, R.F. (1974) 'Amino acid sequence studies on factor XIII and the peptide released during its activation by thrombin', *Biochemistry, 13*, 750-6

Teger-Nilsson, A.C. and Ekelund, H. (1974) 'Fibrinogen to fibrin transformation in umbilical cord blood and purified neonatal fibrinogen', *Thrombosis Research, 5*, 601-12

Tesch, R., Trolp, R. and Witt, I. (1979) 'Electron microsopic studies on the foetal fibrin clot', *Thrombosis Research, 16*, 239-43

Thompson, A.R. (1977) 'Factor IX antigen by radioimmunoassay. Abnormal factor IX protein in patients on warfarin therapy and with hemophilia B', *Journal of Clinical Investigation, 59*, 900-10

Thompson, R.E., Mandle, R. and Kaplan, A.P. (1978) 'Characterization of human high molecular weight kininogen', *Journal of Experimental Medicine, 147*, 488-99

Thunberg, L., Bäckström, G., Grundberg, H., Riesenfeld, J. and Lindahl, U. (1980) 'The molecular size of the antithrombin-binding sequence in heparin', *FEBS Letters, 117*, 203-6

Titani, K., Fujikawa, K., Enfield, D.L., Ericsson, L.H., Walsh, K.A. and Neurath, H. (1975) 'Bovine factor XI (Stuart factor): amino-acid sequence of heavy chain', *Proceedings of the National Academy of Sciences, 72*, 3082-6

Tracy, P.B., Peterson, J.M., Nesheim, M.E., McDuffie, F.C. and Mann, K.G. (1979) 'Interaction of coagulation factor V and factor Va with platelets', *Journal of Biological Chemistry, 254*, 10354-61

Tracy, P.B., Peterson, J.M., Nesheim, M.E., McDuffie, F.C. and Mann, K.G. (1980) 'Platelet interaction with bovine coagulation factor V and factor Va', in K.G. Mann and F.B. Taylor (eds.) *The Regulation of Coagulation*, Elsevier/North Holland, New York, pp. 237-43

Tracy, P.B., Nesheim, M.E. and Mann, K.G. (1981) 'Coordinate binding of factor Va and factor Xa to the unstimulated platelet', *Journal of Biological Chemistry, 256*, 743-51

Tracy, P.B., Eide, L.L., Bowie, E.J.W. and Mann, K.G. (1982) 'Radioimmunoassay of factor V in human plasma and platelets', *Blood, 60*, 59-63

Triantaphyllopoulos, D.C. (1958) 'Anticoagulant effect of incubated fibrinogen', *Canadian Journal of Biochemistry and Physiology, 36*, 249-59

Triantaphyllopoulos, D.C. (1959) 'Nature of the thrombin-inhibiting effect of incubated fibrinogen', *American Journal of Physiology, 197*, 575-9

Triantaphyllopoulos, D.C. and Triantaphyllopoulos, E. (1966) 'Evidence of antithrombic activity of the anticoagulant fraction of incubated fibrinogen', *British Journal of Physiology, 12*, 145-51

Tuszynski, G.P., Bevacqua, S.J., Schmaier, A.H., Colman, R.W. and Walsh, P.N.

(1982) 'Factor XI antigen and activity in human platelets', *Blood, 59,* 1148-56

Ulevitch, R.J., Letchford, D. and Cochrane, C.G. (1974) 'A direct enzymatic assay for the esterolytic activity of activated Hageman factor', *Thrombosis et Diathesis Haemorrhagica, 31,* 30-9

Vannucchi, S., Fibbi, G., Pasquali, F., Del Rosso, M., Cappelletti, R. and Chiarugi, V. (1982) 'Adhesion-dependent heparin production by platelets', *Nature, 296,* 352-3

Vecchione, J. and Zucker, M. (1975) 'Procoagulant activity of platelets in recalcified plasma', *British Journal of Haematology, 31,* 423-8

Vehar, G.A. and Davie, E.W. (1980) 'Preparation and properties of bovine factor VIII (antihemophilic factor)', *Biochemistry, 19,* 401-10

Venneröd, A.H., Laake, K., Solberg, A.K. and Strömland, S. (1976) 'Inactivation and binding of human plasma kallikrein by antithrombin III and heparin', *Thrombosis Research, 9,* 457-66

Vicic, W.J., Ratnoff, O.D., Saito, H. and Goldsmith, G.H. (1979) 'Platelets and surface-mediated clotting activity', *British Journal of Haematology, 43,* 91-8

Villanueva, G. and Danishefsky, I. (1979) 'Conformational changes accompanying the binding of antithrombin III to thrombin', *Biochemistry, 18,* 810-7

Walker, F.J. (1980) 'Regulation of activated protein C by a new protein. A possible function for bovine protein S', *Journal of Biological Chemistry, 255,* 5521-4

Walker, F.J. (1981a) 'Regulation of activated protein C by protein S', *Journal of Biological Chemistry, 256,* 11128-31

Walker, F.J. (1981b) 'Regulation of bovine activated protein C by protein S: the role of the cofactor protein in species specificity', *Thrombosis Research, 22,* 321-7

Walker, F.J., Sexton, P.W. and Esmon, C.T. (1979) 'The inhibition of blood coagulation by activated protein C through the selective inactivation of activated factor V', *Biochimica et Biophysica Acta, 571,* 333-42

Walsh, P.N. (1972a) 'Albumin density gradient separation and washing of platelets and the study of platelet coagulant activities', *British Journal of Haematology, 22,* 205-17

Walsh, P.N. (1972b) 'The role of platelets in the contact phase of blood coagulation', *British Journal of Haematology, 22,* 237-54

Walsh, P.N. (1972c) 'The effects of collagen and kaolin on the intrinsic coagulant activity of platelets. Evidence for an alternative pathway in intrinsic coagulation and requiring factor XII', *British Journal of Haematology, 22,* 393-405

Walsh, P.N. and Griffin, J.H. (1980) 'Contributions of human platelets to the proteolytic activation of blood coagulation factors XII and XI', *Blood, 57,* 106-118

Walz, D.A., Hewett-Emmett, D. and Seegers, W.H. (1977) 'Amino acid sequence of human prothrombin fragments 1 and 2', *Proceedings of the National Academy of Sciences, 74,* 1969-72

Ware, A.G., Fahey, J.L. and Seegers, W.H. (1948) 'Platelet extracts, fibrin formation and interaction of purified prothrombin and thromboplastin', *American Journal of Physiology, 154,* 140-7

Wasteson, A., Glimelius, B., Busch, C., Westermark, B., Heldin, C.-H. and Norling, B. (1977) 'Effect of a platelet endoglycosidase on cell surface associated heparan sulphate of human endothelial and glial cells', *Thrombosis Research, 11,* 309-21

Watada, M., Nakagawa, M., Kitani, T., Okajima, Y., Maeda, Y., Urano, S. and Ijichi, H. (1981) 'Identification of the AT III synthesizing hepatocytes by immunofluorescent technique', *Thrombosis and Haemostasis, 46,* 284

Watt, K.W.K., Takagi, T. and Doolittle, R.F. (1978) 'Amino acid sequence of the β chain of human fibrinogen: Homology with the γ chain', *Proceedings of the National Academy of Sciences, 75,* 1731-5

Watt, K.W.K., Cottrell, B.A., Strong, D.D. and Doolittle, R.F. (1979) 'Amino acid sequence studies on the α chain of human fibrinogen. Overlapping sequences providing the complete sequence', *Biochemistry, 18,* 5410-6

Webster, W.P., Reddick, R.L., Roberts, H.R. and Penick, G.D. (1967) 'Release of factor VIII (anti-haemophilic factor) from perfused organs and tissues', *Nature, 213,* 1146-7

Weidner, N., Ittyerah, T.R., Wochner, R.D. and Sherman, L.A. (1979) 'Investigation of an inflammatory humoral factor as a stimulator of fibrinogen synthesis', *Thrombosis Research, 15,* 651-61

Weiss, H.J. and Hoyer, L.W. (1973) 'Von Willebrand factor: Dissociation from antihemophilic factor procoagulant activity', *Science, 182,* 1149-51

Wiggins, R.C. and Cochrane, C.C. (1979) 'The autoactivation of rabbit Hageman factor', *Journal of Experimental Medicine, 150,* 1122-33

Wiggins, R.C., Bouma, B.N., Cochrane, C.G. and Griffin, J.H. (1977) 'Role of high-molecular weight kininogen in surface-binding and activation of coagulation factor XI and prekallikrein', *Proceedings of the National Academy of Sciences, 74,* 4636-40

Wiggins, R.C., Loskutoff, D.J., Cochrane, C.G. and Griffin, J.H. (1980) 'Activation of rabbit Hageman factor by homogenates of cultured rabbit endothelial cells', *Journal of Clinical Investigation, 65,* 197-206

Wilding, P., Adham, N.F., Mehl, J.W. and Haverback, B.J. (1967) 'Alpha-2-macroglobulin concentrations in human serum', *Nature, 214,* 1226-7

Wilner, G.D., Nossel, H.L. and Leroy, E.C. (1968) 'Activation of Hageman factor by collagen', *Journal of Clinical Investigation, 47,* 2608-15

Witt, I. and Hasler, K. (1972) 'Influence of organically bound phosphorus in foetal and adult fibrinogen on the kinetics of the interaction between thrombin and fibrinogen', *Biochimica et Biophysica Acta, 271,* 357-62

Witt, I. and Müller, H. (1970) 'Phosphorus and hexose content of human foetal fibrinogen', *Biochimica et Biophysica Acta, 221,* 402-4

Witt, I. and Tesch, R. (1979) 'Molecular characterization of human foetal fibrinogen', *Thrombosis and Haemostasis, 42,* 79

Witt, I., Müller, H. and Künzer, W. (1969) 'Evidence for the existence of foetal fibrinogen', *Thrombosis et Diathesis Haemorrhagica, 22,* 101-9

Wuepper, K.D. (1972) 'Precursor plasma thromboplastin antecedent (PTA, clotting factor XI)', *Federation Proceedings, 31,* 624

Wuepper, K.D. (1973) 'Prekallikrein deficiency in man', *Journal of Experimental Medicine, 138,* 1345-55

Wuepper, K.D. and Cochrane, C.G. (1972) 'Plasma prekallikrein: isolation, characterization, and mechanism of action', *Journal of Experimental Medicine, 135,* 1-20

Wuepper, K.D., Miller, D.R. and Lacombe, M.J. (1975) 'Flaujeac trait: deficiency of human plasma kininogen', *Journal of Clinical Investigation, 56,* 1663-72

Yin, E.T., Wessler, S. and Stoll, P.J. (1971) 'Biological properties of the naturally occurring plasma inhibitor to activated factor X', *Journal of Biological Chemistry, 246,* 3703-11

Yin, E.T., Salsgiver, W.J. and Tangen, O. (1979) 'A hitherto undescribed naturally occurring plasma antagonist of activated factor X inhibitor (antithrombin III)', *Thrombosis and Haemostasis, 42,* 122

Young, M.C. and Kolmen, S.N. (1970) 'Recovery of fibrinogen in fibrinolytic dogs', *Thrombosis et Diathesis Haemorrhagica, 23,* 50-7

Zur, M. and Nemerson, Y. (1978) 'The esterase activity of coagulation factor VII. Evidence for intrinsic activity of the zymogen', *Journal of Biological Chemistry*, *253*, 2203-9
Zur, M. and Nemerson, Y. (1980) 'Kinetics of factor IX activation via the extrinsic pathway. Dependence of K_m on tissue factor', *Journal of Biological Chemistry*, *255*, 5703-7

4   The Fibrinolytic Enzyme System

An outline of the fibrinolytic enzyme system was provided in Chapter 1. In this chapter the properties of the individual components of the system are described with an account of their interaction to culminate in the physiological dissolution of fibrin.

Plasminogen

Structure

Human plasminogen is a single-chain glycoprotein with a molecular weight of 92,000 and 791 amino acid residues (Wiman, 1978). The molecule has a carbohydrate content of about two per cent (Wiman and Wallen, 1975a). It was believed to have lysine as the amino-terminal amino acid (Robbins *et al.*, 1967), but it has been found subsequently that native plasminogen has glutamic acid in the amino-terminal position (Wallén and Wiman, 1970). It is now clear that plasminogen with amino-terminal lysine (lys-plasminogen) results from proteolytic degradation of the native form (glu-plasminogen) with removal of peptide material from the amino-terminal portion of the molecule during preparation. Circular dichroism spectra and sedimentation analysis have provided evidence that there are conformational differences between the glu- and lys-forms of plasminogen (Sjöholm *et al.*, 1973; Violand *et al.*, 1975).

Apart from glu- and lys-plasminogen a number of different molecular forms of plasminogen have been demonstrated by the electrophoretic techniques (Wallén and Wiman, 1970; Summaria, *et al.*, 1972). The cause of the microheterogeneity was suggested to lie in differences in the sialic acid content of the different forms (Collen and de Maeyer, 1975). Two major forms of plasminogen can be isolated from plasma by affinity chromatography (Summaria *et al.*, 1976); each form possesses several distinct sub-forms. It has been shown that one of these forms of plasminogen has two oligosaccharide units while the other has only one such unit. Hayes and Castellino (1979a) have shown that the heterogeneity within each unit is due to variable sialic acid content of the oligosaccharide units. The structure of the two oligosaccharide units in plasminogen has been determined (Hayes and Castellino, 1979b, c).

Amino acid sequence analysis has shown that the part of the plasminogen molecule which becomes the heavy chain after activation contains five homologous regions referred to as 'kringles' from their resemblance to a type of Scandinavian biscuit. Similar structures had been identified previously in the prothrombin molecule.

Plasminogen contains regions which specifically bind omega amino acids such as lysine and 6-aminohexanoic acid: these are termed the lysine-binding sites. Plasminogen forms a 1:1 stoichiometric complex with these amino acids (Abiko *et al.*, 1969) with resultant conformational changes in the plasminogen molecule (Sjöholm *et al.*, 1973), changes which also take place on the conversion of glu- to lys-plasminogen. There are a number of lysine-binding sites on plasminogen, one with high affinity and four or five with low affinity for 6-aminohexanoic acid (Markus *et al.*, 1978). The site which binds 6-aminohexanoic acid strongly is located in the amino-terminal part of the molecule (Wiman and Wallén, 1977). It is believed that the conformational change arises from the dissociation of a non-covalent interaction between a site in the amino-terminal part and a lysine-binding site elsewhere in the molecule (Wiman and Wallén, 1975b).

It appears that a number of the kringles contain lysine-binding sites; an exception is kringle 5. There is evidence that the high affinity binding site of plasminogen is located on kringle 1 (Lerch *et al.*, 1980; Vali and Patthy, 1982). Chemical modification of a single tryptophan residue in kringle 4 abolishes lysine-binding activity, suggesting that this residue is at the lysine-binding site (Hochschwender and Laursen, 1981). In other recent studies Arg_{70} and Asp_{56} have been identified as essential residues for the binding of ω-amino-carboxylic acid by the isolated kringle 4 (Trexler *et al.*, 1982): it was suggested that these residues are essential because they provide the complementary charges involved in electrostatic binding of the ligand. The first kringle has been claimed to have an intramolecular interaction with the amino-terminal region of plasminogen which contributes to the maintenance of the compact structure of plasminogen (Lerch *et al.*, 1980).

142

The specificity of plasminogen adsorption to fibrin has been a topic of controversy. Recent evidence supports the view that such adsorption does take place and that the quantities involved, although relatively small, are functionally significant. Lys-plasminogen is adsorbed to fibrin more efficiently than glu-plasminogen (Thorsen, 1975). Rákóczi and associates (1978) have reported that four per cent of native plasma plasminogen binds to fibrin compared with eight per cent for lys-plasminogen. Since the only fragments of plasminogen which adsorb to fibrin are those which also adsorb to lysine-Sepharose it has been proposed that the lysine-binding sites and the sites responsible for the association of plasminogen with fibrin are identical (Wiman and Wallén, 1977). In discord with this view is the finding of Thorsen and colleagues (1981) that the kringle 5-containing fragment of elastase-digested plasminogen can bind to fibrin, but does not contain a lysine-binding site. Fibrinogen appears to possess several sites which can interact with the lysine-binding sites in plasminogen and have been shown to be present in both fragment D and fragment E thus indicating their location in both the amino- and carboxyl-terminal regions of fibrinogen (Wiman *et al.*, 1979). The lysine-binding sites are also involved in the interaction of plasmin with the fast-acting α-antiplasmin (Wiman *et al.*, 1979).

Genetic polymorphism of human plasminogen was reported by Hobart (1979) with two codominant alleles (PLG^1 and PLG^2) on a single autosomal locus. Two common alleles, termed PLGN*A and PLGN*B, were also shown by Raum and associates (1980) together with a number of rare variants. A third common allele (PLG^B) was described in a Japanese population by Nishimukai *et al.* (1981).

Plasma Concentration and Half-life

The plasma concentration of plasminogen has been reported to be around 200 µg/ml using either a radioimmunoassay (Rabiner *et al.*, 1969) or a caseinolytic assay (Cohen *et al.*, 1972). The mean half-life in healthy adults was found to be 2.21 days (Collen *et al.*, 1972), while Harker and colleagues (1972) obtained a shorter time of 1.62 days. The discrepancy may have resulted from differences in the plasminogen preparations used since glu-plasminogen gave a mean half-life of 2.24 days in contrast to less than one day for lys-plasminogen (Collen and Verstraete, 1975).

Synthesis

A number of sites for the synthesis of plasminogen have been proposed over the years. Using a fluorescent antibody technique Barnhart and Riddle (1963) concluded that plasminogen is synthesised in bone marrow eosinophils. All types of human granulocytes have been found subsequently to contain a relatively high concentration of plasminogen (Prokopowicz and Stormorken, 1968), while treatment of leukocytes with cytostatic agents was observed to induce disappearance of the plasminogen suggesting that the granulocytes are a site of synthesis (Prokopowicz *et al.*, 1967). Some data point to the possibility that the kidney may be a source of plasminogen, at least in the cat (Highsmith and Kline, 1971). The liver, shown to be the site of formation of the α- and β-globulins (Miller and Bale, 1954), is likely to be the sole or major source of the β-globulin plasminogen. This has been confirmed by the determination of the genetic types of plasminogen from a donor and a recipient before and after hepatic homotransplantation with change of the genetic type of plasminogen to that of the donor following the transplantation (Raum *et al.*, 1980).

Activation

The activation of plasminogen by urokinase involves the cleavage of a specific sensitive arginyl-valine peptide bond (Arg_{561}-Val_{562}) in the carboxyl-terminal portion of the molecule with the production of a two-chain molecule (Robbins *et al.*, 1967). There is general acceptance of the requirement for hydrolysis of this bond, but subsequent studies suggested that the activation of plasminogen takes place in three steps. It was proposed that a peptide bond in the amino-terminal portion of glu-plasminogen is split in the initial step, leaving Met_{69} as the amino-terminal residue (Wiman and Wallén, 1973). In the second step cleavage of the Arg_{561}-Val_{562} bond takes place. Finally, the Lys_{77}-Lys_{78} bond in the amino-terminal part of the molecule is split completing the formation of lys-plasminogen. More recently it has been shown that the only bond cleaved by urokinase in the presence of plasmin inhibitors is the Arg_{561}-Val_{562} (Summaria *et al.*, 1975, 1977). In the absence of such inhibitors a second bond is split, either Arg_{68}-Met_{69} or Lys_{77}-Lys_{78}. It is now accepted that the cleavages in the amino-terminal part of the plasminogen molecule are caused by plasmin rather than by activator (Wiman, 1978). A schematic representation of the activation of plasminogen is given in Figure 4.1. Cleavage of the Arg_{561}-Val_{562} bond alone in glu-plasminogen provides a plasmin molecule with proteolytic activity, but the activation rate of lys-plasminogen, formed by the cleavage of the plasmin-sensitive bonds in the amino-terminal portion of the mole-

Figure 4.1: Schematic Representation of the Activation of Plasminogen

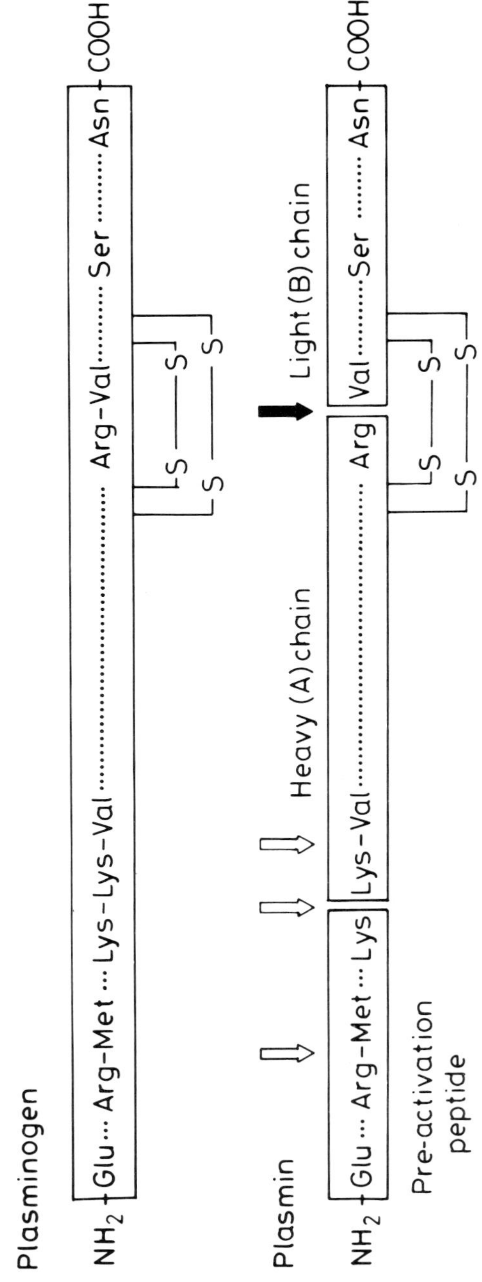

cule, is many times faster (Claeys and Vermylen, 1974). The increased rate of activation of lys-plasminogen is probably determined by the conformational change which follows the release of peptide material (termed the pre-activation peptide) from the amino-terminal. This conformational change is also induced by the addition of omega amino acids. Removal of the pre-activation peptide may, by uncovering fibrin-binding sites, be responsible also for the more efficient adsorption of lys-plasminogen to fibrin.

The pathway leading to the activation of plasminogen probably commences with the cleavage of the Arg_{561}-Val_{562} bond by activator resulting in the formation of catalytic quantities of glu-plasmin from the native glu-plasminogen. The plasmin formed removes the pre-activation peptide with the formation of lys-plasminogen: the accompanying conformational change facilitates the further cleavage of plasminogen by activator with the formation of the two-chain lys-plasmin molecule (Violand and Castellino, 1976). Lys-plasmin is the final form found after activation *in vitro*, but the physiological form of plasmin is uncertain.

Plasmin

Structure and Enzymatic Properties

Plasmin is a proteolytic enzyme with a trypsin-like specificity which cleaves proteins and peptides at arginyl and lysyl bonds (Wallén and Iwanaga, 1968; Groskopf *et al.*, 1968), basic amino esters (Troll *et al.*, 1954; Robbins *et al.*, 1965) and amides (Christensen and Mullertz, 1974). Plasmin appears to have a preference for lysyl bonds in both protein and ester substrates (Weinstein and Doolittle, 1972).

Plasmin consists of two polypeptide chains, the heavy (A) chain and the light (B) chain (Robbins *et al.*, 1967): the chains are connected by two disulphide bridges (Wiman, 1977). The molecular weight of the heavy chain from the amino-terminal part of plasminogen has been estimated at 48,000 (Summaria *et al.*, 1971), 63,000 (Wiman and Wallén, 1973) and 55,000 (Walther *et al.*, 1974). The light chain from the carboxyl-terminal end of the molecule has a weight of about 25,000 (Summaria, *et al.*, 1967). The primary structure of the heavy chain of human plasmin has been established by Sottrup-Jensen and associates (1978), and Wiman (1977) has detailed the primary structure of the light chain.

The active center of plasmin contains a serine residue and a histidine loop and is sited on the light (B) chain of the molecule (Summaria *et al.*, 1967). The histidine loop is homologous to the histidine-containing sequences found in other serine proteases.

Plasmin is a protease of broad specificity and a number of proteins are susceptible to its proteolytic action in addition to the presumed natural substrate fibrin. Other proteins digested by plasmin in puri- fied systems include fibrinogen, factor V (Alagille and Soulier, 1956), factor VIII (Donaldson, 1960; Pasquini and Hershgold, 1973), glucagon and somatotropin (Mirsky *et al.*, 1959), and com- ponents of the complement system (Pillemer *et al.*, 1953).

Fibrinogen Degradation by Plasmin

While the natural substrate for plasmin is fibrin, much of the experi- mental data on the molecular changes which take place in plasmin- induced fibrin dissolution have been obtained from studies on fibrin- ogen: details on fibrinogen degradation have been given in this sec- tion and the differences seen in fibrin degradation are given in the following section. It is unlikely that any significant degradation of fibrinogen takes place through the fibrinolytic enzyme system under physiological circumstances.

On exposure to plasmin the three chains of the fibrinogen mole- cule undergo progressive and asymmetric fragmentation. By the use of ion-exchange chromatography Nussenzweig and associates (1961) distinguished five major end-products of fibrinogen break- down which they termed A, B, C, D and E, but they recognised that intermediary fibrinogen degradation products were formed. These intermediate products were isolated by Marder and his colleagues (1969); their scheme of fibrinogen degradation has been generally accepted,

The initial major fraction to appear in plasmin digests of fibrin- ogen is termed fragment X and its formation is coincident with the appearance of low molecular weight peptides (fragments A, B and C). Fragment X is split further with the formation of two further fragments termed Y and D. The fragment Y is finally cleaved by plasmin to provide an additional fragment D and a fragment termed E (Figure 4.2). Fragments D and E are the plasmin-resistant core products of fibrinogen degradation. It is recognised that the break- down of fibrinogen is a continuous process and that intermediary fragments are formed in addition to those identified and designated as X, Y, D and E (Pizzo *et al.*, 1972).

The molecular changes in the degradation of fibrinogen are represented schematically in Figure 4.3. Fragment X is formed by the removal of peptide material from the carboxyl-terminal end of the Aα chain and, at a slower rate, from the amino-terminal portion of the Bβ chain (Pizzo *et al.*, 1972; Takagi and Doolittle, 1975a). The early cleavage at the carboxyl-terminal end of the Aα chain takes place in the region of the chain containing residues 200 to 250, probably at Lys_{208}-Met_{209}, Lys_{221}-Ser_{222} or Lys_{232}-Ala_{233} (Takagi and Doolittle, 1975b; Doolittle *et al.*, 1977). The early fragment X has fibrinopeptides A and B attached and therefore retains the ability to polymerise on the addition of thrombin. The cleavage of a peptide of around 50,000 daltons from the carboxyl-terminal end of the Aα chain is followed by its degradation into smaller fragments (Hessel, 1975). Plasmin-induced cleavage of the amino-terminal end of the Bβ chain takes place at multiple sites, but the thrombin-sensitive site is not affected. The initial cleavage point on the Bβ chain is between Arg_{42} and Ala_{43} (Takagi and Doolittle, 1975a). Once freed from the parent molecule the $Bβ_{1-42}$ peptide may undergo further breakdown by plasmin action, preferentially at the 21-22 bond to yield $Bβ_{1-21}$ and $Bβ_{22-42}$.

Figure 4.2: Stages in the Degradation of Fibrinogen by Plasmin (approximate molecular weights of fragments in parentheses)

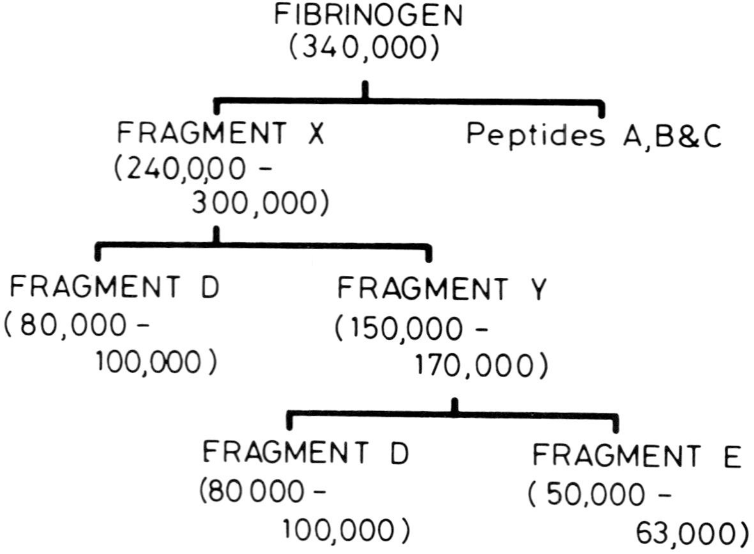

Figure 4.3: Molecular Changes in the Degradation of Fibrinogen by Plasmin (arrows indicate sites of cleavage by plasmin)

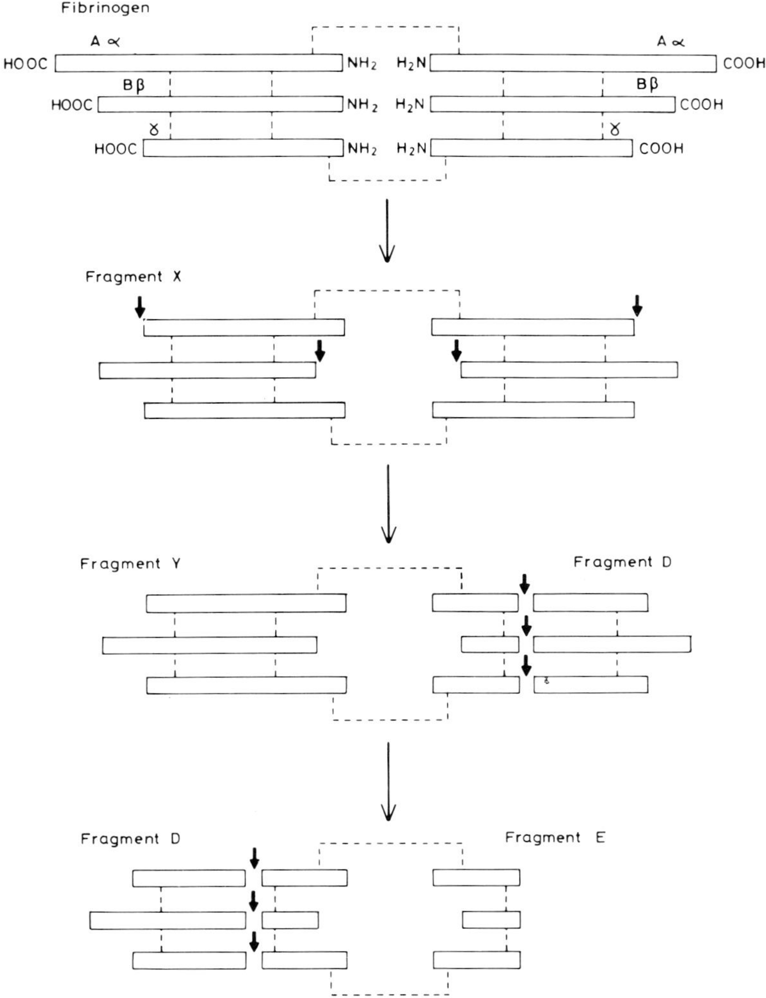

The conversion of fragment X into fragments Y and D involves the cleavage of bonds in each of the Aα-, Bβ- and γ-chains from one side of the dimeric molecule. The portions of the carboxyl-terminal ends of the chains remain linked by disulphide bonds to constitute fragment D. The asymmetric fragment Y is subsequently attacked further by plasmin with removal of a second fragment D, leaving the core fragment E. It is the Aα and Bβ-chain connections between fragment Y and fragment D which are cleaved initially; finally the Lys$_{62}$-Ala$_{63}$ bond of the γ-chain is split to release fragment D (Takagi and Doolittle, 1975a). The removal of the two small peptides leaves Lys$_{53}$ as the stable carboxyl-terminal residue of the γ-chain in fragment E.

Fragment D is heterogeneous in size due to progressive digestion of the carboxyl-terminal portion of the γ-chain (Gaffney and Dubos, 1971; Collen *et al.*, 1975). This molecular attrition is reflected by the various molecular weights reported for fragment D.

Fragment E contains the amino-terminal portions of the polypeptide chains of fibrinogen, but lacks the first 33 amino acid residues of the Bβ-chains, removed in the formation of fragment X. Following the separation of fragment E it undergoes further degradation with removal of both the fibrinopeptide A and the contiguous tripeptide Gly-Pro-Arg. The amino-terminal residues of the γ-chains are, however, conserved in fragment E.

Fibrin Degradation by Plasmin

Fibrin which has not undergone cross-linking provides fragments on exposure to plasmin which are similar to those found after plasmin-fibrinogen interaction (Gaffney, 1973; Pizzo *et al.*, 1973). They may differ, however, in the presence of fibrinopeptide A: fragments X, Y and E from fibrin will not possess the fibrinopeptide whereas it may be present in the fragments from the early digestion of fibrin by plasmin.

The *in vivo* substrate for plasmin is cross-linked fibrin as may be found in fibrinous deposits or intravascular thrombi. Fibrin stabilisation under the influence of factor XIII involves the formation of cross-links between the γ-chains to provide γ dimers and, more slowly, between the α-chains to form α-chain polymers. As a consequence of these cross-links the degradation of cross-linked fibrin by plasmin yields some core fragments which differ from those formed with fibrinogen. In particular, the D dimer (the cross-linked complex of two D fragments) and a complex of D dimer and fragment E.

150

The D dimer contains the cross-linked γ-chain remnants of the original fibrin molecules. The molecular weight of each component of the dimer has been estimated to be 63,000 (Gaffney and Brasher, 1973) or 81,000 (Pizzo *et al.*, 1973), providing a weight for the D dimer of around 160,000 daltons. There is evidence that the stability of the D dimer complex is dependent on calcium ions; it is rapidly converted into monomeric fragment D following removal of calcium (Haverkate and Timan, 1977).

In addition to D dimer, complexes of D dimer with fragment E are found in plasmin digests of cross-linked fibrin (Hudry-Clergeon *et al.*, 1974), although the mechanism of production of such complexes is not established. The complex is maintained by non-covalent bonds which prevent separation into the component fragments (Olexa *et al.*, 1979). Prolonged exposure of the complex to plasmin can result in cleavage of the D dimer from the E fragment with subsequent conversion of the D dimer into monomeric fragment D (Gaffney *et al.*, 1975).

Using chromatography on Sepharose 4B a number of soluble cross-linked fibrin fragments have been obtained after digestion by plasmin (Gaffney *et al.*, 1980): these were characterised as cross-linked X oligomers with around five X fragments bound together by polymerisation and cross-linking sites, and complexes of D dimer-E with fragment Y.

Plasminogen Activators

Activators of plasminogen are widely distributed in blood, tissues and secretions. In the absence of precise characterisation a biochemical classification has not been completed and attempts at classification have been based on their source. It has become clear in recent years that more than one species of activator may be present in a single tissue or secretion. However, since the differentiation of all activators is not yet finalised, they will be considered in this section according to their source.

Tissue Activator

The presence of plasminogen activator in tissues was demonstrated in 1947 by Astrup and Permin. A relatively small proportion of the activator content of tissues is soluble in saline; most appears to be firmly bound to tissue proteins, but can be extracted by the use of potassium thiocyanate (Astrup and Stage, 1952). The concentration of activator extracted from various organs with potassium thiocyanate was examined by Albrechtsen (1957): high concentrations were found in uterus, adrenals and thyroid while spleen, testis and liver had little or no activity.

Properties. Plasminogen activator has been isolated from a variety of mammalian tissues. Heart muscle has been a widely used source for its preparation in a purified form (Bachmann *et al.*, 1964; Rickli and Zaugg, 1970; Wallén and Wiman, 1975; Cole and Bachmann, 1977; Wallén and Rånby, 1977), but ovaries (Kok and Astrup, 1969), parotid glands (Cartwright, 1974) and uterus (Kok, 1979; Rijken *et al.*, 1979) have also been used.

The earlier preparations of tissue activator from heart had estimated molecular weights of around 50,000, close to that found for pig ovary activator (Kok and Astrup, 1969). Using two affinity adsorption steps (fibrin and arginine-Sepharose) Wallén and Rånby (1977) obtained a more highly purified preparation of pig heart activator: the molecular weight was 63,000 daltons, but on reduction the molecular weight had diminished to 31,000. On the basis of heat lability and inhibition pattern with 6-aminohexanoic acid Kok (1979) differentiated three types of activator activity in preparations from human uterus; the major form had a molecular weight of 60,000, was heat stable and was inhibited by 6-aminohexanoic acid. The second form, immunologically similar to the first, had a molecular weight of about 10,000 and was heat labile while the third and minor activity had similarities to urokinase. Rijken and colleagues (1979) also isolated activator from human uterine tissue: their preparation had a molecular weight of around 65,000 and on reduction was separated into two chains of 31,000 and 38,000. The active serine residue was located on the light chain. The uterine activator prepared by Hoylaerts and associates (1982) was also a two-chain molecule and it now appears that tissue activator can be isolated as either a single-chain or a two-chain molecule (Wallén *et al.*, 1981). The activator from melanoma cells in tissue culture can be isolated in single-chain form if the protease inhibitor aprotinin is added to the culture medium whereas a two-chain molecule is obtained in its absence, suggesting that the conversion from the one- into the two-chain form results from proteolytic cleavage (Rijken and Collen, 1981). The binding to fibrin and the fibrinolytic activity of the two forms of activator is very similar (Rijken *et al.*, 1981a).

152

Influence of Fibrin on Plasminogen Activator Activity. Integral to current hypotheses on the mechanism of fibrinolysis is the adsorption of activator to fibrin and the effect of fibrin on the properties of plasminogen activator. It was demonstrated by Camiolo and her colleagues (1971) that the activating properties of tissue activator are greatly enhanced in the presence of fibrin, and this has been confirmed for vascular activator both with fibrin and fibrin monomer (Allen and Pepper, 1981). There is abundant evidence for the adsorption of tissue activator onto fibrin (Thorsen *et al.*, 1972), whereas urokinase binds poorly to fibrin. Avid adsorption to fibrin has also been shown for vascular wall activator (Gurewich *et al.*, 1975) and for post-venous occlusion plasma activator (Ogston *et al.*, 1976; Libeskind *et al.*, 1981). The affinity of tissue activator and vascular wall activator for fibrin is much greater than that for fibrinogen (Kruithof and Bachmann, 1981). The binding site for activator appears to be on the α-chain of fibrin (Lloyd *et al.*, 1981).

Kinetic data have suggested that activation of plasminogen in the presence of fibrin occurs through the binding of an activator molecule to the fibrin surface and the subsequent addition of plasminogen to form a cyclic ternary complex (Hoylaerts *et al.*, 1982): fibrin enhances the activation rate of plasminogen by tissue activator by increasing the affinity of plasminogen for the fibrin-bound activator rather than by increasing the catalytic efficiency of the enzyme. Fibrinogen is much less effective than fibrin in stimulating tissue activator activity (Rånby, 1982).

Vascular Wall Activator

The localisation of plasminogen activator within the endothelial cells of the vascular tree, especially those of veins and venules, was demonstrated by Todd (1959) using his histochemical fibrin slide technique: in this method a thin layer of plasminogen-enriched fibrin is formed over a tissue section and, after incubation, the areas of lysis are related to the histological structures. Pandolfi and his associates (1967) found that the fibrinolytic activity of veins was principally localised to the vasa vasorum of the adventitia. The ability of rabbit vascular endothelial cells to synthesise and secrete activator was later demonstrated by Loskutoff and Edgington (1977).

Properties. The purification of vascular activator has received increasing attention in recent years. The initial attempts were made by Aoki and von Kaulla (1971a) from perfusates of the vascular tree of the lower extremities in cadavers. Their preparation had a molecular weight of 65,000 (Aoki and von Kaulla, 1971b): the more highly purified preparation of Aoki (1974) had a slightly higher molecular weight. Pepper and Allen (1978) purified vascular activator by affinity chromatography on lysine agarose, protamine agarose and gel filtration: this material had a molecular weight of 55,000 and an active center serine was indicated by its inactivation by DFP. There has been differing opinion on whether the activator is a single- or two-chain molecule. An activator purified from saline extracts of canine vascular tissue was found to be a single-chain molecule of 48,000 daltons (Highsmith, 1981), while the preparation by Binder and co-workers (1979), also a single polypeptide chain, had an apparent molecular weight of around 70,000. In contrast, Aasted (1980) purified an activator from vascular perfusates with a molecular weight of 60,000 which, on reduction, gave two bands on SDS gels of 30,000 and 31,000. It is likely, as discussed above, that the two-chain form of activator results from proteolytic cleavage during preparation.

Release from the Vessel Wall. The mechanism of physiological release of activator from the vessel wall into the circulation is not yet clarified. A variety of acute stresses lead to enhanced levels of activator activity in the circulation, events also associated with the release of catecholamines. A link between the two therefore seemed likely. This view was strengthened by the finding that epinephrine injections elicit the release of activator. Some studies, however, have raised doubts on the role of catecholamines in producing activator release since in certain stressful situations, for example, exercise, the rise in the level of catecholamines preceded the release of activator, while in electroconvulsive therapy catecholamine release was reduced in the presence of anesthesia whereas it did not diminish the activator response. In addition, the effect of epinephrine is only partially prevented by β-adrenergic blockade (Cash *et al.*, 1970). Alpha-receptor blockade does not influence the fibrinolytic increase induced by epinephrine (Rosing *et al.*, 1978). This topic is further discussed in Chapter 6.

It has been reported that thrombin can induce the release of plasminogen activator from the isolated perfused dog leg whereas factor Xa had no effect (Hijikata *et al.*, 1980): it was suggested that certain arginine residues of the reactive sites on the vessel wall may participate in the release of activator by proteases. Loskutoff (1981), however, found that thrombin in small quantities decreases the production of plasminogen activator by cultured bovine aortic endothelial cells: the effect was specific and reversible.

The release of activator from the vessel wall in response to venous occlusion is the basis of a widely used test of the ability of the vessel wall to synthesise and release plasminogen activator: this ability has been termed 'fibrinolytic capacity' (Robertson *et al.*, 1972). In this test the activator activity of plasma is measured in blood withdrawn before venous occlusion and, after a 15 to 25 minute period of occlusion induced by a sphygmomanometer cuff on the upper arm inflated to midway between the systolic and diastolic blood pressure, in venous blood from the occluded arm.

Relationship of Vascular Activator to Tissue Activator. All published data point to the identity of vascular wall activator and tissue activator. For example, the tissue activator from human uterus was shown to be immunologically related to the activator obtained by perfusion of cadaver vascular tree (Rijken *et al.*, 1980), while Binder and associates (1981) found similar isolation characteristics, molecular weight, activity on amidolytic substrates, and dependence on fibrin for plasminogen-activating properties for human heart tissue activator and their preparation of cadaver vessel activator.

Circulating Activator

The presence of labile fibrinolytic activity in the plasma of the majority of normal and unstressed individuals was recognised by Fearnley and Tweed (1953) through the use of a technique in which blood was kept cool between withdrawal and assay. Subsequent studies established that plasma fibrinolytic activity in normal subjects and the increase in subjects subjected to such stresses as exercise, electroshock and subcutaneous epinephrine is due to the presence of a plasminogen activator (Sherry *et al.*, 1959; Sawyer *et al.*, 1960).

Assay of Plasma Activator Activity. The most widely used assays of plasma fibrinolytic activity, reflecting the plasma plasminogen activator concentration, are the dilute whole blood or plasma clot lysis time and the euglobulin clot lysis time. Dilution of whole blood or plasma with buffer favours the lysis of clots prepared by the addition of thrombin to such diluted blood or plasma. The alternative technique involves the precipitation of a euglobulin fraction of plasma by acidification and dilution: the bulk of the plasma plasminogen, plas-

155

minogen activator and fibrinogen appear in the euglobulin fraction while the fibrinolytic inhibitors, with the exception of CĪ inactivator and inter-α-trypsin inhibitor (Kluft, 1976), remain in the supernatant fluid. The euglobulin precipitate, redissolved in buffer, may be clotted with thrombin and the lysis time of the clot measured, or the precipitate may be applied to plasminogen-enriched fibrin layers prepared in Petri plates and the area of lysis produced after a standard period of incubation measured.

Properties. The principal problems in the purification of circulating plasminogen activators stem from their low concentration and their lability. A partially purified activator was obtained from post-venous occlusion plasma by gel filtration (Ogston *et al.*, 1976): the activator differed from urokinase in its greater adsorption to forming fibrin and in its lack of inhibition by iodoacetamide. This plasma activator was subsequently shown to be immunologically similar to vascular wall activator (Mackie *et al.*, 1981). Radcliffe and Heinze (1978) used chromatography on lysine-Sepharose to recover activator from human post-exercise plasma: its identity with vascular wall activator was suggested by its similar inhibition profile to that of an activator from homogenised human vein. Husain and colleagues (1981) reported the purification of an activator from human plasma by affinity chromatography on fibrin/Celite followed by gel filtration: the molecular weight was around 63,000 and, because of its low specific activity, it was suggested that it may represent a precursor which is transformed into a more active form by proteases. Using immuno-adsorption chromatography with anti-tissue activator IgG as the immobilised ligand Rijken and co-workers (1980) isolated plasma activator from post-venous occlusion blood. The plasma activator had the same molecular weight as tissue activator prepared from uterus on SDS-polyacrylamide gel electrophoresis, but was slightly larger when examined by gel filtration: it was suggested that this molecular size discrepancy might be the result of a conformational difference related to the transformation of the activator from a tissue form into a soluble form. In the same study Rijken and associates reported that the activator obtained by perfusion of cadaver vascular tree was immunologically related to tissue activator while the activator activity of post-venous occlusion and post-exercise plasma was substantially quenched by anti-tissue activator IgG.

The general conclusion to be drawn from the available data is that one tissue-derived activator is identical or closely similar to the activator obtained from the vessel wall and to that which appears in plasma after such stimuli as exercise and venous occlusion. There is increasing evidence, however, that plasma from resting subjects contains an activator with the immunological and functional properties of urokinase (Dooijewaard *et al.*, 1982; Grasi *et al.*, 1982; Tissot *et al.*, 1982; Wun *et al.*, 1982).

Urokinase

The presence of proteolytic activity in urine was recognised at the end of the last century (Sahli, 1885). Specific fibrinolytic activity in dilute urine was observed by Macfarlane and Pilling (1947) and found to differ from non-specific proteolysis by Williams (1951). Sobel and his associates (1952) attributed the activity to a plasminogen activator and termed it urokinase.

Origin. In early studies cultured monkey and dog kidney cells were observed to accumulate a plasminogen activator with properties similar to those of urokinase (Painter and Charles, 1962). Human kidney cultures produced and released an activator with immunological identity with urokinase: the cells responsible were seen to emerge from small blood vessels (Bernik and Kwaan, 1967). An activator synthesised by cultured human embryo kidney cells was also found to be identical to urokinase (Barlow and Lazer, 1972).

The precise site of urokinase production by the kidney was examined by Astedt (1975) using the fibrin slide histological technique: he found that cultured glomeruli became fibrinolytically active. In further studies (Astedt *et al.*, 1977) early cultures of human fetal kidney were observed to produce a high molecular weight form of urokinase whereas late cultures produced the low molecular weight form. It was suggested that the kidney cells in culture lose the ability to synthesise the light chain of urokinase.

The existence of a proenzyme form of urokinase in cultures of human adult and fetal kidney cells which could be converted into an active form by plasmin or thrombin was suggested by the experiments of Bernik (1973). Nolan and associates (1977) also obtained evidence for a proactivator of urokinase in human embryonic kidney cultures and for its activation by trypsin.

While a renal source for urinary urokinase appears beyond dispute, it has become clear that the ability to produce a urokinase-type activator is not limited to kidney cells, but is a function of many tissues (see below).

Properties. Urokinase prepared in crystalline form was found to be single-chain polypeptide with a molecular weight of 54,000 and an amino-terminal isoleucine (Lesuk *et al.*, 1965, 1967). It contains some six per cent carbohydrate (McLellan *et al.*, 1980). White and co-workers (1966) obtained two separate types of purified urokinase with molecular weights of 31,500 and 54,700: similar molecular weights were reported later by Holmberg *et al.* (1976). It was demonstrated that the lower molecular weight type represents a breakdown product of urokinase which arises during purification (Lesuk *et al.*, 1967). Others have confirmed that the lower molecular weight form is a product of the higher molecular weight form, that conversion takes place catalytically, and that the conversion can be prevented by protease inhibitors (Soberano *et al.*, 1976a, b; Ong *et al.*, 1977). The high molecular weight urokinase preparations of Soberano and colleagues (1976b) comprised a molecule with two polypeptide chains linked by disulphide bonds, and it was suggested that the heavy chain represents the lower molecular weight form. Recent observations have shown that a minichain with a molecular weight of 2426 daltons can be split from the low molecular weight form of urokinase: it probably represents the carboxyl-terminal region of the light chain of the high molecular weight form of urokinase (Henschen and Lottspeich, 1982; Schaller *et al.*, 1982). A tentative scheme for the proteolytic degradation of urokinase is represented in Figure 4.4.

On the basis of its irreversible inhibition by DFP urokinase was designated a serine protease (Landmann and Markwardt, 1970). The active site residues, serine and histidine, are present in the low molecular weight form, and are located in the heavy chain of the high molecular weight form (Ong *et al.*, 1976; Soberano *et al.*, 1976b).

Urokinase activates plasminogen by first-order kinetics, cleaving a single arginyl-bond (see page 144). It appears that the high molecular weight form of urokinase has greater plasminogen-activating activity than the low molecular weight form (Lormeau *et al.*, 1977) although Murano and Aronson (1979) found that the two forms had equivalent activity. Urokinase has proteolytic properties, hydrolysing casein at high concentrations (Alkjaersig *et al.*, 1958). It can also hydrolyse a number of amino acid esters including *p*-toluenesulphonyl-L-arginine methyl ester and lysine ethyl ester (Sherry and Alkjaersig, 1957), *N*-α-acetyl-L-lysine methyl ester (Sherry *et al.*, 1964), *N*-carbo-benzoxy-L-tyrosine *p*-nitrophenyl ester (Lorand and Mozen, 1964), and *N*-α-acetylglycyl-L-lysine methyl ester (Walton, 1967). Urokinase also possesses amidase

Figure 4.4: A Tentative Scheme for the Proteolytic Degradation of Urokinase

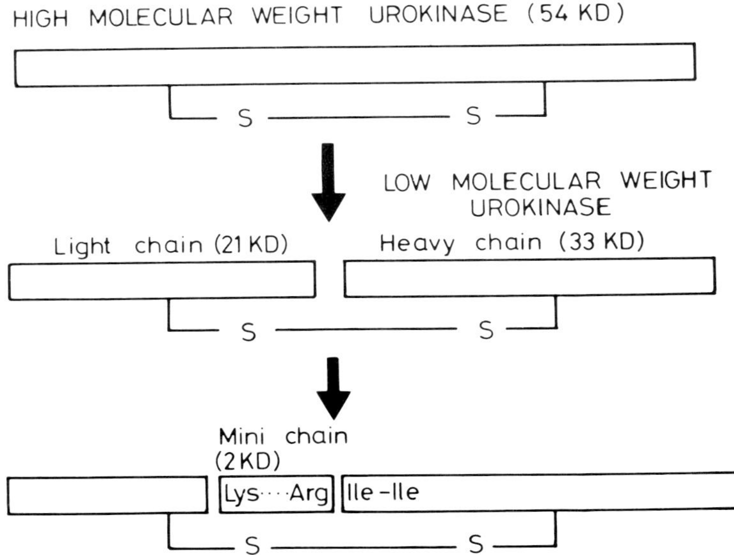

activity on α-acetyl-L-lysine *p*-nitroanilide (Petkov *et al.*, 1973), while the chromogenic substrates Bz-Val-Gly-Arg-*p*-nitroanilide and pyro-Glu-Gly-Arg-*p*-nitroanilide have been developed for the assay of the amidolytic activity of urokinase.

Relationship Between Tissue Activator and Urokinase. Many studies have shown clear differences between urokinase and preparations of tissue activator, vascular activator and the plasma activator which appears after such stimuli as exercise and venous occlusion. Among these were the finding that uterine tissue activator had different substrate requirements from urokinase towards peptide amides (Nieuwenhuizen *et al.*, 1977) and the observation that the two activators are immunologically distinct (Rijken *et al.*, 1981b). A preparation of post-occlusion plasma activator differed from urokinase in its lack of inhibition by iodacetamide (Ogston *et al.*, 1976) and in immunological properties (Mackie *et al.*, 1981), while Astedt (1979) found no increase in anti-urokinase-reacting material in post-venous occlusion plasma with the use of a radioimmunoassay. A number of investigators have also demonstrated that vascular activator can be separated immunologically from urokinase (Aoki and von Kaulla, 1971b; Binder *et al.*, 1979; Kok, 1979; Mackie *et al.*, 1981). Discrepant findings to these were reported by Bernik and colleaues (1974) who claimed that their preparation of human heart activator was immunologically identical to urokinase and that the activity of vascular activator was 80 to 90 per cent quenched by an antiserum to urokinase. The finding, however, was the forerunner of evidence from a number of laboratories that there are two forms of tissue activator, one with functional and immunological similarity to urokinase.

The studies showing the production of a urokinase-type activator by tissues have been performed with cultured human cells. For example, embryonic lung cells were found to produce both a urokinase-type activator and a 73,000 dalton activator resistant to a specific urokinase antiserum whereas embryonic kidney cultures contained only anti-urokinase quenchable activator (Vetterlein *et al.*, 1980). Wilson and associates (1980) also studied a number of tissues in respect of the release of activators by cultured cells susceptible to inhibition by an anti-urokinase serum and found that the activator produced by most of the adult human tissues tested was inhibited. Bernik and colleagues (1979) examined cultures from a number of tissues to determine the type of activator produced: using quenching assays with antisera specific for urokinase and for uterine tissue activator they found that cultures of intestine produced the tissue-type activator only whereas lung and blood vessel cultures produced both types of activator. In a later report by the same investigators (Bernik *et al.*, 1981) both urokinase-type and tissue-type activators were found in cultures of fetal and adult lung, adult spleen, adult thyroid and fetal ureter.

As considered previously, the urokinase-type activator in cultures of kidney cells appears to be in a proenzyme form (Bernik, 1973; Nolan *et al.*, 1977). The pre-urokinase is rapidly activated by plasmin or trypsin, but has been found to be inactivated by thrombin (Wijngaards and Bernik, 1981).

Factor XII-Dependent Activator

The initial observations indicating that the activation of factor XII may lead to the generation of fibrinolytic activity were made by Niewiarowski and Prou-Wartelle (1959). They demonstrated that fibrinolytic activity developed in normal plasma incubated with

160

kaolin, but that this was greatly impaired in factor XII-deficient plasma. The experiments of Iatridis and Ferguson (1962) indicated that factor XII does not activate plasminogen directly, although it has subsequently been reported that amidolytic activity for H-D-valyl-leucyl-L-lysine *p*-nitroanilide evolves in mixtures of plasminogen and factor XII fragments (Goldsmith *et al.*, 1978). After the recognition of the involvement of prekallikrein and high molecular weight kininogen in surface-mediated reactions it was shown that contact-activated fibrinolysis was defective in Fletcher trait plasma (Weiss *et al.*, 1974) and Fitzgerald trait plasma (Saito *et al.*, 1975).

A protein, separate from factor XII, factor XI and plasminogen (termed Hageman factor cofactor), was isolated from plasma and shown to be capable in its activated form of inducing fibrinolytic activity in both normal and Hageman trait plasma (Ogston *et al.*, 1969; Herbert *et al.*, 1972): the material had weak kallikrein activity and did not activate plasminogen directly in purified systems. Preparations of kallikrein purified from human plasma were found to activate plasminogen (Colman, 1969) and it was proposed that prekallikrein acts as a proactivator. This view was supported by the experiments of Laake and Venneröd (1974) who later noted that proactivator could not be demonstrated in Fletcher trait plasma (Venneröd and Laake, 1976). Different findings were reported by Kaplan and Austen (1972): they were able to separate prekallikrein from plasminogen proactivator by the use of SE-Sephadex, Sephadex G-150 and isoelectric focusing, and they proposed that activated factor XII converts proactivator into activator and that the plasmin formed produces factor XII fragments which in a feedback mechanism activate prekallikrein, the kallikrein formed being required for the full activation of factor XII. However, in later studies, Mandle and Kaplan (1979) have found that their plasminogen proactivator is, in fact, prekallikrein while additional proactivator activity was attributed to factor XI. The contribution of factor XI to factor XII-dependent fibrinolytic activity is likely to be small since the plasma of patients with hereditary deficiency of this factor was found to have normal generation of fibrinolytic activity on exposure to kaolin (Ogston *et al.*, 1969). Overall the findings from a number of groups suggests that kallikrein can act as one plasminogen activator, while factor XIIa and factor XIa may also have this property (Figure 4.5). Other agents, however, may be more important final activators in the factor XII-dependent fibrinolytic pathway. A contact-generated activator has been partially purified which has physical properties similar to those of circulating, vascular wall and heart tissue activator, but distinct from those of urokinase (Mackie *et al.*, 1981).

Intrinsic and Extrinsic Activators. One classification of plasma plasminogen activators which has been introduced is a division into 'extrinsic' and 'intrinsic', analogous to that used for the coagulation system. Extrinsic activator refers to that originating from outside the circulation, in particular, from the vascular endothelium, and is therefore synonymous with vascular activator. Intrinsic activators are formed from components within the plasma, the factor XII-dependent activator or activators discussed above being the initial example. The differential inhibition of the intrinsic activators by CĪ inactivator has been used to separate and determine the two classes of activator (Kluft, 1978). The identity of the CĪ inactivator-resistant activity ('extrinsic' activator) with tissue/vascular activator has been shown by its quenching with antibodies against tissue plasminogen activator (Rijken *et al.*, 1980).

In a series of papers Kluft and his associates have postulated the existence of a factor XII-independent intrinsic activator, probably related to that described by Astrup and Rosa (1974). The evidence for such an activator has been indirect, for example, the inhibition of the activity by extracts of *Schistosoma mansoni* (Kluft, 1978; Kluft *et al.*, 1979). From studies using Hageman and Fletcher trait plasmas it was estimated that approximately half of the intrinsic activator activity which can be generated is independent of factor XII (Kluft,

Figure 4.5: Factor XII-Dependent Fibrinolytic Pathways

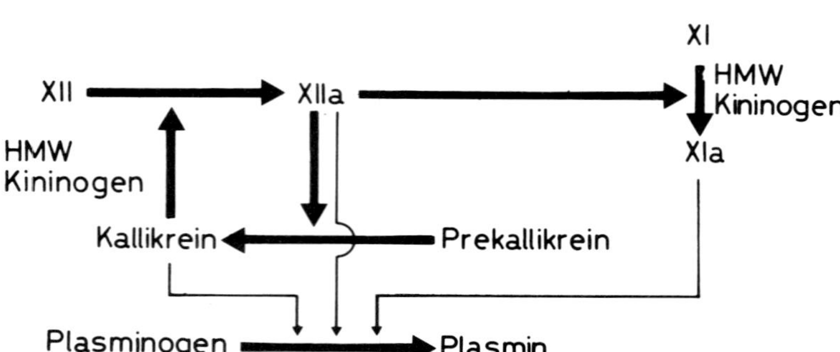

1977, 1979). More recent evidence has been presented to suggest that part at least of the factor XII-independent intrinsic activator is urokinase-related (Kluft *et al.*, 1981). This may be concordant with the detection of urokinase-like activity in human plasma by a number of investigators (see above). For example, Wun and colleagues (1982) purified an agent which had properties which were indistinguishable from a 53,000 dalton component of urinary urokinase in terms of molecular weight, amino acid composition, effect on synthetic peptide substrates, inhibition spectrum and immunological characteristics, and it was suggested that it is present in plasma in a precursor form. The possibility that the precursor form may be converted into the active form by kallikrein or factor XII provides a tentative link to the other intrinsic pathway. At the present time, however, the source of the factor XII-independent intrinsic activator and its relationship to the urokinase-related activator formed in tissue culture is obscure.

Activators in Secretions

Many body secretions and fluids other than blood and urine contain plasminogen activator activity and it has been suggested that the fibrinolytic potential has a function in the maintenance of the patency of small excretory ducts.

Activator activity has been detected in tears (Storm, 1955), milk (Astrup and Sterndorff, 1953), saliva (Albrechtsen and Thayson, 1955/56), seminal fluid (von Kaulla and Shettles, 1953) and aqueous humor (Kwaan and Astrup, 1963; Pandolfi and Kwaan, 1967). Sweat appears to be devoid of activity (Albrechtsen *et al.*, 1958). Albrechtsen and colleagues have also found activator in cerebrospinal fluid, synovial fluid and lymph.

The presence of small quantities of activator in tears was noted by Storm (1955). Using the histochemical fibrin slide technique Pandolfi and Astrup (1967) could not detect activator activity in either the tubules of the lacrimal gland or in its excretory ducts, suggesting that the activator is not secreted by the lacrimal gland. This favors the proposal that the activators in secretions represent circulating plasma activator which has diffused into the lumen of the relevant gland. It is further supported by the finding that the activator in tears is of the tissue activator/vascular wall type, its activity being quenched by antibodies against tissue activator, but not by antiurokinase antibodies (Rijken *et al.*, 1981b).

The activator in milk (galactokinase) has been found by Okamoto and colleagues (1974) to have a molecular weight of 80,000: the activity was higher in colostrum than in mature milk.

Following the original observation (Albrechtsen and Thaysen, 1955) Norén and associates (1975) confirmed the presence of activator activity in saliva. The experiments of Rijken *et al.* (1981b) showed that the activator is, like that in tears, of tissue activator type.

The activator of seminal fluid appears to be derived primarily from Cowper's gland and urethra (Kester, 1971). It has been claimed that seminal fluid contains two closely similar plasminogen activators with molecular weights of 69,000 and 74,000 which are immunologically related to urokinase (Propping *et al.*, 1978). Rijken and co-workers (1981b) found both tissue activator type and urokinase type activators in seminal fluid.

Fibrinolytic activity has been demonstrated in mammalian bile, the activity being attributed to the presence of a plasminogen activator (Oshiba *et al.*, 1969; Norén *et al.*, 1975). King (1972) failed to detect activator activity in bile, but isolated a protease with a molecular weight of around 15,000. Oshiba and Schoenfield (1970) noted that sodium taurocholate induced an increased output of activator into the bile and postulated that this was due to a hypothetical action in transforming a fixed form of tissue activator in liver into a soluble form capable of excretion. On the basis of experiments with extra-corporeal pig liver perfusion Jedrychowski and associates (1972) concluded that bile is the main route for the active excretion of plasma activator. In later studies, however, Oshiba and colleagues (1977) found that the activator activity of human bile, termed bilikinase, was relatively weak and that there was no significant relationship between biliary and plasma fibrinolytic activity.

Activators in Cellular Elements of Blood

Erythrocytes. Fibrinolytic activity was identified in erythrocytes by Kunzer and Haberhausen (1963). A plasminogen activator was subsequently isolated from the stroma-free hemolysate by Semar and colleagues (1969): this activator, termed erythrokinase, could be distinguished from urokinase by differing rates of hydrolysis of synthetic esters and by behavior on acrylamide gel electrophoresis and DEAE-cellulose. The recovery of this activator from red cells was low and its physiological relevance unknown.

Leukocytes. The presence of a plasminogen activator in human granulocytes has been noted by a number of investigators (Goldstein *et al.*, 1971; Granelli-Piperno *et al.*, 1977). Kopitar and associates (1974) isolated an activator from pig leukocytes with a molecular weight of about 30,000.

Platelets. Some early observations suggested that platelets may contain a plasminogen activator, but later investigators have been unable to demonstrate activator activity associated with or derived from platelets (Ekert *et al.*, 1970).

Activators from Macrophages

Unkeless and colleagues (1974) first reported that macrophages can synthesise and secrete plasminogen activators, observing that macrophages from the peritoneal cavity of mice secrete activator *in vitro* following stimulation by thioglycollate. Two activators appeared to be produced; the major macrophage activator had a molecular weight of 48,000 and the other a molecular weight of 28,000. Both were serine proteases. Young cells have been found to be particularly active in producing activator (Vassalli *et al.*, 1978). Exposure to soluble fibrin/fibrinogen complexes has also been reported to stimulate activator production by peritoneal macrophages (Sherman *et al.*, 1976). More recent studies have been unable to detect activator in rabbit alveolar macrophages, the degradation of soluble fibrin complexes being caused by cathepsin B and a cathepsin B-like enzyme (Bang *et al.*, 1981).

Inhibitors of the Fibrinolytic Enzyme System

Plasma Antiplasmins

Alpha$_2$-Antiplasmin

For many years it was accepted that α_2-macroglobulin and α_1-antitrypsin were the major plasma inhibitors of plasmin, the former rapidly-acting and reversible, the latter slowly-acting and irreversible. It is now recognised that the major plasma inhibitor of plasmin is a protein variously termed α_2-antiplasmin, α_2-plasmin inhibitor or primary plasmin inhibitor (Collen, 1976; Moroi and Aoki, 1976; Mullertz and Clemmensen, 1976).

Alpha$_2$-antiplasmin is a single-chain glycoprotein with a carbohydrate content of around 13 per cent and a molecular weight of about 70,000 (Moroi and Aoki, 1976; Wiman and Collen, 1977). Its partial primary structure has been reported (Lijnen *et al.*, 1981c), and some degree of homology with α_1-antitrypsin and antithrombin III shown (Lijnen *et al.*, 1982). It forms a stable 1:1 stoichiometric complex with the B chain of plasmin which has no protease or esterase activity.

The reaction between α_2-antiplasmin and plasmin has been shown by kinetic analysis to take place in two stages. The first is a rapid, reversible second-order reaction involving interactions between the lysine-binding site in the plasmin A chain and corresponding sites in the antiplasmin molecule, and between the active site of plasmin and the corresponding site on antiplasmin. In the second stage there is a slower, irreversible first-order transition (Christensen and Clemmensen, 1977, 1978; Wiman and Collen, 1978a). The lysine-binding sites in plasmin play a dominant role in the rapidity of the reaction with α_2-antiplasmin (Wiman *et al.*, 1978). It appears that the α_2-antiplasmin-binding site is located on the kringle $1 + 2 + 3$ fragment of elastase-digested plasminogen (Thorsen *et al.*, 1981).

It has been proposed that the formation of the stable complex between α_2-antiplasmin and plasmin results from plasmin attack on a specific leucyl-methionyl peptide bond in the carboxyl-terminal portion of the inhibitor: an 8000 dalton peptide is delineated which remains bound to the complex. Following this cleavage a covalent bond is formed between the active serine of plasmin and the carbonyl group of the specific leucyl residue in the antiplasmin (Wiman and Collen, 1979). The complex is sensitive to excess plasmin which splits peptide bonds in the plasmin B chain and in both the amino-terminal and carboxyl-terminal regions of the α_2-antiplasmin.

Evidence has been presented to indicate that two molecular forms of α_2-antiplasmin exist in plasma with differing reactivity towards plasmin and differing ability to bind to fibrin (Clemmensen, 1979; Kluft *et al.*, 1981). The form with affinity for the lysine-binding sites in plasminogen has been estimated to constitute some 65 per cent of the total α_2-antiplasmin (Nilsson *et al.*, 1982); from studies during plasmapheresis it was concluded that the plasminogen-binding form is synthesised primarily while the other type is formed from it.

In addition to its ability to inhibit plasmin, α_2-antiplasmin has been shown to be capable of inhibiting a variety of other proteases participating in blood coagulation and kinin formation. These include factor XII fragments, plasma kallikrein, factor XIa, factor Xa and thrombin (Moroi and Aoki, 1977; Saito *et al.*, 1979). Both urokinase and tissue activator are inhibited slowly by α_2-antiplasmin (Moroi

166

and Aoki, 1976; Korninger and Collen, 1981).

In addition to the rapid inactivation of plasmin, α_2-antiplasmin interferes with the adsorption of plasminogen to fibrin (Aoki *et al.*, 1978). Evidence has also been provided that around 20 per cent of the α_2-antiplasmin is cross-linked to fibrin during blood-clotting, rendering the fibrin clot less susceptible to lysis by plasmin (Sakata and Aoki, 1980). The cross-linking is catalysed by factor XIIIa. Recently reported experiments by Sakata and Aoki (1982) have led them to conclude that the cross-linking of α_2-antiplasmin to fibrin has a significant role in the inhibition of physiological fibrinolysis.

The mean plasma α_2-antiplasmin concentration is approximately 1 μM (Moroi and Aoki, 1976; Mullertz and Clemmensen, 1976; Wiman and Collen, 1977; Aoki and Yamanaka, 1978). The turnover of ^{125}I-labelled α_2-antiplasmin in healthy or atherosclerotic subjects was examined by Collen and Wiman (1979): the plasma half-life was 2.64 ± 0.32 days while the fractional catabolic rate constant was 0.53 ± 0.09 of the plasma pool per day. Synthesis of the inhibitor appears to be by hepatic parenchymal cells (Aoki and Yamanaka, 1978; Högstorp and Saldeen, 1982; Saito *et al.*, 1982). It has been shown that α_2-antiplasmin is susceptible to digestion by neutral proteases from leukocytes (Klingemann *et al.*, 1981).

Alpha$_2$-Macroglobulin

Until the recognition of α_2-antiplasmin α_2-macroglobulin was regarded as the rapidly-acting antiplasmin of plasma. Its role is now recognised as that of the slower reacting inhibitor of plasmin which neutralises excess plasmin after the α_2-antiplasmin has been saturated.

The light chain of plasmin forms a complex in a 1:1 stoichiometry with the 180,000 molecular weight subunit chain of α_2-macroglobulin. Plasmin bound in a complex with α_2-macroglobulin has negligible residual proteolytic activity on such substrates as casein or fibrin; it has been calculated that the fibrinogenolytic activity of the plasmin-α_2-macroglobulin complex is less than one per cent of the uncomplexed plasmin (Harpel and Mosesson, 1973). It does, however, remain esterolytically active (Ganrot, 1967).

It has been proposed that the binding of proteases such as plasmin to α_2-macroglobulin is initiated by a proteolytic attack on the α_2-macroglobulin molecule: this results in a conformational change which irreversibly traps the enzyme within the molecule. The inhibition of proteolytic activity is caused by steric hindrance of access of large substrates to the enclosed enzyme (Barrett and Starkey, 1973). Harpel (1973) has substantiated that complex formation between plasmin and α_2-macroglobulin is accompanied by proteolytic modifications in the subunit structure of the inhibitor. The bond cleaved, as with thrombin, is between the arginine and leucine residue in the 'middle' segment of the polypeptide chain (Sottrup-Jensen *et al.*, 1981).

Alpha₁-antitrypsin

Alpha₁-antitrypsin represents the slow-acting α_1-globulin of Norman and Hill (1958). In purified systems α_1-antitrypsin inhibits plasmin in a time-dependent reaction (Rimon *et al.*, 1966; Gans and Tan, 1967; Crawford and Ogston, 1974). There is evidence that α_1-antitrypsin has two inhibitor sites, one which will bind the trypsin group of proteases and the other which binds the chymotrypsin group of enzymes, and that both locations have a binding site for some aspect of the serine, histidine or aspartic acid residues at the active site of the serine protease (Cohen, 1973).

Antithrombin III

Antithrombin III forms a 1:1 stoichiometric complex with plasmin, and in purified systems it functions as a progressive and time-dependent inhibitor of plasmin. The rate of inactivation is markedly increased in the presence of heparin (Highsmith and Rosenberg, 1974; Crawford and Ogston, 1975). However, it is unlikely that this protease inhibitor has a significant physiological role as an anti-plasmin. The removal of antithrombin III from plasma does not alter the rate or extent of inhibition of plasmin providing normal amounts of α_2-antiplasmin and α_2-macroglobulin are present (Collen, 1976). In a further study it was shown that less than five per cent of the plasma antithrombin was bound to plasmin formed in plasma *in vitro* in the presence of heparin (Semeraro *et al.*, 1978), while only three to eleven per cent of plasmin formed *in vivo* was neutralised by anti-thrombin-heparin complex (Collen *et al.*, 1978).

Cl Inactivator

The ability of Cl inactivator to inhibit the fibrinolytic and proteolytic activities of plasmin in purified systems was first reported by Ratnoff and colleagues (1969), and the observations have subsequently been confirmed by others (Schreiber *et al.*, 1973; Harpel, 1970). Harpel (1970) found that plasmin decreased the inhibitory activity of Cl

168

inactivator in a time-dependent reaction and produced an alteration in its immunoelectrophoretic properties: the conclusion that CÏ inactivator is a plasmin substrate led to the proposal that its inhibitory action on plasmin results from competitive substrate inhibition. In further studies it was shown that plasmin and CÏ inactivator form a 1:1 molar complex with degradation of the inhibitor into several derivatives, one of which retains the ability to complex with plasmin (Harpel and Cooper, 1975). The plasmin-CÏ inactivator complex does not dissociate in the presence of fibrin, but plasmin generated in the presence of fibrin and CÏ inactivator is not inhibited because of its higher affinity for fibrin (Trumpi-Kalshoven, 1978): it is concluded that its influence on fibrinolysis is negligible.

It is now clear that CÏ inactivator has little or no role in the inhibition of plasmin formed in the circulation in the presence of normal concentrations of α_2-antiplasmin and α_2-macroglobulin.

Inter-α-trypsin Inhibitor

Inter-α-trypsin inhibitor was isolated frm human serum by Heide *et al.* (1965) and found to be a labile glycoprotein which readily dissociated into fragments. It has a molecular weight of 177,000 and the normal serum concentration was estimated to be 10-30 mg/100ml. It has been noted by some to have antiplasmin activity in purified systems (Schwick *et al.*, 1967; Steinbuch, 1971), but little or none by others (Gallimore, 1975). It is unlikely to contribute more than a very minor component to the antiplasmin activity of plasma.

Platelet Antiplasmins

The initial descriptions of platelet antifibrinolytic activity was made by Johnson and Schneider (1953) and has been followed by studies demonstrating the association of material inhibitory to fibrinolysis with platelets (Stefanini and Murphy, 1956; Hume, 1958). Alkjaersig (1961) reported that bovine platelets contained an antiplasmin, separable from plasma antiplasmins on kinetic characteristics, which contributed a substantial proportion of the total blood antiplasmin content. Later investigations on human platelets indicated that the platelet contribution to overall blood antiplasmin activity is small (den Ottolander *et al.*, 1967; Ekert *et al.*, 1970). Indeed, Murray and colleagues (1974) failed to obtain any inhibition of plasmin by suspensions of human platelets separated from plasma by gel filtration and used at normal blood concentation.

Progress in the characterisation of platelet antiplasmin activity has been made in recent years, but the precise number and relative importance of these antiplasmins are not finally established. One antiplasmin found in extracts of human platelets is of low molecular weight and is dialysable and heat labile (Ganguly, 1972; Mui *et al.*, 1975). Pig platelets were reported to contain an antiplasmin which could be released by exposure to thrombin, probably from the dense granules; its molecular weight was about 30,000 (Joist *et al.*, 1976). A fraction of human platelets with antiplasmin but not anti-urokinase activity and a molecular weight of around 70,000 was described by Moore and colleagues (1975). Sandjberg Hansen and Clemmensen (1980) have partially purified an inhibitor of plasmin from washed and lysed human platelets: it had an approximate molecular weight of 52,000. While separable from all known plasma protease inhibitors it had functional similarities to α_2-antiplasmin with an equally rapid reaction rate with plasmin.

Established plasma protease inhibitors have been detected in association with platelets. Alpha$_1$-antitrypsin has been identified in platelets and shown to be released on exposure to thrombin (Bagdasarian and Colman, 1978), while both α_2-macroglobulin and α_1-antitrypsin have been shown to be present in membrane and granule fractions of human platelets (Nachman and Harpel, 1976). Alpha$_2$-antiplasmin antigen has been detected in platelets and a mean level of 62 ng/10^9 platelets determined: the protein was released when the platelets were stimulated with thrombin (Plow and Collen, 1981).

It is concluded that platelets contain one or more antiplasmins, incompletely characterised at the present time. While their concentration is low, they may have a regulatory function locally at sites of platelet deposition.

Plasma Antiactivators

A number of the established protease inhibitors have been shown to form inactive complexes slowly with plasminogen activators. These interactions include that between urokinase and α_2-macroglobulin (Ogston *et al.*, 1973), α_2-antiplasmin (Moroi and Aoki, 1976), α_1-antitrypsin (Clemmensen and Christensen, 1976) and antithrombin III (Clemmensen, 1978). Whether separate and specific inhibitors of plasminogen activators exist is a topic of current debate. Such an inhibitor has not yet been isolated and characterised although a variety of experimental evidence suggests their existence. A methodological problem in the investigation of inhibitors of tissue-type plasminogen activators is the lack of a specific and sensitive substrate to assay residual activator activity after incubation with plasma or plasma fractions. In consequence, most studies have used a clot lysis system or fibrin plates to measure residual activator activity. Both methods are theoretically also sensitive to antiplasmin activity.

Using the fibrin plate technique a number of investigators have demonstrated inhibition of urokinase and tissue activator activity by whole plasma (Brakman and Astrup, 1963; Thorsen, 1973; Walker and Ogston, 1982). Urokinase inhibition has also been detected specifically in serum by a clot lysis assay (Bennett, 1967). Others have examined fractions of plasma for inhibitory activity against activators: Hedner and her associates (1970) observed that inhibitory activity against urokinase, measured by a clot lysis technique, eluted in two zones after fractionation of plasma by gel filtration, one in the high molecular weight region and the other just ahead of the albumin peak. Two similar inhibitory fractions of plasma against vascular activator were noted by Aoki and von Kaulla (1971c). One inhibitor of plasminogen activation (Hedner, 1973) was subsequently found to be an inhibitor of activated factor XII (Hedner and Martinsson, 1978). A low molecular weight inhibitor of tissue activator of about 20,000 daltons has been described (Seigel-Ralston and Astrup, 1972).

Evidence has been provided from experiments in which plasminogen-free plasma was fractionated by gel filtration that a number of proteins are capable of inhibiting the activity of urokinase and tissue activator: at least one could be separated from known protease inhibitors (Walker and Ogston, 1982). Inhibition of tissue activator by plasma has also been demonstrated by Kruithof and colleagues (1982): the inhibitory activity was much less in serum prepared from platelet-poor plasma, suggesting that platelets are a major source of this tissue activator inhibitor.

Histidine-rich Glycoprotein

A plasma protein termed histidine-rich 3,8S-α-glycoprotein was described in 1972 (Haupt and Heimburger, 1972; Heimburger *et al.*, 1972). The molecular weight is about 60,000 and it contains two non-covalently linked subunits. It was found later that this protein interacts with the high affinity lysine-binding site of plasminogen and thereby reduces the binding of plasminogen to fibrin (Lijnen *et al.*, 1980). By this mechanism, analogous to that of 6-aminohexanoic acid, histidine-rich glycoprotein has an antifibrinolytic action and may play a role in the regulation of fibrinolysis.

In turnover experiments, using purified ^{125}I-labelled histidine-rich glycoprotein, the plasma radioactivity half-life was found to be 2.93 ± 0.36 days, the fractional catabolic rate constant 0.51 ± 0.06 of the plasma pool per day, and the synthetic rate 1.59 ± 0.35 mg/kg/day. The plasma concentration is around 1.5 μM, and a range of 47 to 147 per cent of a pool of normal donors was found in a group of 99 male and 95 female blood donors (Lijnen *et al.*, 1981a, b).

Platelet Antiactivators

The presence of agents in platelets with inhibitory properties against plasminogen activators has been claimed by a number of investigators. Inhibition of both urokinase and tissue activator by washed platelets was demonstrated in a clot lysis system (den Ottolander *et al.*, 1967, 1969). An agent with inhibitory properties against urokinase and human heart activator and a molecular weight of around 40,000 was partially purified from platelets by Murray and colleagues (1974); a material with anti-urokinase properties was also isolated from platelets by Moore *et al.* (1975). In addition to the inhibitors isolated from platelets, evidence has been provided that a plasma inhibitor of tissue-type activator is derived from platelets (Kruithof *et al.*, 1982). Overall, however, the existence and role of platelet inhibitors of plasminogen activators remains speculative at the present time.

Tissue Inhibitors of Fibrinolysis

Proteins capable of inhibiting plasmin or the activation of plasminogen by urokinase or vascular activator have been isolated from a variety of tissues. Bernik and Kwaan (1971) cultured a number of tissues and examined the supernatant fluid for inhibitory activity: inhibition of plasmin was observed particularly in the supernates of blood vessels whereas lung, kidney, heart, bladder and skin muscle tissue yielded an inhibitor of urokinase. Noordhoek Hegt and

Brakman (1974) detected both antiplasmin and antiactivator activity in the media of human arteries using the fibrinolysis autograph method modified to demonstrate inhibition. Cultured rabbit endothelial cells were shown to contain an inhibitor of fibrinolysis within the cytosol in addition to a membrane-associated plasminogen activator (Losuktoff and Edgington, 1977, 1981). Cultured human endothelial cells may also contain an antiactivator with activity against urokinase and endothelial activator (Loskutoff, 1977). More recently it has been reported that the inhibitor from rabbit endothelial cells blocks the activity of the urokinase-like activator, but not that of the tissue-type activator (Losuktoff and Gerna, 1982). Dosne and associates (1978) have also found that cultures of human umbilical vein endothelial cells prevent activation of plasminogen by urokinase. The antiplasmin activity found in the wall of veins was abolished by an antibody against α_2-antiplasmin (Eriksson and Saldeen, 1980), but the experiments did not indicate whether that inhibitor is adsorbed to the vessel wall from the blood or is produced in the vessel itself.

Inhibitors from the placenta have received considerable attention. Two inhibitors of urokinase activity have been found in extracts of human placenta (Kawano *et al.*, 1968; Uszynski and Abildgaard, 1971). Uszynski and Abildgaard (1971) reported that the major anti-urokinase activity resided in a relatively acidic protein with an estimated molecular weight of 105,000 which inactivated urokinase progressively. The major urokinase inhibitor isolated from placenta by Kawano and colleagues (1970) had a molecular weight of 43,000. A placental inhibitor capable of blocking the action of the activator released from cultured kidney explants was described by Astedt *et al.* (1972), while both vascular activator and tissue activator prepared from human heart were inhibited by the preparation of Aoki and Kawano (1972). A low molecular weight peptide which inhibits urokinase has also been isolated from human placenta (Uszynski, 1980). The mode of action of placental inhibitors is not established, but Holmberg and colleagues (1978) have suggested that the inhibitors complex with the active site serine of urokinase.

Plasminogen-independent Fibrinolysis

The importance of the non-plasmin mediated fibrinolytic activity as a physiological alternative to the plasmin system is presently speculative. Based on the lack of effect of plasminogen removal by affinity chromatography and lack of inhibition by known plasmin inhibitors Moroz and Gilmore (1976) concluded that much of the fibrinolytic activity of normal plasma is independent of plasmin, and they provided evidence pointing to a possible role for the complement system proteases (see Chapter 5). Moroz (1977) later estimated that plasma is responsible for only 20 per cent of the measurable blood fibrinolytic activity, cellular components, in particular, polymorphonuclear leukocytes, accounting for the bulk of the remainder.

Granulocytes

The ability of leukocytes to degrade fibrin has been known since the early years of this century (Rulot, 1904; Opie, 1907). Subsequent studies have shown that leukocytes contain proteases which are capable of digesting casein, fibrin and fibrinogen (Mounter and Atiyeh, 1960; Astrup *et al.*, 1967; Ohlsson, 1971). The leukocyte proteases able to cleave fibrinogen and fibrin have been distinguished from plasmin (Hermann and Miescher, 1965; Ohlsson, 1971; Bilezikian and Nossel, 1977). The neutral proteases in leukocytes include elastases (Ohlsson and Ohlsson, 1974), collagenases (Lazarus *et al.*, 1968) and chymotrypsin-like enzymes (Schmidt and Havemann, 1974). Elastase and cathepsin G have been identified as the major fibrinolytic proteases of leukocytes (Plow, 1980). Elastase appears to be particularly active against fibrinogen (Gramse *et al.*, 1978).

Plow and Edgington (1975) have reported that the fragments produced from fibrinogen by the action of leukocyte protease differ from those produced by plasmin, while Bilezikian and Nossel (1977) found rapid cleavage of all three chains of fibrinogen and noted that the pattern of cleavage of the amino-terminal ends of the Aα- and Bβ-chains of fibrinogen produced by a leukocyte extract was distinct from that seen with plasmin, thrombin and trypsin. Using a more highly purified elastase-like neutral protease, however, Gramse and colleagues (1978) obtained differing results with little degradation of the Bβ- and γ-chains except at high protease concentrations.

Elastase has been shown to be released from leukocytes during clotting, release being induced by events in the early stages of the coagulation pathway (Plow, 1982). Release of leukocyte proteolytic activity may also be induced by lymphokines (Coeugnet, 1979). It has been proposed that the release of proteases from leukocytes constitutes an alternative fibrinolytic system and that their fibrinolytic and anticoagulant properties, due to the fibrinogen degradation

174

products produced, have a physiological role in the coagulation system (Plow and Edgington, 1975). A recent finding is that leukocyte elastase can inactivate the plasmin-inhibitory activities of α_2-antiplasmin and Cl inactivator; this is associated with limited proteolytic cleavage of the inhibitors (Brower and Harpel, 1982).

Mechanism of Physiological Thrombolysis

The mechanism by which fibrin formed in the body is broken down through the action of the fibrinolytic enzyme system has aroused controversy over many years and a number of hypotheses have been advanced. Mullertz (1953) suggested that fibrin has the ability to dissociate plasmin-antiplasmin complexes, and the view that such complexes are central to the fibrinolytic mechanism was elaborated later by Ambrus and Markus (1960). It gained support from experiments reported by their associates (Back *et al.*, 1961), and from the claim that normal plasma contains dissociable plasmin-antiplasmin complexes (Nanninga and Guest, 1964). According to this hypothesis plasmin is formed in the circulation, bound to antiplasmin, but dissociates in the presence of fibrin due to the higher affinity of plasmin for fibrin, allowing thrombolysis to proceed. The lack of significant lysis by plasmin in the presence of excess inhibitor, such as in plasma, makes this hypothesis untenable.

The second important hypothesis was proposed by Sherry and his colleagues (1959) and gave the key role in the lysis of fibrin to plasminogen activator. It was postulated that plasminogen is adsorbed onto fibrin during its formation and that circulating activator diffuses into the fibrin deposit with resulting activation of the intrinsic adsorbed plasminogen (Alkjaersig *et al.*, 1959). The plasmin formed in this site is free to hydrolyse the fibrin, unimpeded by circulating antiplasmins. Any plasmin formed in the circulation is rapidly inactivated by inhibitors so that fibrinogen and other susceptible plasma proteins are protected from proteolysis.

The most recent and generally accepted concept of the mechanism of thrombolysis was propounded by Wiman and Collen (1978b), taking account of the accumulated knowledge on plasminogen and its activation, the discovery of the fast-acting α_2-antiplasmin, and the properties of the vascular wall activator. They proposed that plasminogen is adsorbed onto specific sites on fibrinogen through its lysine-binding sites and is therefore present on the fibrin formed from

175

it. The activator is also strongly adsorbed to fibrin and in the presence of fibrin and its associated plasminogen its activating properties are greatly enhanced. This leads to the rapid and localised formation of plasmin and the consequent cleavage of the fibrin. When associated with fibrin the lysine-binding sites of plasmin are occupied and, as a result, its inactivation by α_2-antiplasmin is very slow. In contrast, any free plasmin released into the plasma will be rapidly neutralised and fibrinogenolysis prevented.

Both the latter two hypotheses emphasise the importance of plasminogen activator in thrombolysis. While activator may diffuse into fibrin deposits from adjacent vascular endothelium, it may reach the fibrin from distant sites in the circulating plasma. The importance of the level of plasminogen activator on thrombolysis highlights the relevance of its assay in physiological studies of fibrinolysis.

References

Aasted, B., (1980) 'Purification and characterization of human vascular plasminogen activator', *Biochimica et Biophysica Acta, 621*, 241-54

Abiko, Y., Iwamoto, M. and Tomikawa, M. (1969) 'Plasminogen-plasmin system. V. A stoichiometric equilibrium complex of plasminogen and a synthetic inhibitor', *Biochimica et Biophysica Acta, 185*, 424-31

Alagille, D. and Soulier, J.P. (1956) 'Action des enzymes protéolytiques sur le sang total "in vitro". Modifications des facteurs de coagulation et du complément', *La Semaine des Hôpitaux de Paris, 32*, 355-9

Albrechtsen, O.K. (1957) 'The fibrinolytic activity of human tissues', *British Journal of Haematology, 3*, 284-91

Albrechtsen, O.K. and Thaysen, J.H. (1955/56) 'Fibrinolytic activity in human saliva', *Acta Physiologica Scandinavia, 35*, 138-45

Albrechtsen, O.K., Storm, O. and Claasen, M. (1958) 'Fibrinolytic activity in some human body fluids', *Scandinavian Journal of Clinical and Laboratory Investigation, 10*, 310-8

Alkjaersig, N. (1961) 'The antifibrinolytic activity of platelets', in S.A. Johnson, R.W. Monto, J.W. Rebuck and R.C. Horn (eds.) *Blood Platelets*, Little, Brown, Boston, pp. 329-36

Alkjaersig, N., Fletcher, A.P. and Sherry, S. (1958) 'The activation of human plasminogen. I. Spontaneous activation in glycerol', *Journal of Biological Chemistry, 233*, 81-5

Alkjaersig, N., Fletcher, A.P. and Sherry, S. (1959) 'The mechanism of clot dissolution by plasmin', *Journal of Clinical Investigation. 38*, 1086-95

Allen, R.A. and Pepper, D.S. (1981) 'Isolation and properties of human vascular plasminogen activator', *Thrombosis and Haemostasis, 45*, 43-50

Ambrus, C.M. and Markus, G. (1960) 'Plasmin-antiplasmin complex as a reservoir of fibrinolytic enzyme', *American Journal of Physiology, 199*, 491-4

Aoki, N. (1974) 'Preparation of plasminogen activator from vascular tree of human cadavers. Its comparison with urokinase', *Journal of Biochemistry, 75*, 731-41

Aoki, N. and von Kaulla, K.N. (1971a) 'The extraction of vascular activator from human cadavers. Some of its properties', *American Journal of Clinical Pathology, 55*, 171-9

Aoki, N. and von Kaulla, K.N. (1971b) 'Dissimilarity of human plasminogen activator and urokinase', *Journal of Laboratory and Clinical Medicine, 78*, 354-62

Aoki, N. and von Kaulla, K.N. (1971c) 'Human serum plasminogen antiactivator: its distinction from antiplasmin', *American Journal of Physiology, 220*, 1137-45

Aoki, N. and Kawano, T. (1972) 'Inhibition of plasminogen activators by naturally occurring inhibitors in man', *American Journal of Physiology, 223*, 1334-7

Aoki, N. and Yamanaka, T. (1978) 'The α_2-plasmin inhibitor levels in liver disease', *Clinica Chimica Acta, 84*, 99-105

Aoki, N., Moroi, M. and Tachiya, K. (1978) 'Effects of α_2-plasmin inhibitor on fibrin clot lysis. Its comparison with α_2-macroglobulin', *Thrombosis and Haemostasis, 39*, 22-31

Astedt, B. (1975) 'Fibrinolysis in cultured cells', *Thrombosis et Diathesis Haemorrhagica, 34*, 644-7

Astedt, B. (1979) 'No crossreaction between circulating plasminogen activator and urokinase', *Thrombosis Research, 14*, 535-9

Astedt, B., Pandolfi, M. and Nilsson, I.M. (1972) 'Inhibitory effect of placenta on plasminogen activation in human organ culture', *Proceedings of the Society for Experimental Biology and Medicine, 139*, 1421-4

Astedt, B., Barlow, G. and Holmberg, L. (1977) 'Time-related release of various molecular forms of urokinase in tissue culture', *Thrombosis Research, 11*, 149-53

Astrup, T. and Permin, P.M. (1947) 'Fibrinolysis in the animal organism', *Nature, 159*, 681-2

Astrup, T. and Rosa, A.T. (1974) 'A plasminogen proactivator-activator system in human blood effective in the absence of Hageman factor', *Thrombosis Research, 4*, 609-13

Astrup, T. and Stage, A. (1952) 'Isolation of soluble fibrinolytic activator from animal tissue', *Nature, 170*, 929

Astrup, T. and Sterndorff, I. (1953) 'A fibrinolytic system in human milk', *Proceedings of the Society for Experimental Biology and Medicine, 84*, 605-8

Astrup, T., Henrichsen, J. and Kwaan, H.C. (1967) 'Protease content and fibrinolytic activity of human leucocytes', *Blood, 29*, 134-8

Bachmann, F., Fletcher, A.P., Alkjaersig, N. and Sherry, S. (1964) 'Partial purification and properties of the plasminogen activator from pig heart', *Biochemistry, 3*, 1578-85

Back, N., Ambrus, J.L. and Mink, I.B. (1961) 'Distribution and fate of I^{131}-labelled components of the fibrinolysin system', *Circulation Research, 9*, 1208-16

Bagdasarian, A. and Colman, R.W. (1978) 'Subcellular localization and purification of platelet α_1-antitrypsin', *Blood, 51*, 139-56

Bang, N.U., Chang, M.L., Mattler, L.E., Burek, P.J., Van Frank, R.M., Zimmermann, R.E., Marks, C.A. and Boxer, L.J. (1981) 'Monocyte/macrophage-mediated catabolism of fibrinogen and fibrin', *Annals of the New York Academy of Sciences, 370*, 568-87

Barlow, G.H. and Lazer, L. (1972) 'Characterization of the plasminogen activator isolated from human embryo kidney cells: comparison with urokinase', *Thrombosis Research, 1*, 201-7

Barnhart, M.I. and Riddle, J.M. (1963) 'Cellular localisation of profibrinolysin (plasminogen)', *Blood, 21*, 306-21

Barrett, A.J. and Starkey, P.M. (1973) 'The interaction of the α_2-macroglobulin with proteinases. Characteristics and specificity of the reaction, and a hypothesis concerning its molecular mechanism', *Biochemical Journal, 133*, 709-24

Bennett, N.B. (1967) 'Inhibitors of plasminogen activation in human serum', *Thrombosis et Diathesis Haemorrhagica, 17*, 12-22

Bernik, M.B. (1973) 'Increased plasminogen activator (urokinase) in tissue culture after fibrin deposition', *Journal of Clinical Investigation, 52*, 823-34

Bernik, M.B. and Kwaan, H.C. (1967) 'Origin of fibrinolytic activity in cultures of the human kidney', *Journal of Laboratory and Clinical Medicine, 70*, 650-61

Bernik, M.B. and Kwaan, H.C. (1971) 'Inhibitors of fibrinolysis in human tissues in culture', *American Journal of Physiology, 221*, 916-21

Bernik, M.B., White, W.F., Oller, E.P. and Kwaan, H.C. (1974) 'Immunological identity of plasminogen activator in human urine, heart, blood vessels, and tissue culture', *Journal of Laboratory and Clinical Medicine, 84*, 546-58

Bernik, M.B., Rijken, D.C. and Wijngaards, G. (1979) 'Production of immunologically distinct plasminogen activators by human tissue in culture', *Thrombosis and Haemostasis, 42*, 414

Bernik, M.B., Wijngaards, G. and Rijken, D.C. (1981) 'Production by human tissues in culture of immunologically distinct, multiple molecular weight forms of plasminogen activators', *Annals of the New York Academy of Sciences, 370*, 592-608

Bilezikian, S. and Nossel, H.L. (1977) 'Unique pattern of fibrinogen cleavage by human leucocyte proteases', *Blood, 50*, 21-8

Binder, B.R., Spragg, J. and Austen, K.F. (1979) 'Purification and characterization of human vascular plasminogen activator derived from blood vessel perfusates', *Journal of Biological Chemistry, 254*, 1998-2003

Binder, B.R., Reissert, G. and Beckmann, R. (1981) 'Isolation and characterization of a plasminogen activator (PA) from human myocardial tissue', *Thrombosis and Haemostasis, 46*, 11

Brakman, P. and Astrup, T. (1963) 'Selective inhibition in human pregnancy blood of urokinase induced fibrinolysis', *Scandinavian Journal of Clinical and Laboratory Investigation, 15*, 603-9

Brower, M.S. and Harpel, P.C. (1982) 'Proteolytic cleavage and inactivation of α_2-plasmin inhibitor and CI inactivator', *Journal of Biological Chemistry, 257*, 9849-54

Camiolo, S.M., Thorsen, S. and Astrup, T. (1971) 'Fibrinogenolysis and fibrinolysis with tissue plasminogen activator', *Proceedings of the Society for Experimental Biology and Medicine, 38*, 277-80

Cartwright, T. (1974) 'Partial purification of a specific plasminogen activator from porcine parotid glands', *Thrombosis et Diathesis Haemorrhagica, 31*, 403-14

Cash, J.D., Woodfield, D.G. and Allan, A.G.E. (1970) 'Adrenergic mechanisms in the systemic plasminogen activator response to adrenaline in man', *British Journal of Haematology, 18*, 487-94

Christensen, U. and Clemmensen, I. (1977) 'Kinetic properties of the primary inhibitor of plasmin from human plasma', *Biochemical Journal, 163*, 389-91

Christensen, U. and Clemmensen, I. (1978) 'Purification and reaction mechanisms of the primary inhibitor of plasmin from human plasma', *Biochemical Journal, 175*, 635-41

Christensen, U. and Mullertz, S. (1974) 'Mechanism of reaction of human plasmin with α-*N*-benzoyl-L-arginine-*p*-nitroanilide', *Biochimica et Biophysica Acta, 334*, 187-98

Claeys, H. and Vermylen, J. (1974) 'Physico-chemical and proenzyme properties of NH_2-terminal glutamic acid and NH_2-terminal lysine human plasminogen', *Biochimica et Biophysica Acta, 343*, 351-9

Clemmensen, I. (1978) 'Inhibition of urokinase by complex formation with human antithrombin III in absence and presence of heparin', *Thrombosis and Haemostasis, 39,* 616-23

Clemmensen, I. (1979) 'Different molecular forms of α_2-antiplasmin', in D. Collen, B. Wiman and M. Verstraete (eds.) *The Physiological Inhibitors of Coagulation and Fibrinolysis,* Elsevier/North Holland, Biomedical Press, Amsterdam, pp. 131-6

Clemmensen, I. and Christensen, U. (1976) 'Inhibition of urokinase by complex formation with human alpha$_1$-antitrypsin', *Biochimica et Biophysica Acta, 429,* 591-9

Coeugnet, E. (1979) 'Lymphokines and thrombosis. Effect of lymphokines on fibrinolysis. Release of fibrinolytic activity by human leucocytes stimulated with lymphokines', *Thrombosis and Haemostasis, 42,* 1240-7

Cohen, A.B. (1973) 'Mechanism of action of α-1-antitrypsin', *Journal of Biological Chemistry, 248,* 7055-9

Cole, E.R. and Bachmann, F.W. (1977) 'Purification and properties of a plasminogen activator from pig heart', *Journal of Biological Chemistry, 252,* 3729-37

Collen, D. (1976) 'Identification and some properties of a new fast-reacting plasmin inhibitor in human plasma', *European Journal of Biochemistry, 69,* 209-16

Collen, D. and de Maeyer, L. (1975) 'Molecular biology of human plasminogen I. Physicochemical properties and microheterogeneity', *Thrombosis et Diathesis Haemorrhagica, 34,* 396-402

Collen, D. and Verstraete, M. (1975) 'Molecular biology of human plasminogen. II. Metabolism in physiological and some pathological conditions in man', *Thrombosis et Diathesis Haemorrhagica, 34,* 403-8

Collen, D. and Wiman, B. (1979) 'Turnover of antiplasmin, the fast-acting plasmin inhibitor of plasma', *Blood, 53,* 313-24

Collen, D., Tytgat, G., Claeys, H., Verstraete, M. and Wallén, P. (1972) 'Metabolism of plasminogen in healthy subjects: effect of tranexamic acid', *Journal of Clinical Investigation, 51,* 1310-8

Collen, D., Kudryk, B., Hessel, B. and Blombäck, B. (1975) 'Primary structure of human fibrinogen and fibrin. Isolation and partial characterization of chains of fragment D', *Journal of Biological Chemistry, 250,* 5808-17

Collen, D., Semeraro, N., Telesforo, P. and Verstraete, M. (1978) 'Inhibition of plasmin by antithrombin-heparin complex. II. During thrombolytic therapy in man', *British Journal of Haematology, 39,* 101-10

Colman, R.W. (1969) 'Activation of plasminogen by human plasma kallikrein', *Biochemical and Biophysical Research Communications, 35,* 273-9

Crawford, G.P.M. and Ogston, D. (1974) 'The influence of α-1-antitrypsin on plasmin, urokinase and Hageman factor cofactor', *Biochimica et Biophysica Acta, 354,* 107-13

Crawford, G.P.M. and Ogston, D. (1975) 'The action of antithrombin III on plasmin and activators of plasminogen', *Biochimica et Biophysica Acta, 391,* 189-92

Donaldson, V.H. (1960) 'Effect of plasmin *in vitro* on clotting factors in plasma', *Journal of Laboratory and Clinical Medicine, 56,* 644-51

Dooijewaard, G., van Iersel, J.J.L., Los, P. and Kluft, C. (1982) 'The intrinsic system of fibrinolysis: identification and partial purification of plasma urokinase', *Haemostasis, 11,* Suppl. 1, 8

Doolittle, R.F., Cassman, K.G., Cottrell, B.A., Friezner, S.J. and Takagi, T. (1977) 'Amino acid sequence on the α-chain of human fibrinogen. Covalent structure of the α-chain portion of fragment D', *Biochemistry, 16,* 1710-5

Dosne, A.M., Dupuy, E. and Bodevin, E. (1978) 'Production of a fibrinolytic inhibitor in cultured endothelial cells derived from human umbilical vein', *Thrombosis Research, 12,* 377-87

Ekert, H., Friedlander, I. and Hardisty, R.M. (1970) 'The role of platelets in fibrinolysis. Studies on the plasminogen activator and anti-plasmin activity of platelets', *British Journal of Haematology, 18,* 575-84

Eriksson, M. and Saldeen, T. (1980) 'Venous macrothromboembolism in an unselected autopsy material and its relation to antiplasmin activity in the vessel wall', *Thrombosis Research, 20,* 555-61

Fearnley, G.R. and Tweed, J.M. (1953) 'Evidence of an active fibrinolytic enzyme in the plasma of normal people with observations on inhibition associated with the presence of calcium', *Clinical Science, 12,* 81-9

Gaffney, P.J. (1973) "Subunit relationships between fibrinogen and fibrin degradation products', *Thrombosis Research, 2,* 201-18

Gaffney, P.J. and Brasher, M. (1973) 'Subunit structure of the plasmin-induced degradation products of crosslinked fibrin', *Biochimica et Biophysica Acta, 295,* 308-13

Gaffney, P.J. and Dubos, P. (1971) 'A structural aspect of human fibrinogen suggested by its plasmin degradation', *FEBS Letters, 15,* 13-6

Gaffney, P.J., Lane, D.A. and Brasher, M. (1975) 'Soluble high-molecular-weight E fragments in the plasmin-induced degradation products of cross-linked fibrin', *Clinical Science and Molecular Medicine, 49,* 149-56

Gaffney, P.J., Joe, F. and Mahmoud, M. (1980) 'Giant fibrin fragments derived from crosslinked fibrin: structure and clinical implication', *Thrombosis Research, 20,* 647-62

Gallimore, M.J. (1975) 'Serum inhibitors of fibrinolysis', *British Journal of Haematology, 31,* 217-31

Ganguly, P. (1972) 'A low molecular weight antiplasmin of human blood platelets', *Clinica Chimica Acta, 39,* 466-8

Ganrot, P.O. (1967) 'Inhibition of plasmin activity by α_2-macroglobulin', *Clinica Chimica Acta, 16,* 328-30

Gans, H. and Tan, B.H. (1967) 'α_1-antitrypsin, an inhibitor for thrombin and plasmin', *Clinica Chimica Acta, 7,* 111-7

Goldsmith, G.H., Saito, H. and Ratnoff, O.D. (1978) 'The activation of plasminogen by Hageman factor (factor XII) and Hageman factor fragments', *Journal of Clinical Investigation, 62,* 54-60

Goldstein, I.M., Wunschmann, B., Astrup, T. and Henderson, E.S. (1971) 'Effects of bacterial endotoxin on the fibrinolytic activity of normal human leukocytes', *Blood, 37,* 447-53

Gramse, M., Bingenheimer, C., Schmidt, W., Egbring, R. and Havemann, K. (1978) 'Degradation products of fibrinogen by elastase-like neutral protease from human granulocytes. Characterization and effects on blood coagulation *in vitro*', *Journal of Clinical Investigation, 61,* 1027-33

Granelli-Piperno, A., Vassalli , J.-D. and Reich, E. (1977) 'Secretion of plasminogen activator by human polymorphonuclear leukocytes. Modulation by glucocorticoids and other effectors', *Journal of Experimental Medicine, 146,* 1693-706

Grasi, B., Jörg, M. and Binder, B.R. (1982) 'Isolation of a plasminogen activator from human plasma by affinity chromatography on anti-urokinase-Sepharose, partial purification of the enzyme', *Haemostasis, 11,* Suppl. 1, 9

Groskopf, W.R., Hsieh, B., Summaria, L. and Robbins, K.C. (1968) 'The specificity of human plasmin on the B-chain of oxidised bovine insulin', *Biochimica et Biophysica Acta, 168,* 376-73

Gurewich, V., Hyde, E. and Lipinski, B. (1975) 'The resistance of fibrinogen and soluble fibrin monomer in blood to degradation by a potent plasminogen activator derived from cadaver limbs', *Blood, 46,* 555-65

Harker, L.A., Schmer, G. and Slichter, S.J. (1972) 'Plasminogen kinetics in man', *Abstracts IIIrd Congress of the International Society of Thrombosis and Haemostasis,* p. 294

Harpel, P.C. (1970) 'Cl inactivator inhibition by plasmin', *Journal of Clinical Investigation, 49,* 568-75

Harpel, P.C. (1973) 'Studies on human plasma α_2-macroglobulin-enzyme interactions. Evidence for proteolytic modification of the subunit chain structure', *Journal of Experimental Medicine, 138,* 508-21

Harpel, P.C. and Cooper, N.R. (1975) 'Studies on plasma Cl inactivator-enzyme interactions. I. Mechanisms on interaction with CIs, plasmin and trypsin', *Journal of Clinical Investigation, 55,* 593-604

Harpel, P.C. and Mosesson, M.W. (1973) 'Degradation of human fibrinogen by plasma α_2-macroglobulin-enzyme complexes', *Journal of Clinical Investigation, 52,* 2175-84

Haupt, H. and Heimburger, N. (1972) 'Humanserumproteine mit holer Affinität zu Carboxymethylcellulose, I', *Hoppe-Seyler's Zeitschrift fur Physiologische Chemie, 353,* 1125-32

Haverkate, F. and Timan, G. (1977) 'Protective effect of calcium in plasmin degradation of fibrinogen and fibrin fragment D', *Thrombosis Research, 10,* 803-12

Hayes, M.L. and Castellino, F.J. (1979a) 'Carbohydrate of the human plasminogen variants. I. Carbohydrate composition, glycopeptide isolation, and characterization', *Journal of Biological Chemistry, 254,* 8768-71

Hayes, M.L. and Castellino, F.J. (1979b) 'Carbohydrate of the human plasminogen variants. II. Structure of the asparagine-linked oligosaccharide unit', *Journal of Biological Chemistry, 254,* 8772-6

Hayes, M.L. and Castellino, F.J. (1979c) 'Carbohydrate of the human plasminogen variants. III. Structure of the *O*-glycosidically linked oligosaccharide unit', *Journal of Biological Chemistry, 254,* 8777-80

Hedner, U. (1973) 'Studies on an inhibitor of plasminogen activation in human serum', *Thrombosis et Diathesis Haemorrhagica, 30,* 414-24

Hedner, U. and Martinsson, G. (1978) 'Inhibition of activated Hageman factor by an inhibitor of the plasminogen activation', *Thrombosis Research, 12,* 1015-23

Hedner, U., Nilsson, I.M. and Jacobsen, C.D. (1970) 'Demonstration of low content of fibrinolytic inhibitors in individuals with high fibrinolytic capacity', *Scandinavian Journal of Clinical and Laboratory Investigation, 25,* 329-36

Heide, K., Heimburger, N. and Haupt, H. (1965) 'An inter-alpha trypsin inhibitor of human serum', *Clinica Chimica Acta, 11,* 82-5

Heimburger, N., Haupt, H., Kranz, T. and Baudner, S. (1972) 'Humanserumproteine mit holer Affinität zu Carboxymethylcellulose, II', *Hoppe-Seyler's Zeitschrift fur Physiologische Chemie, 353,* 1133-40

Henschen, A. and Lottspeich, F. (1982) 'Structural characterisation of different molecular weight forms of urokinase', *Haemostasis, 11,* Suppl. 1, 54

Herbert, R.J., Ogston, D. and Douglas, A.S. (1972) 'Further purification and properties of Hageman factor cofactor', *Biochimica et Biophysica Acta, 271,* 371-7

Hermann, G. and Miescher, P.A. (1965) 'Differentiation of leucocytic fibrinolytic enzymes from plasmin by the use of plasmatic proteolytic inhibitors', *International Archives of Allergy and Applied Immunology, 27,* 346-54

Hessel, B. (1975) 'On the structure of the COOH-terminal part of the Aα-chain of human fibrinogen', *Thrombosis Research, 7,* 75-87

Highsmith, R.F. (1981) 'Isolation and properties of a plasminogen activator derived from canine vascular tissue', *Journal of Biological Chemistry, 256,* 6788-95

Highsmith, R.F. and Kline, D.L. (1971) 'Kidney: primary source of plasminogen after acute depletion in the cat', *Science, 174,* 141-2

Highsmith, R.F. and Rosenberg, R.D. (1974) 'The inhibition of human plasmin by human antithrombin-heparin cofactor', *Journal of Biological Chemistry, 249,* 4335-8

Hijikata, A., Hirata, M. and Kitaguchi, H. (1980) 'Effect of proteases on plasminogen activator release from isolated perfused dog leg', *Thrombosis Research, 20,* 521-31

Hobart, M.J. (1979) 'Genetic polymorphism of human plasminogen', *Annals of Human Genetics, 42,* 419-23

Hochschwender, S.M. and Laursen, R.A. (1981) 'The lysine binding sites of human plasminogen. Evidence for a critical tryptophan in the binding site of kringle 4', *Journal of Biological Chemistry, 256,* 11172-6

Högstörp, H. and Saldeen, T. (1982) 'Rat hepatocytes synthesize alpha$_2$-antiplasmin', *Haemostasis, 11,* Suppl. 1, 46

Holmberg, L., Bladh, B. and Astedt, B. (1976) 'Purification of urokinase by affinity chromatography', *Biochimica et Biophysica Acta, 445,* 215-22

Holmberg, L., Leeander, I., Persson, B. and Astedt, B. (1978) 'An inhibitor from placenta specifically binds urokinase and inhibits plasminogen activator released from ovarian carcinoma in tissue culture', *Biochimica et Biophysica Acta, 544,* 128-37

Hoylaerts, M., Rijken, D.C., Lijnen, H.R. and Collen, D. (1982) 'Kinetics of the activation of plasminogen by human tissue plasminogen activator. Role of fibrin', *Journal of Biological Chemistry, 257,* 2912-9

Hudry-Clergeon, G., Paturel, L. and Suscillon, M. (1974) 'Identification d'un complexe (D-D) . . . E dans les produits de dégradation de la fibrine bovine stailisée par le facteur XIII', *Pathologie et Biologie (Paris), 22,* Suppl., 47-52

Hume, R. (1958) 'An inhibitory effect of platelets on fibrinolysis', *Scottish Medical Journal, 3,* 479-83

Husain, S.S., Lipinski, E. and Gurewich, V. (1981) 'Rapid purification of a high-affinity plasminogen activator from human blood plasma by specific adsorption on fibrin/Celite', *Proceedings of the National Academy of Sciences, 78,* 4625-9

Iatridis, S.G. and Ferguson, J.H. (1962) 'Active Hageman factor: A plasma lysokinase of the human fibrinolytic system', *Journal of Clinical Investigation, 41,* 1277-87

Jedrychowski, A., Parbhoo, S.P. and Hillenbrand, P. (1972) 'Plasminogen activator in bile during extracorporeal perfusion of pig liver', *Gut, 13,* 54-7

Johnson, S.A. and Schneider, C.L. (1953) 'The existence of antifibrinolysin activity in platelets', *Science, 117,* 229-30

Joist. J.H., Niewiarowski, S., Nath, N. and Mustard, J.F. (1976) 'Platelet antiplasmin: its extrusion during the release reaction, subcellular localization, characterization and relationship to antiheparin in pig platelets', *Journal of Laboratory and Clinical Medicine, 87,* 659-69

Kaplan, A.P. and Austen, K.F. (1972) 'The fibrinolytic pathway of human plasma. Isolation and characterization of the plasminogen proactivator', *Journal of Experimental Medicine, 136,* 1378-93

von Kaulla, K.N. and Shettles, L.B. (1953) 'Relationship between human seminal fluid and the fibrinolytic system', *Proceedings of the Society for Experimental Biology and Medicine, 83,* 692-4

Kawano, T., Morimoto, K. and Iemura, Y. (1968) 'Urokinase inhibitor in human placenta', *Nature, 217,* 253-4

Kawano, T., Morimoto, K. and Iemura, Y. (1970) 'Partial purification and properties of urokinase inhibitor from human placenta', *Journal of Biochemistry*, *67*, 333-42

Kester, R.C. (1971) 'The distribution of plasminogen activator in the male genital tract', *Journal of Clinical Pathology*, *24*, 726-31

King, J.B. (1972) 'Fibrinolysis in bile', *Thrombosis et Diathesis Haemorrhagica*, *28*, 299-305

Klingemann, H.-G., Egbring, R., Holst, F., Gramse, M. and Havemann, K. (1981) 'Digestion of α_2-antiplasmin inhibitor by neutral proteases from human leucocytes', *Thrombosis Research*, *24*, 479-83

Kluft, C. (1976) 'Occurrence of Cl inactivator and other proteinase inhibitors in euglobulin fractions and their influence on fibrinolytic activity', *Haemostasis*, *5*, 136-46

Kluft, C. (1977) 'An inventory of plasminogen activators in human plasma', *Thrombosis and Haemostasis*, *38*, 134

Kluft, C. (1978) 'Cl-inactivator-resistant fibrinolytic activity in plasma euglobulin fractions: its relation to vascular activator in blood and its role in euglobulin fibrinolysis', *Thrombosis Research*, *13*, 135-51

Kluft, C. (1979) 'Studies on the fibrinolytic system in human plasma: quantitative determination of plasminogen activators and proactivators', *Thrombosis and Haemostasis*, *41*, 365-83

Kluft, C., Trumpi-Kalshoven, M.M. and Deelder, A.M. (1979) 'Factor XII-independent activator generation in human plasma', in J.F. Davidson, V. Cepelak, M.M. Samama and P.C. Desnoyers (eds.) *Progress in Chemical Fibrinolysis and Thrombolysis*, Vol. 4, Churchill Livingstone, Edinburgh, pp. 362-7

Kluft, C., Los, P. and Jie, A.F.H. (1981) 'Assay and occurrence of two molecular forms of α_2-antiplasmin in plasma', *Thrombosis and Haemostasis*, *46*, 281

Kluft, C., Wijngaards, G. and Jie, A.F.H. (1981) 'The factor XII-independent plasminogen proactivator system of plasma includes urokinase-related activity', *Thrombosis and Haemostasis*, *46*, 343

Kok, P. (1979) 'Separation of plasminogen activators from human uterine tissue and a comparison with activators from human urine and porcine tissue', *Thrombosis and Haemostasis*, *41*, 718-33

Kok, P. and Astrup, T. (1969) 'Isolation and purification of a tissue plasminogen activator and its comparison with urokinase', *Biochemistry*, *8*, 79-86

Kopitar, M., Stegnar, M., Accetto, B. and Lopez, D. (1974) 'Isolation and characterization of plasminogen activator from pig leucocytes', *Thrombosis et Diathesis Haemorrhagica*, *31*, 72-85

Korninger, C. and Collen, D. (1981) 'Inhibition of human tissue plasminogen activator by human plasma: no evidence for a specific antiactivator', *Thrombosis and Haemostasis*, *46*, 280

Kruithof, E.K.O. and Bachmann, F. (1981) 'Relative affinity of vascular and tissue plasminogen activator to fibrin and fibrinogen', *Thrombosis and Haemostasis*, *46*, 163

Kruithof, E.K.O., Ransijn, A. and Bachmann, F. (1982) 'Inhibition of tissue plasminogen activator (TA) by human plasma', *Haemostasis 11*, Suppl. 1, 60

Künzer, W. and Haberhausen, A. (1963) 'Fibrinolytic activity of human erythrocytes', *Nature*, *198*, 396-7

Kwaan, H.C. and Astrup, T. (1963) 'Localisation of fibrinolytic activity in the eye', *Archives of Pathology*, *76*, 595-601

Laake, K. and Venneröd, A.M. (1974) 'Factor XII-induced fibrinolysis: studies on the separation of prekallikrein, plasminogen proactivator and factor XI in human plasma', *Thrombosis Research*, *4*, 285-302

Landmann, H. and Markwardt, F. (1970) 'Irreversible synthetische Inhibitoren der Urokinase', *Experientia, 26,* 145-7

Lazarus, G.S., Daniels, J.R., Brown, R.S., Bladen, H.A. and Fullmer, H.M. (1968) 'Degradation of collagen by a human granulocyte collagenolytic system', *Journal of Clinical Investigation, 47,* 2622-9

Lerch, P.G., Rickli, E.E., Lergier, W. and Gillessen, D. (1980) 'Localization of individual lysine-binding regions in human plasminogen and investigations on their complex-forming properties', *European Journal of Biochemistry, 107,* 7-13

Lesuk, A., Terminiello, L. and Traver, J.H. (1965) 'Crystalline human urokinase: some properties', *Science, 147,* 880-2

Lesuk, A., Terminiello, L., Traver, J.H. and Groff, J.L. (1967) 'Biochemical and biophysical studies of human urokinase', *Thrombosis et Diathesis Haemorrhagica, 18,* 293-4

Libeskind, I.C., Lipinski, B. and Gurewich, V. (1981) 'Binding of blood plasminogen activator to fibrinogen, fibrin monomer and fibrin', *Thrombosis and Haemostasis, 46,* 163

Lijnen, H.R., Hoylaerts, M. and Collen, D. (1980) 'Isolation and characterisation of a glycoprotein with affinity for the lysine binding site (LBS) of plasminogen', *Journal of Biological Chemistry, 255,* 10214-22

Lijnen, H.R., Jacobs, G. and Collen, D. (1981a) 'Histidine-rich glycoprotein in a normal and a clinical population', *Thrombosis Research, 22,* 519-23

Lijnen, H.R., De Cock, F. and Collen, D. (1981b) 'Turnover of human histidine-rich glycoprotein in healthy subjects and during thrombolytic therapy', *Thrombosis Research, 23,* 121-31

Lijnen, H.R., Wiman, B., Van Hoef, B. and Collen, D. (1981c) 'Partial primary structure of human α_2-antiplasmin', *Thrombosis and Haemostasis, 46,* 282

Lijnen, H.R., Wiman, B. and Collen, D. (1982) 'Comparison of the partial primary structure of human α_2-antiplasmin with other plasma protease inhibitors', *Haemostasis, 11,* Suppl. 1, 55

Lloyd, D.A., Cederholm-Williams, S.A. and Sharp, A.A. (1981) 'Binding of plasminogen and vascular plasminogen activator to fibrin and the fibrin alpha-chain', *Thrombosis and Haemostasis, 46,* 163

Lorand, L. and Mozen, M.M. (1964) 'Ester-hydrolysing activity of urokinase preparations', *Nature, 201,* 392-3

Lormeau, J.C., Goulay, J., Vairel, E.G. and Choay, J. (1977) 'The action of urokinases on plasminogens', *Thrombosis and Haemostasis, 38,* 257

Loskutoff, D.J. (1977) 'An inhibitor of plasminogen activator in endothelial cells', *Journal of Cell Biology, 75,* 80a

Loskutoff, D.J. (1981) 'Effect of thrombin on the production of plasminogen activator by endothelial cells', *Thrombosis and Haemostasis, 46,* 82

Loskutoff, D.J. and Edgington, T.S. (1977) 'Synthesis of a fibrinolytic activator and inhibitor by endothelial cells', *Proceedings of the National Academy of Sciences, 74,* 3903-7

Loskutoff, D.J. and Edgington, T.S. (1981) 'An inhibitor of plasminogen activator in rabbit endothelial cells', *Journal of Biological Chemistry, 256,* 4142-5

Loskutoff, D.J. and Gerna, M. (1982) 'An inhibitor in rabbit endothelial cells that recognises urokinase-like but not tissue-type plasminogen activators', *Haemostasis, 11,* Suppl. 1, 48

Macfarlane, R.G. and Pilling, J. (1947) 'Fibrinolytic activity of normal urine', *Nature, 159,* 779

Mackie, M., Booth, N.A. and Bennett, B. (1981) 'Comparative studies on human activators of plasminogen', *British Journal of Haematology, 47,* 77-90

Mandle, R.J. and Kaplan, A.P. (1979) 'Hageman-factor-dependent fibrinolysis:

Generation of fibrinolytic activity in the interaction of human activated factor XI and plasminogen', *Blood, 54*, 850-62

Marder, V.J., Shulman, N.R. and Carroll, W.R. (1969) 'High molecular weight derivatives of human fibrinogen produced by plasmin', *Journal of Biological Chemistry, 244*, 2111-9

Markus, G., De Pasquale, J.L. and Wissler, F.C. (1978) 'Quantitative determination of the binding of ε-aminoproic acid to native plasminogen', *Journal of Biological Chemistry, 253*, 727-32

McLellan, W.L., Vetterlein, D. and Roblin, R. (1980) 'The glycoprotein nature of human plasminogen activators', *FEBS Letters, 115*, 181-4

Miller, L.L. and Bale, W.F. (1954) 'Synthesis of all plasma protein fractions except gamma globulin by the liver. Use of zone electrophoresis and lysine-Σ-C^{14} to define plasma proteins synthesized by the isolated perfused liver', *Journal of Experimental Medicine, 99*, 125-32

Mirsky, I.A., Perisutti, G. and Davis, N.C. (1959) 'Destruction of glucagon, adrenocorticotropin and somatotropin by human blood plasma', *Journal of Clinical Investigation, 38*, 14-20

Moore, S., Pepper, D.S. and Cash, J.D. (1975) 'The isolation and characterisation of a platelet-specific β-globulin (β-thromboglobulin) and the detection of anti-urokinase and antiplasmin released from thrombin-aggregated washed human platelets', *Biochimica et Biophysica Acta, 379*, 360-9

Moroi, M. and Aoki, N. (1976) 'Isolation and characterization of α_2-plasmin inhibitor from human plasma', *Journal of Biological Chemistry, 251*, 5956-65

Moroi, M. and Aoki, N. (1977) 'Inhibition of proteases in coagulation, kinin-forming and complement systems by α_2-plasmin inhibitor', *Journal of Biochemistry, 82*, 969-72

Moroz, L.A. (1977) 'Increased blood fibrinolytic activity after aspirin ingestion', *New England Journal of Medicine, 296*, 525-9

Moroz, L.A. and Gilmore, N.J. (1976) 'Fibrinolysis in normal plasma and blood: evidence for significant mechanisms independent of the plasminogen-plasmin system', *Blood, 48*, 531-45

Mounter, L.A. and Atiyeh, W. (1960) 'Proteases of human leucocytes', *Blood, 15*, 52-9

Mui, P.T.K., James, H.L. and Ganguly, P. (1975) 'Isolation and properties of a low molecular weight antiplasmin of human blood platelets and serum', *British Journal of Haematology, 29*, 627-37

Mullertz, S. (1953) 'The action of plasmin on fibrin and fibrinogen in blood', *Acta Physiologica Scandinavica, 28*, 29-40

Mullertz, S. and Clemmensen, I. (1976) 'The primary inhibitor of plasmin in human plasma', *Biochemical Journal, 159*, 545-53

Murano, G. and Aronson, D.L. (1979) 'High and low molecular weight urokinase', *Thrombosis and Haemostasis, 42*, 1066-8

Murray, J., Crawford, G.P.M., Ogston, D. and Douglas, A.S. (1974) 'Studies on an inhibitor of plasminogen activators in human platelets', *British Journal of Haematology, 26*, 661-5

Nachman, R.L. and Harpel, P.C. (1976) 'Platelet α_2-macroglobulin and α_1-antitrypsin', *Journal of Biological Chemistry, 251*, 4514-21

Nanninga, L.B. and Guest, M.M. (1964) 'On the interaction of fibrinolysin (plasmin) with the inhibitors antifibrinolysin and soy bean trypsin inhibitor', *Archives of Biochemistry and Biophysics, 108*, 542-51

Nieuwenhuizen, W., Wijngaards, G. and Groenvald, E. (1977) 'Synthetic substrates and the discrimination between urokinase and tissue plasminogen activator activity', *Thrombosis Research, 11*, 87-9

Niewiarowski, S. and Prou-Wartelle, O. (1959) 'Rôle du facteur contact (facteur

Hageman) dans la fibrinolyse', *Thrombosis et Diathesis Haemorrhagica, 3,* 593-603

Nilsson, I., Cedergren, B. and Wiman, B. (1982) 'On the significance of the two forms of human α_2-antiplasmin with different affinity for the lysine-binding sites in plasminogen', *Haemostasis, 11,* Suppl. 1, 62

Nishimukai, H., Kera, Y., Sakata, K. and Yamasawa, K. (1981) 'Genetic polymorphism of plasminogen: A new basic variant (PLG B) and population study in Japanese', *Vox Sanguinis, 40,* 422-5

Nolan, C., Hall, L.S., Barlow, G.H. and Tribby, I.I.E. (1977) 'Plasminogen activator from human embryonic kidney cell cultures', *Biochimica et Biophysica Acta, 496,* 384-400

Noordhoek Hegt, V and Brakman, P. (1974) 'Histochemical study of an inhibitor of fibrinolysis in the human arterial wall', *Nature, 248,* 75-6

Norén, I., Ramström, G. and Wallén, P. (1975) 'Fibrin plate method with reagents purified by affinity chromatography and its use for determination of fibrinolytic and other proteolytic activity in saliva, bile and plasma', *Haemostasis, 4,* 110-24

Norman, P.S. and Hill, B.M. (1958) 'Studies on the plasmin system. III. Physical properties of the two plasmin inhibitors in plasma', *Journal of Experimental Medicine, 108,* 639-49

Nussenweig, V., Seligmann, M., Pelmont, J. and Grabar, P. (1961) 'Les produits de dégradation du fibrinogene humain par la plasmin', *Annales de l'institut Pasteur, 100,* 377-89

Ogston, D., Ogston, C.M., Ratnoff, O.D. and Forbes, C.D. (1969) 'Studies on a complex mechanism for the activation of plasminogen by kaolin and by chloroform: the participation of Hageman factor and additional cofactors', *Journal of Clinical Investigation, 48,* 1786-801

Ogston, D., Bennett, B., Herbert, R.J. and Douglas, A.S. (1973) 'The inhibition of urokinase by α_2-macroglobulin', *Clinical Science, 44,* 73-9

Ogston, D., Bennett, B. and Mackie, M. (1976) 'Properties of a partially purified preparation of a circulating plasminogen activator', *Thrombosis Research, 8,* 275-84

Ohlsson, K. (1971) 'Properties of leucocyte protease', *Clinica Chimica Acta, 32,* 399-405

Ohlsson, K. and Olsson, I. (1974) 'The neutral proteases of human granulocytes. Isolation and partial characterization of granulocyte elastases', *European Journal of Biochemistry, 42,* 519-27

Okamoto, U., Nagamatsu, Y. and Matsumoto, M. (1974) 'Studies on a plasminogen activating system in human milk. I. Isolation of plasminogen activator and proactivator from human milk, and estimation of their molecular weights', *Acta Haematologica Japonica, 37,* 121-31

Olexa, S.A., Budzynski, A.Z. and Marder, V.J. (1979) 'Modification of high molecular weight plasmic degradation products of human crosslinked fibrin', *Biochimica et Biophysica Acta, 576,* 39-50

Ong, E.B., Johnson, A.J. and Schoellmann, G. (1976) 'Identification of an active site histidine in urokinase', *Biochimica et Biophysica Acta, 429,* 252-7

Ong, E.B., Soberano, M.E., Johnson, A.J. and Schoellmann, G. (1977) 'Studies on the biochemistry of urokinase', *Thrombosis and Haemostasis, 38,* 801-8

Opie, E.L. (1907) 'Experimental pleurisy. Resolution of a fibrinous exudate', *Journal of Experimental Medicine, 9,* 391

Oshiba, S. and Schonfield, L.J. (1970) 'Plasminogen activator in bile stimulated by sodium taurcholate in isolated hamster livers', *Proceedings of the Society for Experimental Biology and Medicine, 133,* 89-92

Oshiba, S., Hata, S. and Okamoto, S. (1969) 'A plasminogen activator in mammalian bile', *Japanese Journal of Physiology, 19*, 212-9

Oshiba, S., Ariga, T., Sawai, H. and Imai, H. (1977) 'Some aspects of biliary fibrinolysis', *Thrombosis and Haemostasis, 38*, 273

den Ottolander, G.J.H., Leijnse, B. and Cremer-Elfrink, H.M.J. (1967) 'Plasmatic and platelet antiplasmins and anti-activators', *Thrombosis et Diathesis Haemorrhagica, 18*, 404-15

den Ottolander, G.J.H., Leijnse, B. and Cremer-Elfrink, H.M.J. (1969) 'Plasmatic and thrombocytic antiplasmins and anti-activators. II', *Thrombosis et Diathesis Haemorrhagica, 21*, 26-34

Painter, R.H. and Charles, A.F. (1962) 'Characterization of a soluble plasminogen activator from kidney cell cultures', *American Journal of Physiology, 202*, 1125-30

Pandolfi, M. and Astrup, T. (1967) 'A histochemical study of the fibrinolytic activity. Cornea, conjunctiva, and lacrimal gland', *Archives of Ophthalmology, 77*, 258-64

Pandolfi, M. and Kwaan, H.C. (1967) 'Fibrinolysis in the anterior segment of the eye', *Archives of Ophthalmology, 77*, 99-104

Pandolfi, M., Nilsson, I.M. and Robertson, B. (1967) 'Fibrinolytic activity of human veins', *Lancet, ii*, 127-8

Pasquini, R. and Hershgold, E.J. (1973) 'Effects of plasmin on human factor VIII (AHF)', *Blood, 41*, 105-11

Pepper, D.S. and Allen, R. (1978) 'Isolation and characterization of human cadaver vascular endothelial activator', in J.F. Davidson, R.M. Rowan, M.M. Samama and P.C. Desnoyers (eds.) *Progress in Chemical Fibrinolysis and Thrombolysis*, Vol. 3, Raven Press, New York, pp. 91-8

Petkov, D., Christova, E. and Karadjova, M. (1973) 'Amidase activity of urokinase I. Hydrolysis of α-N-acetyl-L-lysine p-nitroanilide', *Thrombosis et Diathesis Haemorrhagica, 29*, 276-85

Pillemer, L., Ratnoff, O.D., Blum, L. and Lepow, I.H. (1953) 'The inactivation of complement and its components by plasmin', *Journal of Experimental Medicine, 97*, 573-89

Pizzo, S.V., Schwartz, M.L., Hill, R.L. and McKee, P.A. (1972) 'The effect of plasmin on the subunit structure of human fibrinogen', *Journal of Biological Chemistry, 247*, 636-45

Pizzo, S.V., Taylor, L.M., Schwartz, M.L., Hill, R.L. and McKee, P.A. (1973) 'Subunit structure of fragment D from fibrinogen and crosslinked fibrin', *Journal of Biological Chemistry, 248*, 4584-90

Plow, E.F. (1980) 'The major fibrinolytic proteases of human leucocytes', *Biochimica et Biophysica Acta, 630*, 47-56

Plow, E.F. (1982) 'Leukocyte elastase release during blood coagulation. A potential mechanism for activation of the alternative fibrinolytic pathway', *Journal of Clinical Investigation, 69*, 564-72

Plow, E.F. and Collen, D. (1981) 'Presence and release of α_2-antiplasmin from human platelets', *Blood, 58*, 1069-74

Plow, E.F. and Edgington, T.S. (1975) 'An alternative pathway for fibrinolysis. I. The cleavage of fibrinogen by leucocyte proteases at physiologic pH', *Journal of Clinical Investigation, 56*, 30-8

Prokopowicz, J. and Stormorken, H. (1968) 'Fibrinolytic activity of leucocytes in smears of bone marrow and peripheral blood', *Scandinavian Journal of Haematology, 5*, 129-37

Prokopowicz, J., Rejniak, L. and Niewiarowski, S. (1967) 'Influence of cytostatic agents on fibrinolytic and proteolytic enzymes and on phagocytosis of guinea-pig leucocytes', *Experientia, 23*, 813-4

Propping, D., Zaneveld, L.J.D., Tauber, P.F. and Schumacher, G.F.B. (1978)

'Purification of plasminogen activators from human seminal plasma',
Biochemical Journal, 171, 435-44

Rabiner, S.F., Goldfine, I.D., Hart, A., Summaria, L. and Robbins, K.C. (1969)
'Radioimmunoassay of human plasminogen and plasmin', *Journal of
Laboratory and Clinical Medicine, 74*, 265-73

Radcliffe, R. and Heinze, T. (1978) 'Isolation of plasminogen activator from
human plasma by chromatography on lysine-Sepharose', *Archives of
Biochemistry and Biophysics, 189*, 185-94

Rákóczi, I., Wiman, B. and Collen, D. (1978) 'On the biological significance of the
specific interaction between fibrin, plasminogen and antiplasmin', *Biochimica et
Biophysica Acta, 540*, 295-300

Rånby, M. (1982) 'Studies on the kinetics of plasminogen activation by tissue
plasminogen activator', *Biochimica et Biophysica Acta, 704*, 461-9

Ratnoff, O.D., Pensky, J., Ogston, D. and Naff, G.B. (1969) 'The inhibition of
plasmin, plasma kallikrein, plasma permeability factor and the C'1r
subcomponent of the first component of complement by C'1 esterase inhibitor',
Journal of Experimental Medicine, 129, 315-31

Raum, D., Marcus, D. and Alper, C.A. (1980) 'Genetic polymorphism of human
plasminogen', *American Journal of Human Genetics, 32*, 681-9

Raum, D., Marcus, D., Alper, C.A., Levey, R., Taylor, P.D. and Starzl, T.E.
(1980) 'Synthesis of human plasminogen by the liver', *Science, 208*, 1036-7

Rickli, E.E. and Zaugg, H. (1970) 'Isolation and purification of highly enriched
tissue plasminogen activator from pig heart', *Thrombosis et Diathesis
Haemorrhagica, 23*, 64-76

Rijken, D.C. and Collen, D. (1981) 'Purification and characterization of the
plasminogen activator secreted by human melanoma cells in culture', *Journal of
Biological Chemistry, 256*, 7035-41

Rijken, D.C., Wijngaards, G., Zaal-De Jong, M. and Welbergen, J. (1979)
'Purification and partial characterization of plasminogen activator from human
uterine tissue', *Biochimica et Biophysica Acta, 580*, 140-53

Rijken, D.C., Wijngaards, G. and Welbergen, J. (1980) 'Relationship between
tissue plasminogen activator and the activators in blood and vascular wall',
Thrombosis Research, 18, 815-30

Rijken, D.C., Hoylaerts, M. and Collen, D. (1981a) 'On the fibrinolytic properties
of single-chain and two-chain human tissue plasminogen activator', *Thrombosis
and Haemostasis, 46*, 12

Rijken, D.C., Wijngaards, G. and Welbergen, J. (1981b) 'Immunological
characterization of plasminogen activator activities in human tissues and body
fluids', *Journal of Laboratory and Clinical Medicine, 97*, 477-86

Rimon, A., Shamash, Y. and Shapiro, B. (1966) 'The plasmin inhibitor of human
plasma. IV. Its action on plasmin, trypsin, chymotrypsin and thrombin', *Journal
of Biological Chemistry, 241*, 5102-7

Robbins, K.C., Summaria, L., Elwyn, D. and Barlow, G.H. (1965) 'Further studies
on the purification and characterization of human plasminogen and plasmin',
Journal of Biological Chemistry, 240, 541-50

Robbins, K.C., Summaria, L., Hsieh, B. and Shah, R.J. (1967) 'The peptide chains
of human plasmin. Mechanism of activation of human plasminogen', *Journal of
Biological Chemistry, 242*, 2333-42

Robertson, B.R., Pandolfi, M. and Nilsson, I.M. (1972) ' "Fibrinolytic capacity" in
healthy volunteers as estimated from effect of venous occlusion of arms', *Acta
Chirurgica Scandinavica, 138*, 429-36

Rosing, D.R., Redwood, D.R., Brakman, P., Astrup, T. and Epstein, S.E. (1978)
'The fibrinolytic response of man to vasoactive drugs measured in arterial
blood', *Thrombosis Research, 13*, 419-28

188

Rulot, H. (1904) 'Intervention des leucocytes dans l'autolyse de la fibrine (fibrinolyse de Dastre)', *Archives Internationales de Physiologie, 1*, 152-8

Sahli, W. (1885) 'Ueber das Vorkommen von Pepsin und Trypsin im normalen menschlichen Hara', *Pflugers Archiv fur Physiologie, 36*, 209-29

Saito, H., Ratnoff, O.D., Waldmann, R. and Abraham, J.P. (1975) 'Fitzgerald trait: deficiency of a hitherto unrecognized agent, Fitzgerald factor, participating in surface-mediated reactions of clotting, fibrinolysis, generation of kinin and the property of diluted plasma enhancing vascular permeability (PF/Dil)', *Journal of Clinical Investigation, 55*, 1082-9

Saito, H., Goldsmith, G.H., Moroi, M. and Aoki, N. (1979) 'Inhibitory spectrum of α_2-plasmin inhibitor', *Proceedings of the National Academy of Sciences, 76*, 2013-7

Saito, H., Goodenough, L.T., Knowles, B.B. and Aden, D.P. (1982) 'Synthesis and secretion of α_2-plasmin inhibitor by established human liver cell lines', *Proceedings of the National Academy of Sciences, 79*, 5684-7

Sakata, Y. and Aoki, N. (1980) 'Cross-linking of α_2-plasmin inhibitor to fibrin by fibrin-stabilizing factor', *Journal of Clinical Investigation, 65*, 290-7

Sakata, Y. and Aoki, N. (1982) 'Significance of cross-linking of α_2-plasmin inhibitor to fibrin in inhibition of fibrinolysis and in hemostasis', *Journal of Clinical Investigation, 69*, 536-42

Sandjberg Hansen, M. and Clemmensen, I. (1980) 'Partial purification and characterization of a new fast-acting plasmin inhibitor from human platelets', *Biochemical Journal, 187*, 173-80

Sawyer, W.D., Fletcher, A.P., Alkjaersig, N. and Sherry, S. (1960) 'Studies on the thrombolytic activity of human plasma', *Journal of Clinical Investigation, 39*, 426-34

Schaller, J., Nick, H., Rickli, E., Gillessen, D., Lergier, W. and Studer, R.O. (1982) 'The two chain structure of human low molecular weight (LMW) urinary urokinase, evidenced by sequence analysis', *Haemostasis, 11*, Suppl. 1, 54

Schmidt, W. and Havemann, K. (1974) 'Isolation of elastase-like and chymotrypsin-like neutral proteases from human granulocytes', *Hoppe-Seyler's Zeitschrift fur Physiologische Chemie, 355*, 1077-82

Schreiber, A.D., Kaplan, A.P. and Austen, K.F. (1973) 'Inhibition by Cl 1NH of Hagemann factor fragment activation of coagulation, fibrinolysis and kinin generation', *Journal of Clinical Investigation, 52*, 1402-9

Schwick, H.G., Heimburger, N. and Haupt, H. (1967) 'Purification and chemical-physical properties of some proteinase inhibitors of plasma', *Thrombosis et Diathesis Haemorrhagica, 18*, 302

Seigel-Ralston, A. and Astrup, T. (1972) 'A previously undescribed fibrinolysis inhibitor in human blood', *Nature, 234*, 180-1

Semar, M., Skoza, I. and Johnson, A.J. (1969) 'Partial purification and properties of a plasminogen activator from human erythrocytes', *Journal of Clinical Investigation, 48*, 1777-85

Semeraro, N., Colucci, M., Telesforo, P. and Collen, D. (1978) 'The inhibition of plasmin by antithrombin-heparin complex. I. In human plasma *in vitro*', *British Journal of Haematology, 39*, 91-9

Sherman, L.A., Lee, J. and Stewart, C.C. (1977) 'Release of fibrinolytic enzymes by macrophages in response to soluble fibrin', *Thrombosis and Haemostasis, 38*, 46

Sherry, S. and Alkjaersig, N. (1957) 'Studies on the fibrinolytic enzyme of human plasma', *Thrombosis et Diathesis Haemorrhagica, 1*, 264-76

Sherry, S., Lindemeyer, R.I., Fletcher, A.P. and Alkjaersig, N. (1959) 'Studies on enhanced fibrinolytic activity in man', *Journal of Clinical Investigation, 38*,

810-22

Sherry, S., Alkjaersig, N. and Fletcher, A.P. (1964) 'Assay of urokinase preparations with the synthetic substrate acetyl-L-lysine methyl ester', *Journal of Laboratory and Clinical Medicine, 64*, 145-53

Sjöholm, I., Wiman, B. and Wallén, P. (1973) 'Studies on the conformational changes of plasminogen induced during activation to plasmin and by 6-amino-hexanoic acid', *European Journal of Biochemistry, 39*, 471-9

Sobel, G.W., Mohler, S.R., Jones, N.W., Dowdy, A.B.C. and Guest, M.M. (1952) 'Urokinase: an activator of plasma profibrinolysin extracted from urine', *American Journal of Physiology, 171*, 768-9

Soberano, M.E., Ong, E.B. and Johnson, A.J. (1976a) 'The effects of inhibitors on the catalytic conversion of urokinase', *Thrombosis Research, 9*, 675-81

Soberano, M.E., Ong, E.B., Johnson, A.J., Levy, M. and Schoellmann, G. (1976b) 'Purification and characterization of two forms of urokinase', *Biochimica et Biophysica Acta, 445*, 763-73

Sottrup-Jensen, L., Claeys, H., Zajdel, M., Petersen, T.E. and Magnusson, S. (1978) 'The primary structure of human plasminogen: isolation of two lysine-binding fragments and one "mini"-plasminogen (MW 38,000) by elastase-catalyzed-specific limited proteolysis', in J.F. Davidson, R.M. Rowan, M.M. Samama and P.C. Desnoyers (eds.) *Progress in Chemical Fibrinolysis and Thrombolysis*, Vol. 3, Raven Press, New York, pp. 191-209

Sottrup-Jensen, L., Hansen, H.F., Mortensen, S.B., Petersen, T.E., Magnusson, S. and Jörnvall, H. (1981) 'Mechanism of proteinase-binding to α_2-macroglobulin', *Thrombosis and Haemostasis, 46*, 87

Stefanini, M. and Murphy, I.S. (1956) 'Studies on platelets. XIV. Human platelets as a source of antifibrinolysin', *Journal of Clinical Investigation, 35*, 355-61

Steinbuch, M. (1971) 'Les antiprotéases du plasma', *Revue Française de Transfusion, 14*, 61-82

Storm, O. (1955) 'Fibrinolytic activity in human tears', *Scandinavian Journal of Clinical and Laboratory Investigation, 7*, 55-8

Summaria, L., Hsieh, B., Groskopf, W.R. and Robbins, K.C. (1967) 'The isolation and characterisation of the *S*-carboxymethyl β (light) chain derivative of human plasmin', *Journal of Biological Chemistry, 242*, 5046-52

Summaria, L., Robbins, K.C. and Barlow, G.H. (1971) 'Isolation and characterization of the *S*-carboxymethyl heavy chain derivative of human plasmin', *Journal of Biological Chemistry, 246*, 2143-6

Summaria, L., Arzadon, L., Berabe, P. and Robbins, K.C. (1972) 'Studies on the isolation of the multiple molecular forms of human plasminogen and plasmin by isoelectric focusing methods', *Journal of Biological Chemistry, 247*, 4691-702

Summaria, L., Arzadon, L., Berabe, P. and Robbins, K.C. (1975) 'The activation of plasminogen to plasmin by urokinase in the presence of the plasmin inhibitor Trasylol', *Journal of Biological Chemistry, 250*, 3988-95

Summaria, L., Spitz, F., Arzadon, L., Boreisha, G. and Robbins, K.C. (1976) 'Isolation and characterization of the affinity chromatography forms of human glu- and lys-plasminogens and plasmins', *Journal of Biological Chemistry, 251*, 3693-9

Summaria, L., Boreisha, I.G., Arzadon, L. and Robbins, K.C. (1977) 'Activation of human glu-plasminogen to glu-plasmin by urokinase in presence of plasmin inhibitors', *Journal of Biological Chemistry, 252*, 3945-51

Takagi, T. and Doolittle, R.F. (1975a) 'Amino acid sequence studies on plasmin-derived fragments of human fibrinogen: amino-terminal sequences of intermediate and terminal fragments', *Biochemistry, 14*, 940-6

Takagi, T. and Doolittle, R.F. (1975b) 'Amino acid studies on the α-chain of human fibrinogen. Location of four plasmin attack points and a covalent

190

cross-linking site', *Biochemistry, 14*, 5149-56

Thorsen, S. (1973) 'The inhibition of tissue plasminogen activator and urokinase-induced fibrinolysis by some natural proteinase inhibitors and by plasma and serum from normal and pregnant subjects', *Scandinavian Journal of Clinical and Laboratory Investigation, 31*, 57-9

Thorsen, S. (1975) 'Difference in the binding to fibrin of native plasminogen and plasminogen modified by proteolytic degradation. Influence of ω-aminocarboxylic acids', *Biochimica et Biophysica Acta, 393*, 55-65

Thorsen, S., Glas-Greenwalt, P. and Astrup, T. (1972) 'Differences in the binding to fibrin of urokinase and tissue plasminogen activator', *Thrombosis et Diathesis Haemorrhagica, 28*, 65-74

Thorsen, S., Clemmensen, I., Sottrup-Jensen, L. and Magnusson, S. (1981) 'Adsorption to fibrin of native fragments of known primary structure from human plasminogen', *Biochimica et Biophysica Acta, 668*, 377-87

Tissot, J.-D., Schneider, Ph. and Bachmann, F. (1982) 'Isolation from human plasma of a plasminogen activator (PA) identical to urinary high Mr-urokinase', *Haemostasis, 11*, Suppl. 1, 7

Todd, A.S. (1959) 'The histological localisation of fibrinolysin activator', *Journal of Pathology and Bacteriology, 78*, 281-3

Trexler, M., Vali, Z. and Patthy, L. (1982) 'Structure of the ω-amino-carboxylic binding sites of human plasminogen. Arginine 70 and aspartic acid 56 are essential for binding of ligand by kringle 4', *Journal of Biological Chemistry, 257*, 7401-6

Troll, W., Sherry, S. and Wachman, J. (1954) 'The action of plasmin on synthetic substrates', *Journal of Biological Chemistry, 208*, 85-93

Trumpi-Kalshoven, M.M. (1978) 'The relevance of Cl inhibitor in the inhibition of the fibrinolytic activity of plasmin', in J.F. Davidson, R.M. Rowan, M.M. Samama and P.C. Desnoyers (eds.) *Progress in Chemical Fibrinolysis and Thrombolysis*, Vol. 3, Raven Press, New York, pp. 257-67

Unkeless, J.C., Gordon, S. and Reich, E. (1974) 'Secretion of plasminogen activator by stimulated macrophages', *Journal of Experimental Medicine, 139*, 834-50

Uszynski, M. (1980) 'Isolation of peptides with anti-urokinase activity from the human placenta', *Thrombosis and Haemostasis, 42*, 1411-6

Uszynski, M. and Abildgaard, U. (1971) 'Separation and characterization of two fibrinolytic inhibitors from human placenta', *Thrombosis et Diathesis Haemorrhagica, 25*, 580-9

Vali, Z. and Patthy, L. (1982) 'Location of the intermediate and high affinity ω-aminocarboxylic acid-binding sites in human plasminogen', *Journal of Biological Chemistry, 257*, 2104-10

Vassalli, J.-D., Hamilton, J. and Reich, E. (1976) 'Macrophage plasminogen activator: modulation of enzyme production by anti-inflammatory steroids, mitotic inhibitors, and cyclic nucleotides', *Cell, 8*, 271-81

Venneröd, A.M. and Laake, K. (1976) 'Prekallikrein and plasminogen proactivator: absence of plasminogen proactivator in Fletcher factor deficient plasma', *Thrombosis Research, 8*, 519-22

Vetterlein, D., Young, P.L., Bell, T.E. and Roblin, R. (1979) 'Immunological characterization of multiple molecular weight forms of human cell plasminogen activators', *Journal of Biological Chemistry, 254*, 575-8

Vetterlein, D., Bell, T.E., Young, P.L. and Roblin, R. (1980) 'Immunological quantitation and immunoadsorption of urokinase-like plasminogen activators secreted by human cells', *Journal of Biological Chemistry, 255*, 3665-72

Violand, B.N. and Castellino, F.J. (1976) 'Mechanism of the urokinase-catalyzed activation of human plasminogen', *Journal of Biological Chemistry, 251*,

3906-12

Violand, B.N., Sodetz, J.M. and Castellino, F.J. (1975) 'The effect of ϵ-amino caproic acid on the gross conformation of plasminogen and plasmin', *Archives of Biochemistry and Biophysics, 170*, 300-5

Walker, J.E. and Ogston, D. (1982) 'The inhibition of tissue activator and urokinase by human plasma', *Thrombosis and Haemostasis, 47*, 265-8

Wallén, P. and Iwanaga, S. (1968) 'Differences between plasmic and tryptic digests of human *S*-sulpho-fibrinogen', *Biochimica et Biophysica Acta, 154*, 414-7

Wallén, P. and Rånby, M. (1977) 'Purification and some properties of tissue plasminogen activator', *Thrombosis and Haemostasis, 38*, 61

Wallén, P. and Wiman, B. (1970) 'Characterization of human plasminogen. I. On the relationship between different molecular forms of plasminogen demonstrated in plasma and found in purified preparations', *Biochimica et Biophysica Acta, 221*, 20-30

Wallén, P. and Wiman, B. (1975) 'Purification of tissue activator using affinity chromatography. Effect of fibrin on the enzymatic properties of the activator', *Thrombosis et Diathesis Haemorrhagica, 34*, 609

Wallén, P., Rånby, M., Bergsdorf, N. and Kok, P. (1981) 'Purification and characterization of tissue plasminogen activator. On the occurrence of two different forms and their enzymatic properties', in J.F. Davidson, I.M. Nilsson and B. Astedt (eds.) *Progress in Chemical Fibrinolysis and Thrombolysis*, Vol. 5, Churchill Livingstone, Edinburgh, pp. 16-23

Walther, P.J., Steinmann, H.H., Hill, R.M. and McKee, P.A. (1974) 'Activation of human plasminogen by urokinase. Partial characterization of a pre-activation peptide', *Journal of Biological Chemistry, 249*, 1173-81

Walton, P.L. (1967) 'The hydrolysis of α-N-acetylglycyl-L-lysine methyl ester by urokinase', *Biochimica et Biophysica Acta, 132*, 104-14

Weinstein, M.J. and Doolittle, R.F. (1972) 'Differential specificities of thrombin, plasmin and trypsin with regard to synthetic and natural substrates and inhibitors', *Biochimica et Biophysica Acta, 255*, 577-90

Weiss, A.S., Gallin, J.I. and Kaplan, A.P. (1974) 'Fletcher factor deficiency. A diminished rate of Hageman factor activation caused by absence of prekallikrein with abnormalities of coagulation, fibrinolysis, chemotactic activity, and kinin generation', *Journal of Clinical Investigation, 53*, 622-33

White, W.F., Barlow, G.H. and Mozen, M.M. (1966) 'The isolation and characterization of plasminogen activators (urokinase) from human urine', *Biochemistry, 5*, 2160-9

Wijngaards, G. and Bernik, M.B. (1981) 'Activation and "inactivation" of a plasminogen proactivator (preurokinase) in human tissue culture media', *Thrombosis and Haemostasis, 46*, 10

Williams, J.R.B. (1951) 'The fibrinolytic activity of urine', *British Journal of Experimental Pathology, 32*, 530-7

Wilson, E.L., Becker, M.L.B., Hoal, E.G. and Dowdle, E.B. (1980) 'Molecular species of plasminogen activators secreted by normal and neoplastic human cells', *Cancer Research, 40*, 933-8

Wiman, B. (1977) 'Primary structure of the B-chain of human plasmin', *European Journal of Biochemistry, 76*, 129-37

Wiman, B. (1978) 'Biochemistry of the plasminogen to plasmin conversion', in P.J. Gaffney and S. Balkuv-Ulutin (eds.) *Fibrinolysis: Current Fundamental and Clinical Concepts*, Academic Press, London, pp. 47-60

Wiman, B. and Collen, D. (1977) 'Purification and characterization of human antiplasmin, the fast-acting plasmin inhibitor', *European Journal of Biochemistry, 78*, 19-26

Wiman, B. and Collen, D. (1978a) 'On the kinetics of the reaction between human

antiplasmin and plasmin', *European Journal of Biochemistry, 84*, 573-8
Wiman, B. and Collen, D. (1978b) 'Molecular mechanism of physiological fibrinolysis', *Nature, 272*, 549-50
Wiman, B. and Collen, D. (1979) 'On the mechanism of the reaction between human α_2-antiplasmin and plasmin', *Journal of Biological Chemistry, 254*, 9291-7
Wiman, B. and Wallén, P. (1973) 'Activation of human plasminogen by an insoluble derivative of urokinase. Structural changes of plasminogen in the course of activation to plasmin and demonstration of a possible intermediate compound', *European Journal of Biochemistry, 36*, 25-31
Wiman, B. and Wallén, P. (1975a) 'On the primary structure of human plasminogen and plasmin. Purification and characterization of cyanogen-bromide fragments', *European Journal of Biochemistry, 57*, 387-94
Wiman, B. and Wallén, P. (1975b) 'Structural relationships between "glutamic acid" and "lysine" forms of human plasminogen and their interaction with the NH_2-terminal activation peptide as studied by affinity chromatography', *European Journal of Biochemistry, 50*, 489-94
Wiman, B. and Wallén, P. (1977) 'The specific interaction between plasminogen and fibrin. A physiological role of the lysine binding site in plasminogen', *Thrombosis Research, 10*, 213-22
Wiman, B., Boman, L. and Collen, D. (1978) 'On the kinetics of the reaction between human antiplasmin and a low-molecular-weight form of plasmin', *European Journal of Biochemistry, 87*, 143-6
Wiman, B., Lijnen, H.R. and Collen, D. (1979) 'On the specific interaction between the lysine-binding sites in plasmin and complementary sites in α_2-antiplasmin and in fibrinogen', *Biochimica et Biophysica Acta, 579*, 142-54
Wun, T.-C., Schleuning, W.-D. and Reich, E. (1982) 'Isolation and characterization of urokinase from human plasma', *Journal of Biological Chemistry, 257*, 3276-83

5 Interactions Between the Hemostatic and Complement Systems

It would be anticipated that the various systems of the body involved in the defensive responses to injury and the invasion by micro-organisms would function in an integrated manner. These systems include the hemostatic, complement and kinin-forming mechanisms. This chapter reviews the various points of interaction between the hemostatic and complement systems although definitive conclusions on their functional importance is premature.

Outline of the Complement System

The complement system of man comprises a group of proteins which interact sequentially to induce irreversible damage of the target membrane and, additionally, produce a number of biologically active products. The activation of the complement system takes place through two major pathways, the classical pathway and the alternative (properdin) pathway (Figure 5.1). A detailed account of the pathways and their constituent components may be obtained in the review by Reid and Porter (1981).

The Classical Pathway

In the classical pathway activation of the first component of complement (Cl) is initiated by the union of an antibody of the IgM or IgC classes with its corresponding antigen. The first component of complement is a macromolecular complex of three proteins (Clq, Clr and Cls). The Clq subunit, with a molecular weight of some 400,000 and a mixed globular and fibrillar structure, has close structural resemblances to collagen including a triple helical formation in its fibrillar region. Clr and Cls are both single polypeptide chains of molecular weight around 83,000.

On the union of antibody with antigen there is binding of the Fc

194

Figure 5.1: The Complement Pathways

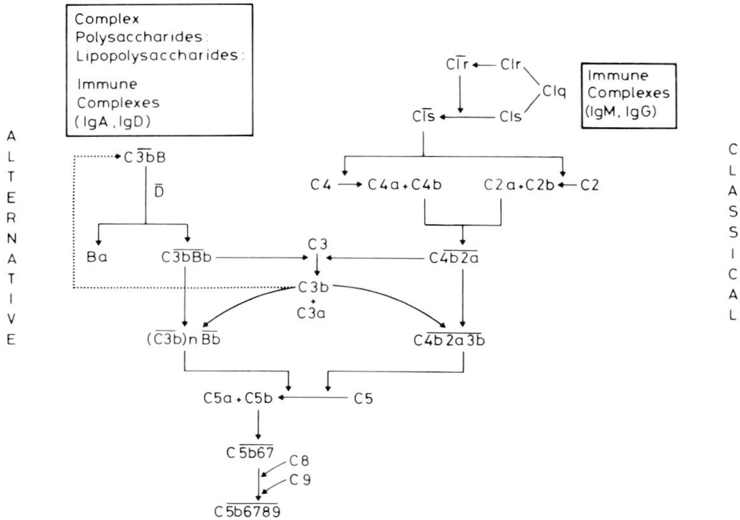

region of the immunoglobulin to the Clq subunit. Such binding results in a conformational change in the Clq and there is resultant activation of the Clr subunit: in its activated form (C̄lr) this subunit can cleave the single-chain Cls subunit with the formation of the active two-chain serine protease C̄ls. Cls hydrolyses the fourth and second complement components. The fourth component of complement (C4), composed of three disulphide-linked polypeptide chains, is cleaved with the formation of a small peptide (C4a) and a larger fragment (C4b). The second component (C2), a single-chain glycoprotein, is split into a small peptide (C2b) and the active fragment C2a. The C4b and C2a fragments bind together to form a complex C̄4b2a (C3 convertase) which possesses a proteolytic site in the C2a portion and is capable of cleaving C3. The proteolysis of C3 provides a small fragment (C3a) and the major fragment C3b which has the ability to bind to immune complexes and a variety of surfaces. Its presence on the cell surface changes the specificity of adjacent C̄4b2a complexes to that of a C5 convertase (C̄4b2a3b). Cleavage of C5 takes place with the production of a small fragment (C5a) which is released into the fluid phase, and a larger fragment (C5b) which forms a macromolecular complex with the terminal complement components C6 and C7 and, finally, with C8 and C9. The macromolecular complex induces cellular injury and lysis through the formation of ultrastructural lesions in the membrane.

The fragments C3a and C5a, with molecular weights of some 9000 and 11,000 respectively, are termed anaphylatoxins. They have the properties of releasing histamine from mast cells and basophils, contracting smooth muscle and increasing capillary permeability. C5a is also chemotactic for leukocytes. It is the anaphylatoxins which are responsible for the biological effects consequent on the activation of the complement system additional to the death and lysis of the antibody-bearing cells.

A number of proteins have a role in the regulation of the complement system. $\overline{\text{Cl}}$ inactivator has an important function in the classical pathway through its ability to inhibit Cls and thereby limit the cleavage of C4 and C2. The C4-binding protein specifically binds to the C4b fragment and contributes to the regulation of activation of complement through the classical pathway. BlH globulin (H) binds to C3b in an equimolar ration, preventing its interaction with other proteins and rendering it susceptible to proteolytic cleavage by C3b inactivator (C3bINA): through the combined action of these agents the level of C3b in the serum is finely controlled.

The Alternative Pathway

The alternative pathway is initiated by the interaction of certain serum components with complex polysaccharides, for example, inulin and zymosan; lipopolysaccharides; and some immune complexes containing antibodies of the IgA or IgD classes. This pathway bypasses Cl, C4 and C2, but produces a C3 convertase which leads to the assembly of the terminal lytic complex ($\overline{\text{C5b-9}}$). The serum components contributing to the alternative pathway include factor B (C3 proactivator), factor D (C3 proactivator convertase) and properdin (P). The activated forms of both factor B and factor D are serine proteases; it appears that most of the factor D in plasma is in the activated form ($\overline{\text{D}}$). Properdin is a glycoprotein which is present in plasma in a native form functionally different from its activated form.

It is likely that a low level of C3b formation takes place continuously in plasma, but it is normally inactivated by the action of BlH globulin and C3b inactivator. Binding to certain activators of the alternative pathway protect C3b from such inactivation. In the presence of C3b factor B is cleaved by factor D to yield fragments Ba and

196

Bb and the larger fragment Bb, in combination with C3b, form the alternative pathway C3 convertase ($\overline{C3bBb}$). Properdin functions by preventing spontaneous dissociation of the $\overline{C3bBb}$ complex and its subsequent inactivation. The $\overline{C3bBbP}$ complex cleaves C3 into the fragments C3a and C3b and can be converted into a C5 convertase by the deposition of further C3b on the cell surface. The C3b liberated by the action of C3 convertase can be used to generate more $\overline{C3bBb}$ in a positive feedback mechanism. The formation of the C5 convertase initiates the assembly of the terminal $\overline{C5b\text{-}9}$ complex.

Complement and Platelets

Complement Factors Associated with Platelets

The selective association of components of complement with platelets has been sought by a number of investigators. On the basis of aggregation of washed platelets by antisera to C4 and C3 Taylor and Müller-Eberhard (1970) suggested that these components were bound specifically to the platelet membrane. C4 activity in platelets was also found by Nagaki *et al* (1965). In contrast, Nachman (1968) could detect no C3 or C4 protein in platelets, and Mueller-Eckhardt and Lüscher (1968) found no active complement component on washed human platelets. A Cl hemolytic activity was found to be associated with platelets (Wautier *et al.*, 1974) and the same workers (Wautier *et al.*, 1976) confirmed the presence of Cl on gel filtered platelets by the precipitation reaction with an anti-Clq monospecific serum or immunodiffusion, and platelet aggregation induced by this antiserum.

The Interactions Between Immune Complexes, Complement and Platelet Aggregation and Secretion

Humphrey and Jaques (1955) demonstrated that the addition of antigen and antibody to the blood of the non-immune rabbit resulted in platelet aggregation and the release of vasoactive amines. The observation that the platelets of rabbit and other mammalian species undergo aggregation and/or release of amines on exposure to antigen-antibody complexes has been repeatedly confirmed (Barbaro, 1961; Des Prez *et al.*, 1961; Gocke and Osler, 1965; Henson and Cochrane, 1969; Henson, 1970; Marney, 1971; Pfueller and Lüscher, 1972; Siraganian, 1972).

Marney (1971) showed that the aggregation induced by soluble antigen and antibody is diphasic. The first wave of aggregation requires complement and divalent cations and precedes the release of ADP and amines: the second wave is accompanied by release of the platelet ADP. Henson (1970) described two mechanisms for the release of vasoactive amines. The first is the adherence of platelets to the immune complexes which requires the action of C3, and lysis of the platelets which requires C6 and which is accompanied by the release of the vasoactive amines. The second mechanism of vasoactive amine release was shown to be non-lytic and was seen if particulate antigen had fixed complement components up to C3.

In contrast to the requirement for complement in the aggregation of rabbit platelets in the presence of antigen-antibody complexes, such complexes can induce washed human platelets to secrete and aggregate as a result of direct interaction of IgG with platelets and do not require complement activation (Pfueller and Lüscher, 1972). This species difference depends on the phenomenon termed immune adherence. The platelets of the rabbit, as described by Henson (1970), guinea-pig and dog are immune adherence positive; they are attracted to sites of fixation of the third component of complement on biological membranes. There is adherence of the platelets to the immune complexes and, as a result, release of vasoactive amines and nucleotides. In the second phase there is lysis of the platelets due to generation of the lytic terminal complement complex C5-9 close to the platelet membrane with consequent bystander lysis of the platelets. In this lytic phase there is liberation of further platelet contents including coagulant material. The platelets of man are immune-adherence negative.

In addition to the complement-independent aggregation of human platelets by immune complexes, evidence has been provided that human platelets aggregate and undergo a release reaction by a complement-dependent process: release precedes aggregation which is dependent on both complement and ADP released during the first phase (Pfueller and Lüscher, 1974a).

Zymosan, which activates complement through the alternative pathway, induces platelet aggregation and a release reaction in human platelet-rich plasma: the reactions are magnesium-dependent and plasma treated with hydrazine or cobra venom factor is inactive (Zucker and Grant, 1974). Both the alternative pathway of complement activation and fibrinogen appear to be involved in the generation of platelet aggregating and secretion-promoting activity by zymosan (Zucker *et al.*, 1974; Pfueller and Lüscher, 1974b).

Other alternative pathway activators (inulin, endotoxin) do not develop platelet aggregating activity when incubated with human plasma.

Endotoxin, Platelets and Complement

The interactions between platelets and the complement system have been incriminated in the induction of the changes of disseminated intravascular coagulation by endotoxin. Endotoxin is associated with thrombocytopenia and disseminated intravascular coagulation in a number of animal species including the rabbit (Beller *et al.*, 1969; Brown and Lachmann, 1974), guinea-pig (Kane *et al.*, 1973) and dog (Garner *et al.*, 1974). Endotoxic lipopolysaccharide has a similar action to antigen-antibody complexes in triggering platelet aggregation and secretion *in vitro* (Des Prez *et al.*, 1961; Des Prez and Bryant, 1966) and there is abundant evidence that endotoxin activates complement (Gilbert and Braude, 1962; Gewurz *et al.*, 1968).

A role for complement in the thrombocytopenia induced by endotoxin infusion is suggested by the sparing of platelets in cobra venom factor-treated rabbits (Fong and Good, 1971; Brown and Lachmann, 1973; Fong *et al.*, 1974) or dogs (Garner *et al.*, 1974), and the protection from thrombocytopenia in C4-deficient guinea-pigs challenged with endotoxin. Müller-Berghaus and Lohmann (1974), however, were unable to find any difference in the fall in platelet numbers between normal and C6-deficient rabbits following endotoxin infusion. It appears that endotoxin requires the classical pathway for the development of thrombocytopenia, but endotoxin also activates the alternative pathway (Kane *et al.*, 1973; Fong *et al.*, 1974).

The relevance of these observations to human disease is uncertain. Endotoxemia is associated with thrombocytopenia in man (Cohen and Gardner, 1966; Corrigan *et al.*, 1968), but while endotoxins from a variety of sources cause platelet aggregation and serotonin secretion when added to rabbit platelet-rich plasma, there is no aggregation and only slow release of serotonin from human platelets (Nagayama *et al.*, 1971).

Role of Platelets in the Complement-Coagulation Interaction

The suspicion that complement might be involved in the enhancement of blood coagulation was raised by experiments showing that the clotting of whole rabbit blood, but not platelet-poor blood, is accelerated by antigen-antibody complexes (Robbins and Stetson,

1959) and by endotoxin (McKay and Shapiro, 1958). Inulin, an alternative pathway activator, was found to shorten the clotting time of normal platelet-rich rabbit plasma, but not that of C6-deficient rabbits' platelet-rich plasma or normal rabbits' platelet-free plasma (Zimmerman and Müller-Eberhard, 1971; Zimmerman *et al.*, 1971a). Siraganian (1972) reported that the platelet release reaction in response to inulin, zymosan and antigen-antibody aggregates is absent in C6-deficient rabbit plasma: he suggested that the linkage between the complement and clotting systems was the activation of platelet factor 3, shown by Horowitz and colleagues (1962) to be the result of the action of endotoxin or immune aggregates on rabbit platelets. It has also been concluded by Zimmerman (1974) that the enhancement of coagulant activity by complement-activating agents in some animal species is the result of a complement-mediated effect on the platelet with the release of procoagulant material from the immune-adherent platelets. The evolution of platelet coagulant activity appears to precede the complement-induced lysis of the platelet (Zimmerman, 1974).

C6-deficient human subjects have no demonstrable abnormality in coagulation (Heusinkveld *et al.*, 1974). Zimmerman (1974) has attributed this species difference to the relative insensitivity of human platelets to complement, presumably secondary to their immune adherence negativity. In contrast to rabbit platelets, human platelets do not respond with significant clot-promoting activity after complement activation by inulin or endotoxin. Zucker and Grant (1974), however, have shown that activation of the alternative pathway by zymosan can cause human platelets to release serotonin. Zimmerman (1974) has reported that activation of complement through the classical pathway with rabbit antihuman platelet antibody results in some clot-promoting activity developing in C6-deficient platelet-rich plasma: after the addition of C6 marked enhancement in the development of clot-promoting activity and serotonin release takes place. He suggested, therefore, that the effect of complement on human platelets is analogous to that seen with rabbit platelets.

Complement and Normal Platelet Function

In addition to the platelet aggregation and secretion resulting from exposure to antigen-antibody complexes with complement activation, there have been a number of reported observations to suggest that complement may have a role in normal platelet function.

200

Complement and Collagen-induced Platelet Aggregation. The close chemical and structural similarities between collagen and Clq has already been noted. These include, in addition to the triple helical structure, linkage of glucose-galactose disaccharides with the hydroxyl group of hydroxylysine, the presence of collagen-like amino acid sequences in each of the three polypeptide chains of Clq and susceptibility to collagenase (Calcott and Müller-Eberhard, 1972; Reid, 1974).

The possibility that complement might be involved in collagen-induced platelet aggregation was raised by the finding that such aggregation is inhibited by anti-complement drugs such as chloroquine (Jobin and Tremblay, 1969). It was found later that decomplemented dogs lost their ability to aggregate with collagen (Garner *et al.*, 1974), while Chater (1976) reported that decomplementation with cobra venom factor *in vitro* in man, dog and rabbit abolished the response of platelets to collagen, and that incubation of gel filtered platelets with antibodies to Cl, C3 and C5 inhibited their collagen response. Microscopic evidence was presented that the cobra venom factor does not affect adhesion of platelets to collagen. These results led Chater to suggest that components of the complement system are involved in the induction of the platelet release reaction by collagen. Tiffany and Penner (1980) found that C4 was required in guinea-pigs for platelet aggregation in response to low concentrations of insoluble collagen whereas adhesion of platelets to collagen was not affected and there was no change in the response of the platelets to ADP or ristocetin.

It has been shown that collagen and Clq are capable of binding to the same sites on sensitised sheep erythrocytes and human platelets (Suba and Csako, 1976), and that Clq inhibits collagen-induced platelet aggregation and secretion (Cazenave *et al.*, 1976; Suba and Csako, 1976). Platelet adherence to collagen is also inhibited by Clq (Cazenave *et al.*, 1976). It has been demonstrated that it is the collagen-like structure of Clq which is responsible for its inhibitory effect on platelet aggregation and adhesion to collagen (Wautier *et al.*, 1977). It has been proposed, therefore, that Clq becomes associated with the sites on the platelet surface which normally interact with collagen and in this way inhibits the platelet reactions induced by collagen. Since the inhibitory concentration of Clq is less than that which is found in plasma it has been suggested that Clq might have an important influence on normal hemostasis by regulating platelet reactivity to collagen. More recently it has been reported that the Cls subcomponent also inhibits platelet aggregation and adhesion to collagen, perhaps by interacting with collagen (Wautier *et al.*, 1981).

Factor $\bar{D}$ and Thrombin-induced Platelet Aggregation. Factor $\bar{D}$, with chemical and immunological similarities to thrombin, has been shown to be capable of inhibiting thrombin-induced platelet aggregation without affecting collagen or arachidonic acid-induced aggregation (Davis and Kenney, 1979). Factor D was found to inhibit thrombin binding to platelets and it was concluded that this component of the alternative pathway acts as a competitive inhibitor of thrombin binding to platelets, thereby influencing thrombin-induced platelet aggregation.

Formation of C5-9 Complex on Platelets. Using radiolabelled complement components Zimmerman and Kolb (1976) demonstrated that normal human platelets can initiate the formation and binding of the C5-9 complex onto the platelet membrane in a process which is not dependent on either the classical or alternative pathways of complement activation. While the C5-9 complex is the mediator of complement-directed membrane damage, the absence of lysis was ascribed to the relative resistance of human platelets to the lytic effect of the complex. It was speculated that modulation of the platelet membrane by complement is required for normal platelet function.

An identical or closely related finding is that the C5-9 complex is formed on the platelet surface during thrombin activation and greatly enhances the aggregation induced by thrombin (Polley and Nachman, 1978). The suggested sequence of reactions is the interaction of thrombin with its receptor on the platelet membrane, the activation of complement on the platelet surface, the formation of the C5-9 complex, and enhancement of platelet aggregation. It was proposed that the C3 convertase generated by the interaction of thrombin with the platelet membrane is distinct from that resulting from activation of the classical or alternative pathways. The precise mechanism of the platelet aggregation enhancement is unknown, but it was suggested by Polley and Nachman (1979) that it is mediated through the arachidonic acid-endoperoxide pathway since it was inhibited by cyclo-oxygenase inhibitors.

Platelet Chemotactic Factor

It has been reported that a protein, extractable from the lysosomal granules of human platelets, can generate chemotactic activity for polymorph leukocytes after incubation with C5 (Weksler and Coupal, 1973). It was found that this factor is released during collagen or epinephrine-induced platelet aggregation and during blood coagulation. Such a mechanism might promote the invasion of thrombi by leukocytes and thereby assist in their dissolution through the action of the leukocyte proteases.

Anaphylatoxin-induced Platelet Aggregation

An anaphylatoxin generated from rat plasma by dextran or cobra venom, considered to be a cleavage product of C5, was found to be capable of inducing aggregation in platelet-rich plasma from cats and guinea-pigs, but it did not affect the platelets of man, dog, rabbit or, indeed, rat (Benner *et al.*, 1975). The C3a anaphylatoxin from the guinea-pig has been shown to cause a specific and rapid release of serotonin from guinea-pig platelets (Becker *et al.*, 1978).

Platelet Function in Patients with Hereditary Deficiency of Complement Components

Platelet function has been studied in a small number of patients with hereditary deficiency of a single component of complement. In the patient with homozygous C3 deficiency described by Gomperts and Rabson (1976) platelet aggregation in response to ADP, collagen, thrombin, epinephrine and ristocetin was normal as were the bleeding time and platelet factor 3 availability. The two C5-deficient subjects studied by Breckenridge and colleagues (1977) had normal platelet counts, bleeding times and platelet factor 3 release and normal aggregation in response to ADP, thrombin and collagen, and a variable response to ristocetin. The platelet-rich plasma of one of the patients, who suffered from systemic lupus erythematosus, failed to respond to epinephrine, but abnormal platelet aggregation has been reported in some patients with systematic lupus erythematosus in the absence of C5 deficiency (Regan *et al.*, 1974). Platelet function studies in a C6-deficient subjects revealed a normal platelet count, normal platelet factor 3 release and a normal aggregation response to ADP, epinephrine and collagen (Heusinkveld *et al.*, 1974): the aggregation response to thrombin was not examined. Two patients with C6 deficiency have been reported to show reduced platelet aggregability to thrombin (Wautier *et al.*, 1979), perhaps due to an inability to form the C5-9 complex. Formal platelet function studies were not reported from the patient with C7 deficiency, but the platelet count and bleeding times were normal (Boyer *et al.*, 1975).

Complement and the Coagulation System

The complement system and the coagulation system both involve the sequential formation of a series of proteases. This analogy and the contribution each makes to homeostasis have led to a series of studies to ascertain whether the complement and coagulation systems are interrelated through the influence of one on the other.

Relationship Between Complement and Coagulation

It was considered at one time that prothrombin and the first component of complement are identical (Bordet, 1920), but this was later shown to be fallacious (Boulanger and Rice, 1951). The lack of a factor common to both systems was supported by the finding that there is no difference between the total complement levels in normal serum and serum from patients with hereditary deficiencies of factors V, VIII and IX and from patients treated with dicoumarol-like drugs (Sharp, 1957).

Interest in the relationship between complement and coagulation was reawakened by the finding that C6-deficient rabbits had a coagulation abnormality characterised by a prolonged whole blood clotting time and decreased prothrombin consumption, abnormalities which could be corrected by the addition of purified C6 (Zimmerman *et al.*, 1971). It was further shown that activation of the alternative pathway by the polysaccharide inulin shortened the clotting time of normal platelet-rich rabbit plasma, but not that of the C6-deficient plasma (Zimmerman and Müller-Eberhard, 1971a). The complement-mediated action of inulin on blood coagulation in rabbits was shown to be due primarily to an effect on blood platelets, through the action of the terminal components of complement in enhancing platelet coagulant activity (Zimmerman and Müller-Eberhard, 1971b; Zimmerman, 1974).

Coagulation Tests in Patients with Hereditary Deficiency of Complement Components

The opportunity to study the hemostatic mechanism in human C6 deficiency was afforded by the recognition of a young woman with a hereditary deficiency of this component (Heusinkveld *et al.*, 1974): all standard tests of coagulability provided results which lay within the normal range. These tests included the whole blood clotting time, prothrombin consumption, activated partial thromboplastin time, prothrombin time, and individual clotting factor levels. A further patient has been described in whom tests of coagulability were also normal (Wautier *et al.*, 1979).

Studies in patients with hereditary deficiency of other complement components have been reported subsequently. A patient with C3 deficiency had neither a bleeding tendency nor a history of thrombotic episodes (Gomperts and Rabson, 1976), and all coagulation tests and the level of individual clotting factors were within normal. Two siblings with hereditary C5 deficiency were studied by Breckenridge and associates (1977): the coagulation tests were again normal. The patients with C7 deficiency have had limited coagulation studies performed but no abnormalities were found in the partial thromboplastin time, prothrombin time or thrombin time (Boyer *et al.*, 1975; Delage *et al.*, 1977).

C4-deficient guinea-pigs have been found to have normal clotting times (Kane *et al.*, 1973), suggesting that C4 is not required for normal clotting in this species. Dodds and colleagues (1975) reported, however, that their C4-deficient guinea-pigs had prolonged partial thromboplastin times and reduced factor VIII activity compared to normal guinea-pigs. Coagulation studies in human C4 deficiency are not available.

Influence of Components of Complement on Coagulation

Factor B and the Activation of Prothrombin. Fragment Bb from factor B of the alternative pathway has been shown to be capable of activating prothrombin in the absence of factor X (Fair *et al.*, 1981):

Figure 5.2: The Activation of Prothrombin by Fragment Bb

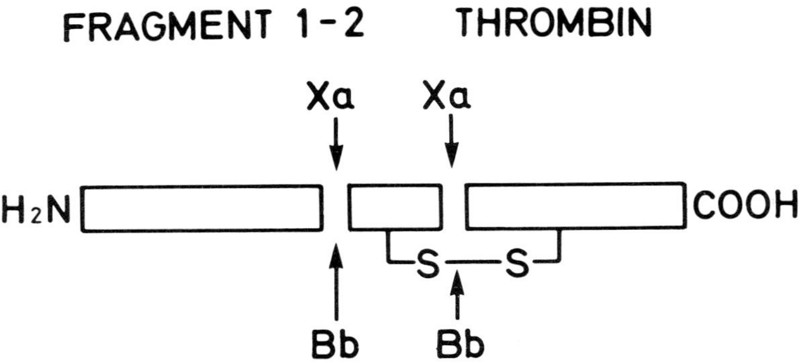

the prothrombin cleavage products produced by fragment Bb were similar to those obtained with factor Xa as analysed by SDS-poly-acrylamide gel electrophoresis. A diagrammatic representation of this reaction is provided in Figure 5.2.

Complement and Tissue Factor Activity. A further mechanism by which the complement system is linked to the coagulation system is through the influence of products of complement activation on the production of tissue factor activity by leukocytes. A variety of stimuli have been shown to induce leukocytes, particularly monocytes, to generate tissue factor activity: these include endotoxin (Rivers *et al.*, 1975; Niemetz *et al.*, 1977), antigen-antibody complexes (Roth-berger *et al.*, 1977), and platelets or platelet membranes (Niemetz and Marcus, 1974). Evidence that products of complement activation could also stimulate the production of tissue factor activity by leukocytes came initially from Prydz and colleagues (1977) who showed that guinea-pig C3b increased markedly the tissue factor

Figure 5.3: Induction of Leukocyte Generation of Tissue Factor Activity

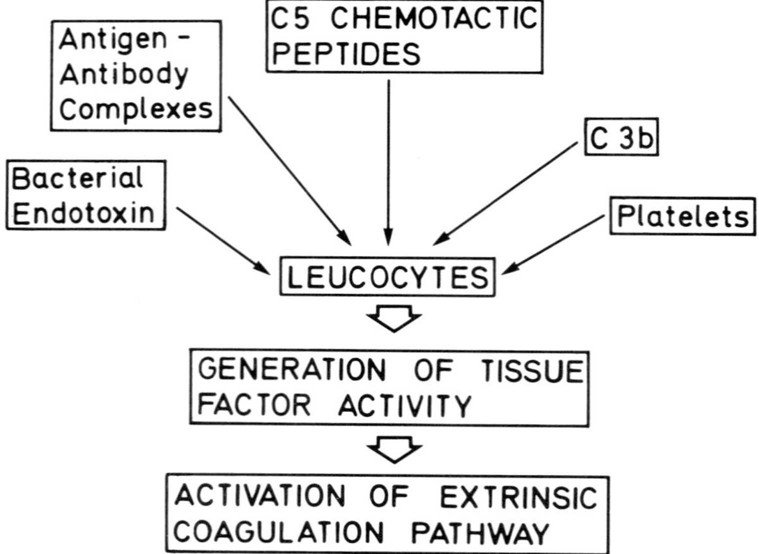

activity of cultured human monocytes. Muhlfelder and co-workers (1979) later reported that leukocytes incubated with human plasma after complement activation by cobra venom factor, zymosan or inulin, or with the C5 chemotactic fragment, generated increased amounts of tissue factor. In contrast, C3 chemotactic fragments had only very weak stimulatory effects on tissue factor activity generation. No activity was generated in C5-deficient plasma. Stimuli inducing leukocytes to generate tissue factor activity are summarised in Figure 5.3.

Influence of Coagulation Factors on the Complement System

Factor XII and Complement. Factor XII has been tentatively linked with the complement system. The possibility that factor XII might have a role in the activation of complement was raised by the finding of Donaldson (1968) that Cl esterase activity was induced in hereditary angioedema plasma, deficient in Cl inactivator, by the addition of purified activated factor XII in high concentration. Factor XII may influence the complement system only indirectly, however, since it mediates the production of kallikrein and plasmin which may themselves activate complement (Gigli *et al.*, 1968).

Eisen and Loveday (1972) found that the activation of Cls induced by the addition of factor XII activators to C$\bar{\text{I}}$ inactivator-deficient serum was not related to their potency as factor XII activators and that the spontaneous generation of Cl esterase activity (C$\bar{\text{l}}$s) was only slightly delayed in such serum depleted of factor XII. The development of Cl esterase activity did not take place in normal sera either spontaneously or after the addition of factor XII-activating agents. Although these findings do not point to an important direct role for factor XII in the activation of Cls, Eisen and Loveday suggest that factor XII, or the proteases formed under its influence, may participate in the spontaneous generation of Cl esterase activity in hereditary angioedema. The two phases proposed for this generation are the conversion of small amounts of Cls into its active form by factor XII or associated proteases, and the autocatalytic activation of the greater part of the Cls by the C$\bar{\text{l}}$s formed. Evidence has been provided more recently that the 28,000 dalton factor XII fragment can activate Cl directly rather than through the formation of kallikrein or plasmin (Ghebrehiwet *et al.*, 1981).

A further role for factor XII in the complement sequence has been proposed by Wooldridge and co-investigators (1975). They suggested that the magnitude of activation of the alternative pathway by zymosan is modulated by the availability if factor XII and early components of the classical pathway: this proposal was based on the finding that C3 inactivation by zymosan was impaired in C2 or factor XII-deficient serum, with repair of the abnormality by reconstitution of the serum with purified C2 or factor XII fragments. It was proposed that the contribution of factor XII to the formation of C3 convertase of the classical pathway is through the generation of plasmin or other undefined factors which can activate or substitute for complement components. The C3 convertase formed generates C3b which amplifies the production of the alternative pathway C3 convertase.

Thrombin and Factor D. Thrombin has been reported to be capable of replacing factor D in a hemolytic assay (Davis *et al.*, 1978), but full details of this reaction are not available.

Cold-promoted Activation of Complement. Plasma stored at low temperature has been shown to develop increased factor VII activity through a pathway involving factor XII, kallikrein and high molecular weight kininogen (Gjønnaess, 1972a, 1972b; Saito and Ratnoff, 1975; Saito *et al.*, 1975). Kondo and associates (1976) reported that, in comparison with plasma, the serum of some patients with hepatic cirrhosis have reduced levels of complement after storage at 4°C: they suggested that cold-promoted activation of complement is the result of blood coagulation although the precise mechanism of such activation is uncertain. Factor VII was not considered to be a factor in the cold activation of complement since Trasylol and soy bean trypsin inhibitor blocked the cold activation of factor VII, but not that of complement. Although the activation of factor VII on the addition of thrombin resulted in activation of complement at 37°C, this did not take place at 4°C: factor VII and thrombin were accordingly not considered to be responsible for the phenomenon of cold-promoted activation of complement.

Complement and the Fibrinolytic Enzyme System

Complement has been linked with fibrinolysis in two ways: firstly, in hypotheses to explain the lysis of clots made from dilute whole blood or plasma and thrombin and, secondly, in the influence of plasmin on components of the complement system.

Complement and the Lysis of Clots

Starting with a report in 1967 Taylor and his colleagues have built up a hypothesis implicating complement to explain the lysis of dilute whole blood clots prepared by the addition of thrombin. They found that platelets were required for the lysis of such dilute whole blood clots and that specific antisera to plasminogen, C3, C4 and γM-globulin inhibited the lysis of the blood clots. It was suggested that the platelets adsorb γM-globulin with consequent triggering of a complement reaction involving at least Cl, C2, C3 and C4. The platelets, incorporated in the fibrin network, induce activation of plasminogen either by enzyme activity associated with the components of complement or by the release of their intrinsic enzymes. In a subsequent report (Taylor and Müller-Eberhard, 1970) the non-hemolytic form of C8 (C8i) was claimed to accelerate clot lysis markedly and it was concluded that it behaves as a plasminogen activator. This hypothesis has been extended to couple the fibrinolytic and coagulation mechanisms in a 'hemostatic coagulolytic system' (Taylor *et al.*, 1973). It was proposed that thrombin or activated factor X modifies the platelet membrane coincidentally with the polymerisation of fibrinogen to fibrin. Recruitment of γM-globulin and components of complement onto the platelet membrane follows, resulting in platelet fusion and clot retraction. Consequent on the activation of complement and a further modification of the platelet membrane there is release of procoagulant material and activation of plasminogen on the surface of the platelet by C8i, the plasmin formed hydrolysing fibrin. An alternative suggestion is that the platelet membrane change results in the release of platelet activators of plasminogen: there is, however, little evidence that plasminogen activator activity is associated with or derived from platelets (Reid and Silver, 1964; Ekert *et al.*, 1970).

A different hypothesis to implicate complement in the fibrinolytic mechanism was put forward subsequently. Using a thrombin-clotted dilute whole plasma system Schreiber and Austen (1974) observed that the plasma of two patients with acquired C3 deficiency and plasma depleted of C3 by cobra venom factor or treatment with zymosan did not sustain clot lysis. Normal clot lysis took place in plasma deficient in C2, indicating that the reaction is not dependent on the classic C3 convertase. There was also a normal response in C6-deficient plasma implying that the terminal component complex (C5b-9) is not a prerequisite. These experiments, therefore, implicated C3 as being an essential component in the lysis of thrombin-induced clots in plasma, but the mechanism of fibrinolytic activation is obscure. In contrast to the results of Taylor and his associates, platelets were not essential for lysis although they did augment it to some extent. It may be noted that the euglobulin clot lysis time was found to be essentially normal in a patient with hereditary C3 deficiency (Gomperts and Rabson, 1976).

A further hint of possible relationship between fibrinolysis and complement has come from experiments reported by Moroz and Gilmore (1976). They found that prior heating to 56°C for 30 minutes or pretreatment of plasma with zymosan decreased the spontaneous fibrinolytic activity of the plasma and suggest, therefore, that components of the complement activation sequence contribute to normal fibrinolysis. They postulate further, on the basis of a stimulatory effect of DFP, *p*-toluene-sulphonyl-L-arginine methyl ester (TAMe). salicylaldoxine and hydrazine, that there is, additionally, a fibrinolytic role for a protease inhibitor involved in the control of the complement pathway, perhaps C3b inactivator. Moroz and Gilmore emphasise that the fibrinolytic activity induced is independent of plasmin formation.

A further mechanism by which complement activation may induce the formation of plasmin has been proposed recently (Sundsmo and Wood, 1981). Using purified materials the Bb fragment was shown to be capable of cleaving and activating plasminogen (Figure 5.4).

Plasmin and the Complement System

Since many of the steps in the complement sequence involve a limited proteolytic reaction with the formation of an active enzyme by the cleavage of a precursor protein it could be expected that proteases associated with the coagulation or fibrinolytic systems might interact with components of the complement sequence. In the case of plasmin this has been shown to be so in a number of ways, at least in *in vitro* studies.

Initial reports of the inactivation of human complement by streptokinase-activated plasma (Pillemer *et al.*, 1953; Lepow *et al.*, 1953) with inactivation of C2 and C4, but not C3, were followed by observations demonstrating that this phenomenon results from the conversion of Cl into an active enzyme (Lepow *et al.*, 1954). Confirmation that plasmin is capable of activating the Cls subcomponent of the first component of complement was provided by Ratnoff and Naff (1967) who demonstrated that the concentration of plasmin

Figure 5.4: The Activation of Plasminogen by Bb

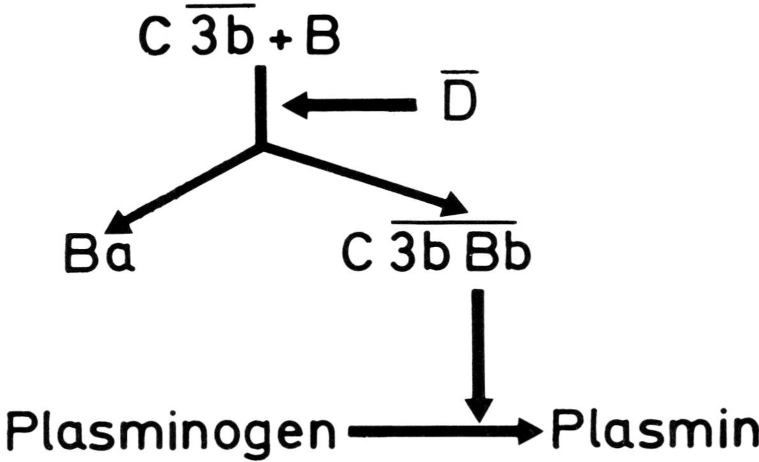

which can achieve this activation is comparable to that which can be generated from plasma plasminogen. More recently it has been shown that plasmin cleaves the polypeptide chain of Cls at a site near to or identical with that attacked by C̄lr, and also at another distant site (Cooper, 1979): Clr too may be activated by plasmin in purified systems (Cooper *et al.*, 1979). It is theoretically possible, therefore, that plasmin could activate the complement system by bypassing the remainder of the recognition unit. Once generated the esterase properties of C̄ls are diminished by further incubation with plasmin (Ratnoff and Naff, 1967).

It has been reported that a number of proteolytic enzymes including plasmin, can replace C̄ls in the generation of C3 convertase (Loos, 1977). Older observations by Laurell and her colleagues (1965) provide some evidence that C4 may be directly inactivated by plasmin.

One report has suggested that proteolytic enzymes, including plasmin and trypsin, can replace factor D̄ in the alternative pathway (Brade *et al.*, 1974). The physiological contribution of such an interaction is obscure, but it implies that plasmin could activate the alternative pathway.

Plasmin was reported by Ward (1967) to be able to cleave a small fragment from the C3 molecule which had leukocyte chemotactic activity, but which differed from C3a in that it had no anaphylatoxin activity. However, Bokisch *et al.* (1969) could obtain C3a by treatment with C3 convertase, plasmin or trypsin. They suggested that the chemotactic C3 fragment described by Ward, which was devoid of smooth muscle-contracting activity, is a degradation product of C3a.

A precursor form of C4 (pro-C4) has been identified in guinea-pig plasma and within macrophages and in human plasma (Gigli, 1978). Plasmin has been shown to be able to convert pro-C4 in cell lysates of guinea-pig macrophages into a protein with the subunit structure of native C4 (Goldberger and Colten, 1980), raising the possibility that plasmin converts pro-C4 into C4 *in vivo.*

The physiological relevance of these various points of possible interaction of plasmin with the complement system is far from clear. It is relevant to recall that any plasmin formed in plasma is very rapidly bound to and neutralised by α_2-antiplasmin. Unless a mechanism exists analogous to the lysine-binding sites of plasminogen which localise plasmin to its fibrin substrate where it is protected from neutralisation by α_2-antiplasmin, it is difficult to understand how plasmin-complement interactions could take place under physiological circumstances.

References

Barbaro, J.F. (1961) 'The release of histamine from rabbit platelets by means of antigen-antibody precipitations. I. The participation of the immune complex in histamine release', *Journal of Immunology, 86,* 369-76

Becker, S., Meuer, S., Hadding, U. and Bitter-Suermann, D. (1978) 'Platelet activation: a new biological activity of guinea-pig C3a anaphylatoxin', *Scandinavian Journal of Immunology, 7,* 173-80

Beller, F.K., Graeff, H. and Gorstein, F. (1969) 'Disseminated intravascular coagulation during the continuous infusion of endotoxin in rabbit', *American Journal of Obstetrics and Gynecology, 103,* 544-54

Benner, K.V., Schumacher, K.-A. and Classen, H.-G. (1975) 'Platelet aggregation induced by anaphylatoxin and its inhibition', *Arzneimittel-Forschung, 25,* 1635-8

Bokisch, V.A., Müller-Eberhard, J. and Cochrane, C.G. (1969) 'Isolation of a fragment (C3a) of the third component of human complement containing anaphylatoxin and chemotactic activity and description of an anaphylatoxin inactivator of human serum', *Journal of Experimental Medicine, 129,* 1109-30

Bordet, J. (1920) 'Considérations sur les théories de la coagulation du sang', *Annales de l'Institut Pasteur, 34,* 561-95

Boulanger, P. and Rice, C.E. (1951) 'Parallel studies of complement and coagulation: relative effects of dicoumarol', *Canadian Journal of Medical Sciences 29*, 5-12

Boyer, J.T., Gall, E.F., Norman, M.E., Nilsson, U.R. and Zimmerman, T.S. (1975) 'Hereditary deficiency of the seventh component of complement', *Journal of Clinical Investigation, 56*, 905-13

Brade, V., Nicholson, A., Bitter-Suermann, D. and Hadding, U. (1974) 'Formation of the C3-cleaving properdin enzyme on zymosan. Demonstration that factor D is replaceable by proteolytic enzymes', *Journal of Immunology, 113*, 1735-43

Breckenridge, R.T., Rosenfeld, S.I., Graff, K.S. and Leddy, J.P. (1977) 'Hereditary C5 deficiency in man. III. Studies of hemostasis and platelet responses to zymosan', *Journal of Immunology, 118*, 12-6

Brown, D.L. and Lachmann, P.J. (1973) 'The behaviour of complement and platelets in lethal endotoxin shock in rabbits', *International Archives of Allergy and Applied Immunology, 45*, 193-205

Brown, D.L. and Lachmann, P.J. (1974) 'The relationship of complement-mediated platelet damage to intravascular coagulation: comparison between endotoxin and inulin in the rabbit', *Advances in the Biosciences, 12*, 300-10

Calcott, M.A. and Müller-Eberhard, H.J. (1972) 'Clq protein of human complement', *Biochemistry, 11*, 3443-50

Cazenave, J.-P., Assimah, S.N., Painter, R.H., Packham, M.A. and Mustard, J.F. (1976) 'Clq inhibition of the interaction of collagen with human platelets', *Journal of Immunology, 116*, 162-3

Chater, B.V. (1976) 'The role of membrane bound complement in the aggregation of mammalian platelets by collagen', *British Journal of Haematology, 32*, 515-24

Cohen, P. and Gardner, F.H. (1966) 'Thrombocytopenia as laboratory sign and complication of gram-negative bacteramic infection', *Archives of Internal Medicine, 117*, 113-24

Cooper, N.R. (1979) 'Interrelationships between the complement system and the coagulation, kinin-forming and fibrinolytic systems', *Thrombosis and Haemostasis, 42*, 262

Cooper, N.R., Miles, L.A. and Griffin, J.H. (1979) 'Activation of purified Cl, the first complement component, by purified plasma kallikrein and plasmin', *Thrombosis and Haemostasis, 42*, 251

Corrigan, J.J., Ray, W.L. and May, N. (1968) 'Changes in blood coagulation system associated with septicemia', *New England Journal of Medicine, 279*, 851-6

Davis, A.E. and Kenney, D.M. (1979) 'Properdin factor D̄. Effects on thrombin-induced platelet aggregation', *Journal of Clinical Investigation, 64*, 721-8

Davis, A.E., Rosenberg, R.D., Fenton, J.W., Bing, D.H., Rosen, F.S. and Alper, C.A. (1978) 'Is factor D of the alternative pathway a fragment of thrombin?' *Journal of Immunology, 120*, 1771

Delage, J.M., Bergeron, P., Simard, J., Lehner-Netsch, G. and Prochazka, E. (1977) 'Hereditary C7 deficiency. Diagnosis and HLA studies in a French-Canadian family', *Journal of Clinical Investigation, 60*, 1061-9

Des Prez, R.M. and Bryant, R.E. (1969) 'Two mechanisms of immunologically-induced injury to rabbit platelets', *Journal of Immunology, 102*, 241-52

Des Prez, R.M., Horowitz, H.I. and Hook, E.W. (1961) 'Effects of bacterial endotoxin on rabbit platelets. I. Platelet aggregation and release of platelet factors *in vitro*', *Journal of Experimental Medicine, 114*, 857-73

Dodds, W.J., Raymond, S.L., Moynihan, A.C., Pickering, R.J. and Amiraian, K.

(1975) 'Coagulation studies in C4-deficient guinea pigs', *Thrombosis et Diathesis Haemorrhagica, 34,* 602

Donaldson, V.H. (1968) 'Mechanisms of activation of C′1 esterase in hereditary angioneurotic edema plasma *in vitro.* The role of Hageman factor, a clot-promoting agent', *Journal of Experimental Medicine, 127,* 411-29

Eisen, V. and Loveday, C. (1972) 'Activation of arginine and tyrosine esterase in serum from patients with hereditary angio-oedema', *British Journal of Pharmacology, 46,* 157-66

Ekert, H., Friedlander, I. and Hardisty, R.M. (1970) 'The role of platelets in fibrinolysis. Studies on the plasminogen activator and antiplasmin activity of platelets', *British Journal of Haematology, 18,* 575-84

Fair, D.S., Sundsmo, J.S., Schwartz, B.S., Edgington, T.S. and Müller-Eberhard, H.J. (1981) 'Prothrombin activation by factor B (Bb) of the alternative pathway of complement', *Thrombosis and Haemostasis, 46,* 301

Fong, J.S.C. and Good, R.A. (1971) 'Prevention of the localized and generalized Schwartzman reaction by an anticomplementary agent, cobra venom factor', *Journal of Experimental Medicine, 134,* 642-55

Fong, J.S.C., White, J.G. and Good, R.A. (1974) 'Inhibition of endotoxic lipopolysaccharide-mediated platelet aggregation by cobra venom anticomplementary factor', *Blood, 44,* 399-409

Garner, R., Chater, B.V. and Brown, D.L. (1974) 'The role of complement in endotoxin shock and disseminated intravascular coagulation: experimental observations in the dog', *British Journal of Haematology, 28,* 393-401

Gewurz, H., Shin, H.S. and Mergenhagen, S.E. (1968) 'Interactions of the complement system with endotoxic lipopolysaccharide: consumption of each of the six terminal complement components', *Journal of Experimental Medicine, 128,* 1049-57

Ghebrehiwet, B., Silverberg, M. and Kaplan, A.P. (1981) 'Activation of the classical pathway of complement by Hageman factor fragment', *Journal of Experimental Medicine, 153,* 665-76

Gigli, I. (1978) 'A single chain precursor of C4 in human serum', *Nature, 272,* 836-7

Gigli, I., Mason, J.W., Colman, R.W. and Austen, K.F. (1968) 'Interaction of kallikrein with C′1 esterase inhibitor (C′1aINH)', *Journal of Immunology, 101,* 814

Gilbert, V.E, and Braude, A.L. (1962) 'Reduction of serum complement in rabbits after injection of endotoxin', *Journal of Experimental Medicine, 116,* 477-90

Gjønnaess, H. (1972a) 'Cold-promoted activation of factor VII. III. Relation to the kallikrein system', *Thrombosis et Diathesis Haemorrhagica, 28,* 182-93

Gjønnaess, H. (1972b) 'Cold-promoted activation of factor VII. IV. Relation to the coagulation system', *Thrombosis et Diathesis Haemorrhagica, 28,* 194-205

Gocke, D.J. and Osler, A.G. (1965) '*In vitro* damage to rabbit platelets by an unrelated antibody reaction. I. General characteristics of the reaction', *Journal of Immunology, 94,* 236-46

Goldberger, G. and Colten, H.R. (1980) 'Precursor complement protein (pro-C4) is converted *in vitro* to native C4 by plasmin', *Nature, 286,* 514-6

Gomperts, E.D. and Rabson, A.R. (1976) 'Coagulation studies in a patient with homozygous C3 deficiency', *British Journal of Haematology, 35,* 165-6

Henson, P.M. (1970) 'Release of vasoactive amines from rabbit platelets induced by sensitized mononuclear leucocytes and antigen', *Journal of Experimental Medicine, 131,* 287-304

Henson, P.M. and Cochrane, C.G. (1969) 'Immunological induction of increased vascular permeability. II. Two mechanisms of histamine release from rabbit platelets involving complement', *Journal of Experimental Medicine, 129,*167-83

214

Heusinkveld, R.S., Leddy, J.P., Klemperer, M.R. and Breckenridge, R.T. (1974) 'Hereditary deficiency of the sixth component of complement in man. II. Studies of hemostasis', *Journal of Clinical Investigation, 53,* 554-8

Horowitz, H.I., Des Prez, R.M. and Hook, E.W. (1962) 'Effects of bacterial endotoxin on rabbit platelets', *Journal of Experimental Medicine, 115,* 619-33

Humphrey, J.M. and Jaques, R. (1955) 'The release of histamine and 5-hydroxytryptamine (serotonin) from platelets by antigen-antibody reactions (*in vitro*)', *Journal of Physiology, 128,* 9-27

Jobin, F. and Tremblay, F. (1969) 'Platelet reactions and immune processes. II. The inhibition of platelet aggregation by complement inhibitors', *Thrombosis et Diathesis Haemorrhagica, 22,* 466-81

Kane, M.A., May, J.E. and Frank, M.M. (1973) 'Interactions of the classical and alternative complement pathways with endotoxin lipopolysaccharide', *Journal of Clinical Investigation, 52,* 370-6

Kondo, M., Hosokawa, K. and Masuda, M. (1976) 'Cold activation of complement. I. Presence of coagulation-related activator', *Journal of Immunology, 117,* 486-90

Laurell, A.-B., Lundh, B. and Malmquist, J. (1965) 'Inability of a highly purified streptokinase preparation to inactivate complement in serum', *Acta Pathologica et Microbiologica Scandinavia, 64,* 318-28

Lepow, I.H., Pillemer, L. and Ratnoff, O.D. (1953) 'The influence of calcium ions on the inactivation of human complement and its components by plasmin', *Journal of Experimental Medicine, 98,* 277-89

Lepow, I.H., Wurz, L., Ratnoff, O.D. and Pillemer, L. (1954) 'Studies on the mechanism of inactivation of human complement by plasmin and by antigen-antibody aggregates', *Journal of Immunology, 73,* 146-58

Loos, M. (1977) 'Generation of EAC$\overline{42}$ (C3 convertase) by proteolytic enzymes', *Federation Proceedings, 36,* 1209

Marney, S.R. (1971) 'Rabbit platelet injury by soluble antigen and antibody. II. Platelet aggregation in heparinized plasma and citrated plasma, relationship to ADP and inhibition by cobra factor', *Journal of Immunology, 106,* 82-90

McKay, D.G. and Shapiro, S.S. (1958) 'Alterations in the blood coagulation system induced by bacterial endotoxin. I. *In vivo* (generalized Schwartzman reaction)', *Journal of Experimental Medicine, 107,* 353-67

Moroz, L.A. and Gilmore, N.J. (1976) 'Fibrinolysis in normal plasma and blood: evidence for significant mechanisms independent of the plasminogen-plasmin system', *Blood, 48,* 531-45

Mueller-Eckhardt, C.H. and Lüscher, E.F. (1968) 'Immune reactions of human blood platelets. A comparative study on the effects of heterologous antiplatelet anti-serum of antigen-antibody complexes of aggregated gamma-globulin and thrombin on platelets', *Thrombosis et Diathesis Haemorrhagica, 20,* 155-67

Mulfelder, T.W., Niemetz, J., Kreutzer, D., Beebe, D., Ward, P.A. and Rosenfeld, S.I. (1979) 'C5 chemotactic fragment induces leukocyte production of tissue factor activity. A link between complement and coagulation', *Journal of Clinical Investigation, 63,* 147-50

Müller-Berghaus, G. and Lohmann, E. (1974) 'The role of complement in endotoxin-induced disseminated intravascular coagulation: studies in congenitally C6-deficient rabbits', *British Journal of Haematology, 28,* 403-18

Nachman, R.L. (1968) 'Platelet proteins', *Seminars in Hematology, 5,* 18-31

Nagaki, K., Fujikawa, K. and Inai, S. (1965) 'Studies on the fourth component of complement. II. The fourth component of complement in guinea pig and human platelets', *Biken Journal, 8,* 129-41

Nagayama, M., Zucker, M.B. and Beller, F.K. (1971) 'Effects of a variety of endotoxins on human and rabbit platelet function', *Thrombosis et Diathesis*

Haemorrhagica, 26, 467-73

Niemetz, J. and Marcus, A.J. (1974) 'The stimulating effect of platelets and platelet membranes on the procoagulant activity of leucocytes', *Journal of Clinical Investigation, 54,* 1437-43

Niemetz, J., Muhlfelder, T., Chierego, M.E. and Troy, B. (1977) 'Procoagulant activity of leucocytes', *Annals of the New York Academy of Sciences, 283,* 208-17

Pfueller, S.L. and Lüscher, E.F. (1972) 'The effect of aggregated immunoglobulins on human blood platelets in relation to their complement-fixing abilities', *Journal of Immunology, 109,* 517-25

Pfueller, S.L. and Lüscher, E.F. (1974a) 'Studies of the mechanisms of the human platelet release reaction induced by immunologic stimuli. I. Complement-dependent and complement-independent reactions', *Journal of Immunology, 112,* 1204-10

Pfueller, S.L. and Lüscher, E.F. (1974b) 'Studies of the mechanisms of the human platelet release reaction induced by immunologic stimuli. II. Effects of zymosan', *Journal of Immunology, 112,* 1211-18

Pillemer, L., Ratnoff, O.D. and Lepow, I.H. (1953) 'The inactivation of complement and its components by plasmin', *Journal of Experimental Medicine, 97,* 573-89

Polley, M.J. and Nachman, R. (1978) 'The human complement system in thrombin-mediated platelet function', *Journal of Experimental Medicine, 147,* 1713-26

Polley, M.J. and Nachman, R.L. (1979) 'Normal complement in thrombin-mediated platelet function. Uptake of the C5b-9 complex', *Journal of Experimental Medicine, 150,* 633-45

Prydz, H., Allison, A.C. and Schorlemmer, H.U. (1977) 'Further link between complement activation and blood coagulation', *Nature, 270,* 173-4

Ratnoff, O.D. and Naff, G.B. (1967) 'The conversion of C'1s to C'1 esterase by plasmin and trypsin', *Journal of Experimental Medicine, 125,* 337-58

Regan, M.G., Lackner, H. and Karpatkin, S. (1974) 'Platelet function and coagulation profile in lupus erythematosus. Studies in 50 patients', *Annals of Internal Medicine, 81,* 462-8

Reid, K.B.M. (1974) 'A collagen-like amino acid sequence in a polypeptide chain of human Clq (a subcomponent of the first component of complement)', *Biochemical Journal, 141,* 189-203

Reid, K.B.M. and Porter, R.R. (1981) 'The proteolytic activation systems of complement', *Annual Review of Biochemistry, 50,* 433-64

Reid, W.O. and Silver, M.J. (1964) 'Activation of the blood fibrinolytic enzyme system by platelets', *American Journal of Physiology, 206,* 1255-61

Rivers, R.P.A., Hathaway, W.E. and Weston, W.L. (1975) 'The endotoxin-induced coagulant activity of human monocytes', *British Journal of Haematology, 30,* 311-6

Robbins, J. and Stetson, C.A. (1959) 'An effect of antigen-antibody interaction on blood coagulation', *Journal of Experimental Medicine, 109,* 1-8

Rothberger, H., Zimmerman, T.S., Spiegelberg, H.L. and Vaughan, J.H. (1977) 'Leucocyte procoagulant activity. Enhancement of production *in vitro* by IgG and antigen-antibody complexes', *Journal of Clinical Investigation, 59,* 549-57

Saito, H. and Ratnoff, O.D. (1975) 'Alteration of factor VII activity by activated Fletcher factor (a plasma kallikrein): a potential link between the intrinsic and extrinsic blood-clotting systems', *Journal of Laboratory and Clinical Medicine, 85,* 405-15

Saito, H., Ratnoff, O.D., Waldmann, R. and Abraham, J.P. (1975) 'Fitzgerald trait: deficiency of a hitherto unrecognized agent, Fitzgerald factor,

216

participating in surface-mediated reactions of clotting, fibrinolysis, generation of kinin and the property of diluted plasma enhancing vascular permeability (PF/Dil)', *Journal of Clinical Investigation, 55*, 1082-9

Schreiber, A.D. and Austen, K.F. (1974) 'Hageman factor-independent fibrinolytic pathway', *Clinical and Experimental Immunology, 17*, 587-600

Sharp, A.A. (1957) 'Relationship of complement to blood coagulation', *Nature, 179*, 632-3

Siraganian, R.P. (1972) 'Platelet requirement in the interaction of the complement and clotting systems', *Nature New Biology, 239*, 208-10

Suba, E.A. and Csako, G. (1976) 'Clq (Cl) receptor on human platelets: inhibition of collagen-induced platelet aggregation by Clq (Cl) molecules', *Journal of Immunology, 117*, 304-9

Sundsmo, J.S. and Wood, L.M. (1981) 'Activated factor B (Bb) of the alternative pathway of complement activation cleaves and activates plasminogen', *Journal of Immunology, 127*, 877-80

Taylor, F.B. and Müller-Eberhard, H.J. (1967) 'Factors influencing lysis of whole blood clots', *Nature, 216*, 1023-5

Taylor, F.B. and Müller-Eberhard, H.J. (1970) 'Qualitative description of factors involved in the retraction and lysis of dilute whole blood clots and in the aggregation and retraction of platelets', *Journal of Clinical Investigation, 49*, 2068-88

Taylor, F.B., Nilsson, U.R., Creech, R.H., Carroll, E.T. and Beisswenger, J.G. (1973) 'Coagulolysis: mechanism of formation and lysis of dilute whole blood clots and application of this assay in the study of ceratin hypercoagulable states', *Series Haematologica, VI*, 4, 529-48

Tiffany, M.L. and Penner, J.A. (1980) 'Effect of complement on collagen-induced platelet aggregation', *Journal of Laboatory and Clinical Medicine, 96*, 796-802

Ward, P.A. (1967) 'A plasmin-split fragment of C'3 as a new chemotactic factor', *Journal of Experimental Medicine, 126*, 189-206

Wautier, J.L., Tobelem, G.M., Pelter, A.P. and Caen, J.F. (1974) 'Evidence for Cl on human platelets', *Haemostasis, 2*, 281-6

Wautier, J.L., Tobelem, G.M., Peltier, A.P. and Caen, J.P. (1976) 'Cl and human platelets. II. Detection by immunological methods and role', *Immunology, 30*, 459-65

Wautier, J.L., Souchon, H., Reid, K.B.M., Peltier, A.P. and Caen, J.P. (1977) 'Studies on the mode of reaction of the first component of complement with platelets; interaction between collagen-like portion of Clq and platelets', *Immunochemistry, 14*, 763-6

Wautier, J.L., Peltier, A.P. and Caen, J.P. (1979) 'Complement (C6) and platelet activation by thrombin in humans', *Thrombosis Research, 15*, 589-91

Wautier, J.L., Legrand, Y.J., Fauvel, F. and Caen, J.P. (1981) 'Inhibition of platelet collagen interactions by the Cls subcomponent of the first component of complement', *Thrombosis Research, 21*, 3-12

Weksler, B.B. and Coupal, C.E. (1973) 'Platelet-dependent generation of chemotactic activity in serum', *Journal of Experimental Medicine, 137*, 1419-30

Wooldridge, D., Spragg, J. and Austen, K.F. (1975) 'Interaction of Hageman factor dependent systems with the complement sequence', *Life Sciences, 16*, 802-3

Zimmerman, T.S. (1974) 'The platelet in complement-coagulation interaction', *Advances in Biosciences, 12*, 291-8

Zimmerman, T.S. and Kolb, W.P. (1976) 'Human platelet-initiated formation and uptake of the C5-9 complex of human complement', *Journal of Clinical Investigation, 57*, 203-11

Zimmerman, T.S. and Müller-Eberhard, H.J. (1971a) 'Blood coagulation initiation by a complement-mediated pathway', *Journal of Experimental Medicine, 134,* 1601-7

Zimmerman, T.S. and Müller-Eberhard, H.J. (1971b) 'Initiation of coagulation by complement activation: generation of platelet associated clot-promoting activity', *Blood, 38,* 791

Zimmerman, T.S., Arroyave, C.M. and Müller-Eberhard, H.J. (1971) 'A blood coagulation abnormality in rabbits deficient in the sixth component of complement (C6) and its correction by purified C6', *Journal of Experimental Medicine, 134,* 1591-600

Zucker, M.B. and Grant, R.A. (1974) 'Aggregation and release reaction induced in human blood platelets by zymosan', *Journal of Immunology, 112,* 1219-30

Zucker, M.B., Grant, R.A., Alper, C.A., Goodkofsky, I. and Lepow, I.H. (1974) 'Requirement for complement components and fibrinogen in the zymosan-induced release reaction of human blood platelets', *Journal of Immunology, 113,* 1744-51

6 Hormonal Influences on the Hemostatic Mechanism

A variety of observations have pointed to possible hormonal influences on the physiological control of platelet function, coagulation and fibrinolysis. In addition, there are some tantalising data available to suggest that the fibrinolytic system may be involved in the modulation of hormone production or action. This chapter reviews some of the available information on the relationships between hormones and hemostatic function and components. At the present time definitive conclusions on these associations are lacking. This arises in part from the complex interrelationships between the various endocrine glands which militates against precise interpretations of the influence of individual hormones on hemostasis. It can be readily appreciated that the differentiation of the changes which result from direct induction of synthesis, release or inhibition of hemostatic factors by a single hormone from those which are secondary to the effect of the hormone on other tissues or organs, including other endocrine glands, may be extremely difficult. A further problem in interpretation is that the majority of reported studies have examined changes in components of the hemostatic mechanism after the administration of large and unphysiological quantities of hormone, or have measured such components in patients with clinical syndromes resulting from gross hormonal deficiency or excess.

Catecholamines

The adrenal medulla secretes epinephrine (adrenaline) and norepinephrine (noradrenaline) in a ratio of about four to one. Both are synthesised from tyrosine and, together with dopamine, are collectively termed catecholamines. Norepinephrine acts as the transmitter substance at sympathetic postganglionic nerve terminals and dopamine is a neurotransmitter in the brain.

The release of catecholamines from the adrenal medulla is provoked by a variety of stimuli including hypoxia, hypothermia, hypoglycemia, acidemia, exercise and such emotional reactions as fear and pain. On their release from the adrenals they are removed very rapidly from the circulation either by uptake into nerve terminals and other tissues or by degradation by catechol-*O*-methyl-transferase and monoamine oxidase.

The effects of catecholamines on target organs follow their combination with specific receptor sites on the cells. These adrenergic receptors are divided into two main types, the α-receptors which serve mainly excitatory functions such as vasoconstriction and which respond preferentially to norepinephrine, and the β-receptors which serve predominantly inhibitory functions such as vasodilation and which respond principally to epinephrine. Many β effects are mediated by activation of adenylate cyclase which catalyses the conversion of ATP into $3',5'$-cyclic adenosine monophosphate (cyclic AMP). Most α effects are associated with inhibition of adenylate cyclase. Two classes of α-receptor have been delineated and termed α_1- and α_2-receptors: these correspond to an earlier terminology of postsynaptic and presynaptic respectively. There are also two types of β-receptor, designated β_1 and β_2 on the basis of their response to synthetic agonists and antagonists. Beta$_1$-receptors are present in heart muscle and are stimulated by both epinephrine and norepinephrine. Beta$_2$-receptors are present in vascular, bronchial and other smooth muscles and are stimulated by epinephrine; norepinephrine has a relatively weak effect of β_2-receptors.

Influence on Platelet Numbers and Function

Epinephrine infusions result in an increase in the peripheral platelet count (McClure *et al.*, 1965). The source of the platelets is the exchangeable platelet pool (Branehög *et al.*, 1973). It appears that α-receptor stimulation causes the release of platelets from the spleen whereas β-receptor stimulation has an opposite effect with trapping of platelets by the spleen; this latter effect is mediated by both β_1- and β_2-receptors (Olsson *et al.*, 1976; Fredén *et al.*, 1978). The α stimulation effect predominates after infusion of epinephrine, explaining the rise in the platelet count.

Catecholamines induce platelet aggregation *in vitro* (Clayton and Cross, 1963; O'Brien, 1963; Mitchell and Sharp, 1964): both epinephrine and norepinephrine induce two waves of aggregation, but norepinephrine has been found to be ten-fold less active, and isoprenaline to be inactive (O'Brien, 1963). In addition, the effect of ADP on platelet aggregation is increased by epinephrine, norepinephrine and dopamine, even at low concentration (Ardlie *et al.*, 1966): this potentiation shows itself both as an acceleration of primary aggregation and as diminution in the ADP concentration required to initiate the second phase of aggregation. Epinephrine also potentiates aggregation by thrombin (Thomas, 1967), but this may represent, at least in part, potentiation by the ADP which this agent induces platelets to secrete. Trace amounts of thrombin have been found to be required for epinephrine-induced aggregation by human platelets (Cazanave *et al.*, 1981).

Infused catecholamines also influence platelet function. Daily injections of epinephrine over a 14-day period were found to shorten platelet survival in the rabbit (Ozge *et al.*, 1963), and norepinephrine caused intravascular platelet aggregation in the dog (Haft *et al.*, 1972).

The human platelet possesses receptor sites for epinephrine; McMillan and co-workers (1979) estimated that the number of binding sites averaged 105,000 per platelet. Earlier studies showing inhibition of epinephrine-induced aggregation by α-adrenergic blockers such as phentolamine led to the conclusion that the receptor was α-adrenergic (Mills and Roberts, 1967; Bygdeman and Johnson, 1969). More recent studies indicate that human platelets possess only α_2-adrenergic receptors (Hoffman *et al.*, 1979), but of two types, one with higher affinity for agonists (Lynch and Steer, 1981).

The mechanism by which epinephrine induces platelet aggregation is not entirely clear. It binds to the receptors on the surface of the platelet and inhibits the platelet adenylate cyclase by an action on an α-receptor with a resultant reduction in the platelet cyclic AMP. Cyclic AMP probably acts on a protein kinase to phosphorylate a protein which facilitates the removal of calcium from the platelet cytoplasm and its storage in the dense tubular system (Käser-Glanzmann *et al.*, 1977): the reduction in cyclic AMP resulting from the epinephrine-receptor interaction allows the cytoplasmic calcium concentration to rise and stimulate such processes as the actin-myosin interaction and the activation of platelet phospholipases A_2 and C.

Fibrinogen is essential for the induction of platelet aggregation and granule secretion by ADP and epinephrine, and acts as a bridge between adjacent platelets in an aggregate. Epinephrine, as well as ADP, induces the appearance of specific high affinity sites for fibrinogen on platelets (Bennett and Vilaire, 1979).

Both epinephrine and norepinephrine, in common with other amines, are taken up by the dense granules of platelets by a glucose-dependent active-transport mechanism (Sano *et al.*, 1959; Bull and Zucker, 1965).

Influence on the Coagulation System

The first recorded observation that parenteral epinephrine shortens the coagulation time of blood was made by Vosburgh and Richards in 1903. This finding in dogs was confirmed in cats (Cannon and Gray, 1914; Grabfield, 1916), and it was shown that epinephrine had no effect on coagulation when added directly to blood (Wiggers, 1909).

Ingram (1961) reported that infusions of epinephrine result in an increase in factor VIII procoagulant activity; norepinephrine had no effect. This observation has been confirmed repeatedly (Egeberg, 1963; Gader *et al.*, 1973a). The increase in factor VIII appears very rapidly, indicating that it is released from preformed stores rather than synthesised. The epinphrine-induced rise in factor VIII can be prevented by the β-adrenergic blocker propranolol, but not by the α-blocker phentolamine (Ingram and Vaughan Jones, 1966), leading to the conclusion that the effect of epinephrine on factor VIII is mediated by β-receptors. In later experiments it was found that infusions of isoprenaline and the β_2-antagonist salbutamol increased the factor VIII level indicating that it could be classified as a β_2 effect,

The plasma level of factor VIII rises rapidly in response to various physiological stimuli. In the case of exercise this rise can be prevented by the prior administration of β-blockers, demonstrating an adrenergic mediation for the increase in factor VIII (Cohen *et al.*, 1968). In other stress situations, however, β-adrenergic blockade does not prevent the factor VIII increase: examples are pneumoencephalography (Mannucci *et al.*, 1971) and venous occlusion (Ponari *et al.*, 1973).

Epinephrine infusions do not affect the plasma level of factors V, X, IX, XI, XII, XIII or fibrinogen (Barrow *et al.*, 1969; Mannucci *et al.*, 1975; Ingram *et al.*, 1977).

Influence on the Fibrinolytic System

Biggs and her colleagues (1947) first described the increase in blood fibrinolytic activity following the injection of epinephrine and showed that the addition of epinephrine to blood did not influence its fibrinolytic activity. This effect of parenteral epinephrine has been confirmed in many studies (Truelove, 1951; Kwaan *et al.*, 1957; Tanser and Smellie, 1964; Cash and Allan, 1967). The increase in blood fibrinolytic activity was shown to be due to the release of a plasminogen activator (Sherry *et al.*, 1959).

The effect of norepinephrine on blood fibrinolytic activity is less conclusive. Truelove (1951) obtained no increase in activity with dl-norepinephrine whereas Genton and co-workers (1961) reported that intravenous epinephrine and norepinephrine produced equivalent decreases in the euglobulin clot lysis time. More recent studies strongly suggest that norepinephrine induces little or no activator release (Gader *et al.*, 1973a).

Experiments with different adrenergic agonists and antagonists have been carried out to elucidate the type of receptor responsible for the release of plasminogen activator after epinephrine infusion. The ability of the β_2-agonist salbutamol to induce activator release led to the proposal that β_2-adrenergic receptors are involved (Gader *et al.*, 1973a). Studies with a number of β-adrenergic antagonists have indicated that the release mechanism is of greater complexity with either several different β-adrenergic receptor sites or multifocal components on one site (Gader *et al.*, 1974a).

Catecholamine release is a common factor in the general response to stress and it has been proposed that it is responsible for the increase in plasma plasminogen activator activity seen in such stresses as physical exertion, perhaps through its vasoactive properties. Evidence against this view has come from a number of sources. In the experiments of Pina-Cabral and Rodriques (1974) electroconvulsive therapy resulted in a large increase in the blood level of catecholamines and of plasminogen activator whereas modified electroconvulsive therapy, where barbiturate anesthesia combined with curarisation was administered immediately before the electroshock, was associated with a substantial reduction in catecholamine release, but with a similar increase in plasma activator activity. Changes in blood catecholamine levels and plasma fibrinolytic activity resulting from exercise and exposure to heat have been shown to correlate poorly and it has been concluded that the release of activator is not the direct result of changes in the plasma epinephrine concentration (Britton *et al.*, 1975; Hawkey *et al.*, 1975). Further evidence that the rise in plasminogen activator in stress situations is not mediated by stimulation of β-adrenergic receptors comes from studies showing that β-adrenergic blockade does not influence the fibrinolytic response to such stresses. For example, the non-selective β-blocker

propranolol or propranolol together with the α-blocker phentolamine did not prevent the increased fibrinolytic activity in response
to venous occlusion (Ponari *et al*, 1973), and β-blockade did not
influence the exercise-induced rise in fibrinolysis (Cohen *et al.*,
1968).

Corticosteroids

The influence of corticosteroids on the hemostatic mechanism has
been assessed from measurements on patients with adrenal cortex
hyperactivity, on patients with a variety of disorders receiving
corticosteroid preparations or, of greater significance, on healthy
volunteers given a natural or synthetic corticosteroid.

Influence on Platelet Numbers and Function

Patients with a history of a bleeding tendency given prednisone in a
dosage of 45 mg/day showed an increase in the platelet count and a
shortening of the bleeding time, but there was no change in platelet
aggregation in response to ADP, collagen or epinephrine and no
alteration in platelet adhesiveness in glass bead columns (McKenna
et al., 1981). Prednisone (20 mg daily for 11 days) did not affect
either the platelet count or platelet adhesiveness in healthy subjects
(Isacson, 1970), and intravenous hydrocortisone did not influence
platelet aggregation (Hutton *et al.*, 1980). Of tenuous physiological
significance is the finding that the addition of large amounts of hydrocortisone sodium succinate to canine blood prevented platelet
aggregation in response to ADP, thrombin and collagen (Nelson and
Taylor, 1975).

Influence on the Coagulation System

A number of early observations suggested that corticosteroids shortened the clotting time of whole blood. For example, Menczel and
Dreyfuss (1960) found acceleration of the clotting time in siliconised
tubes two and four hours after a single oral dose of 10 mg prednisone
in patients receiving dicoumarol for thromboembolic disease.
Coagulation assays in a patient with Cushing's syndrome revealed a
short one-stage prothrombin time, a short partial thromboplastin
time and raised factor VIII procoagulant levels, changes reversed by
bilateral adrenalectomy and mimicked by administration of prednisone to patients (Ozsoylu *et al.*, 1962). An increase in the level of

factor VIII in Cushing's syndrome was also found by Sjöberg and associates (1976) who additionally noted increased levels of factors IX and V. An increase in factor VIII procoagulant activity was seen in healthy volunteers given 20 mg prednisone daily for eleven days, but in this study there was no change in the prothrombin time or in the level of factor V and fibrinogen (Isacson, 1970). A significant increase in factor VIIIR:Ag was observed in patients with collagen or hematological diseases treated with prednisone (Jørgensen *et al.*, 1982): an increase in antithrombin III and a decrease in fibrinogen were also seen in these patients. The conclusion from all these observations on the influence of corticosteroids on the coagulation system is that relatively high quantities of these steroids induce an increase in the plasma level of factor VIII.

Influence on the Fibrinolytic System

Studies on the influence of corticosteroids on the fibrinolytic system have not provided uniform conclusions. Early observations using insensitive techniques suggested that therapeutic doses of ACTH and cortisone did not affect plasma fibrinolytic activity (Truelove, 1952). A later investigation showed that ACTH and corticosteroids increased the blood fibrinolytic activity when given to patients with rheumatoid arthritis and other inflammatory diseases (Chakrabarti *et al.*, 1964): five patients with non-inflammatory illnesses also showed increased fibrinolytic activity after treatment with ACTH. In contrast to these findings, ACTH or cortisone was found to decrease pathologically high fibrinolytic activity in patients with hepatic cirrhosis and cryptogenic splenomegaly (Kwaan and McFadzean, 1956; Kwaan *et al.*, 1957). In the study of Isacson (1970) where 12 healthy volunteers took 20 mg prednisone daily for 11 days there was a decrease in circulating activator activity, perhaps related to the finding of a decrease in the content of activator in the wall of superficial veins. There was no alteration in the plasma levels of plasminogen or antiplasmin activity. In a group of patients treated with prednisone Jørgensen *et al.* (1982) found no definite changes in fibrinolysis as assessed by the euglobulin clot lysis time, but there was a fall in the plasminogen concentration.

There is evidence that the fibrinolytic activity of cultured cells is sensitive to hormonal control. The activity of human embryonic lung cells was found to decrease when low concentrations of glucocorticoids were included in the culture medium (Rifkin, 1978). Such an effect could be the result of decreased production of activator or stimulation of inhibitor production. Using cultures of human fibroblasts Crutchley and colleagues (1981) observed that corticosteroids induce the production of an inhibitor directed against plasminogen activator, both urokinase and the activator produced by untreated cells.

Thyroxine/Triiodothyronine

Influence on Platelet Numbers and Function

Platelet numbers and platelet aggregation have been reported to be normal in patients with hypothyroidism while platelet adhesiveness, measured by a glass bead column technique, was found to be low in such patients (Edson *et al.*, 1975).

Influence on the Coagulation System

In a number of early studies it was noted that the vitamin K-dependent coagulation factors, as determined by the one-stage prothrombin time or the Owren P & P method, were diminished in a substantial proportion of patients with hyperthyroidism (Lord and Andrus, 1941; Bechgaard, 1946; Lamberg and Gordin, 1954). A reduction in the individual vitamin K-dependent factors was confirmed by Ikkala and colleagues (1962). Later studies on smaller numbers of patients, however, have shown normal prothrombin times and unaltered levels of the individual vitamin K-dependent factors in hyperthyroidism (Egeberg, 1963; Simone *et al.*, 1965; Rennie *et al.*, 1978). Hepatic dysfunction is a well-recognised complication of severe hyperthyroidism and may have been the cause of the reduced vitamin K-dependent factors in the earlier studies.

An elevation in factor VIII coagulant activity has been reported in a variable proportion of patients with hyperthyroidism (Egeberg, 1963; Simone *et al.*, 1965; Farid *et al.*, 1976), attributed to increased sensitivity to catecholamines (Simone *et al.*, 1965). The levels of factors V, XI and XII are unchanged in thyrotoxic patients (Egeberg, 1963; Rennie *et al.*, 1978). Fibrinogen levels are also normal (Simone *et al.*, 1965; Rennie *et al.*, 1978).

Alpha$_1$-antitrypsin and CĪ inactivator have been found to be significantly increased in patients with hyperthyroidism whereas the mean α_2-macroglobulin level was not different from control subjects (Rennie *et al.*, 1978).

The reduced circulating thyroxine and triiodothyronine in hypo-

226

thyroidism has been found to be associated with reduced levels of factor VIII (Egeberg, 1963; Simone *et al.*, 1965; Edson *et al.*, 1975; Farid *et al.*, 1976), while Farid and colleagues (1976) found a positive relationship between the factor VIII concentration and serum thyroxine. Other coagulation factors have been reported to be reduced in hypothyroid patients (Simone *et al.*, 1965; Edson *et al.*, 1975; Rennie *et al.*, 1978; van Oosterom *et al.*, 1979), perhaps a reflection of the general decrease in protein synthesis found in hypothyroidism (Crispell *et al.*, 1956). Van Oosterom and associates (1979) found that the disappearance and reappearance rates for the vitamin K-dependent factors after blocking their synthesis with coumarin drugs was slower in hypothyroid patients. Where coagulation factors have been reduced in patients with hypothyroidism, treatment with thyroxine has induced a return of their level to normal.

Significantly increased α_2-macroglobulin and decreased antithrombin III levels have been observed in hypothyroid patients (Rennie *et al.*, 1978).

Influence on the Fibrinolytic System

There is general agreement that blood fibrinolytic activity, assessed in most studies by the euglobulin clot lysis technique, is diminished in patients with hyperthyroidism (Hume, 1965; Bennett *et al.*, 1967; Farid *et al.*, 1976; Rennie *et al.*, 1978). Control of thyroid overactivity with methylthiouracil was accompanied by a progressive shortening of the lysis time (Bennett *et al.*, 1967). In one study fibrinolytic capacity was not significantly different in the hyperthyroid patients (Rennie *et al.*, 1978), suggesting that synthesis and release of plasminogen activator from the vascular wall is not impaired. The relationship between indices of thyroid overactivity and fibrinolytic activity was not close, making it unlikely that the reduced fibrinolysis is a direct result of an altered hormone level.

The plasma plasminogen concentration is lowered in hyperthyroidism (Bennett *et al.*, 1967; Rennie *et al.*, 1978), but in one study no relationship was seen between the level of plasminogen and indices of liver function (Rennie *et al.*, 1978). Increases in both the total antiplasmin activity and the level of a serum inhibitor of plasminogen activation were reported in patients with hyperthyroidism (Bennett *et al.*, 1967); information on the level of α_2-antiplasmin or histidine-rich glycoprotein is not available.

The changes in the fibrinolytic system in hypothyroidism are the converse of those seen in hyperthyroidism. Most have found an increase in circulating plasminogen activator levels (Bennett *et al.*, 1967; Rennie *et al.*, 1978). In contrast, Farid and co-workers (1976) reported a statistically significant prolongation of the euglobulin clot lysis time, those with the lowest thyroxine levels tending to have the least fibrinolytic activity. Plasminogen concentrations have been reported to be slightly higher in hypothyroid patients (Bennett *et al.*, 1967), but the difference from control subjects was not statistically significant in the later study of Rennie *et al.* (1978).

Estrogens and Progestogens

The sex hormones fall into three broad functional categories — estrogens, progestogens and androgens which are produced in varying amounts in the ovaries, testes and adrenal glands: all are steroidal in structure. Estrogens may be defined as substances which are capable of producing characteristic changes in the reproductive tract, in particular, the vaginal epithelium. The principal active estrogen secreted by the ovary is estradiol; estrone and estriol are other naturally occurring estrogens. While the term progestogen implies its ability to maintain pregnancy, it may be more specifically defined as a substance which induces secretory changes in the estrogen-primed endometrium. Naturally-occurring progestogens include progesterone and 17-hydroxyprogesterone.

The epidemiological association between the use of oral contraceptives and thromboembolic disorders has provided the stimulus for a multitude of studies on the effect of estrogen/progestogen combinations on the hemostatic system. Indeed, our knowledge on the effects of estrogens and progestogens on hemostasis comes almost entirely from studies on women receiving oral contraceptives, although it is likely that the changes in hemostatic components associated with pregnancy and the menstrual cycle are mediated principally through hormonal variations. Relatively few studies have been published on the effect of estrogens or progestogens on the hemostatic mechanism of men.

Many discrepancies are apparent in the results of different studies on the influence of estrogen/progestogen preparations on the hemostatic system. Methodological differences may account for some of the variation. Additionally, in many of the studies a group of women receiving oral contraceptive preparations were compared with a group not on such drugs; in other studies measurements were made on the same women in consecutive cycles with alternate use of the drugs. It is clear that the subjects of the former type of study had usually received the contraceptive preparation for considerably longer periods and it is in these studies that the changes in hemostatic components have been most prominent. A further factor which may have contributed to the varying conclusions is the possibility that responses may vary according to the nature of the estrogen used. Consistent with this possibility is the finding by Davies and colleagues (1976) that different responses in a range of hemostatic components follow the use of estriol succinate compared to those with ethinyl-estradiol.

Influence on Platelet Numbers and Function

Findings on the blood platelet count in women receiving combined estrogen/progestogen preparations have been inconsistent. A number of earlier studies concluded that a rise in the platelet count was induced (Margulis *et al.*, 1965; Miller *et al.*, 1965; Ygge *et al.*, 1969). Others have been unable to find any significant change in platelet numbers (Amris and Starup, 1967; Nilsson and Kullander, 1967; Dawson *et al.*, 1976; Meade *et al.*, 1976; Tsakok, 1979).

Conclusions on the effect of estrogens and progestogens on platelet function have also been discrepant; it is likely that this is due, at least in part, to the varied methods used in assessment. While Ygge and co-workers (1969) found decreased platelet adhesiveness, measured by the retention of platelet to glass bead columns after the addition of ADP to platelet-rich plasma, in women receiving combined estrogen/progestogen preparations, others using the Hellem technique have not detected any alteration in platelet adhesiveness (Meade *et al.*, 1976). Farbiszewski and Kuroska (1970) did find some increase in adhesiveness, but this did not attain statistical significance.

The effect of estrogen in platelet adhesiveness has been examined in men: 3 mg of estriol daily for 14 days failed to alter adhesiveness as assessed by the glass bead column technique or by the forearm-cut method of Borchgrevink (Bennett *et al.*, 1966). Adhesiveness measured by the rotating bulb method was also found to be unaffected by estrogen administration to men (Elkeles *et al.*, 1968).

Platelet aggregation in response to ADP, collagen and epinephrine and serotonin release have been reported to be unchanged in women taking estrogen/progestogen contraceptive pills (Poller *et al.*, 1969; Carvalho *et al.*, 1977). Others have found increased platelet sensitivity to ADP in women on combined preparations as assessed by electrophoretic mobility behavior (Bolton *et al.*, 1968) or by optical density changes (Farbiszewski and Kurowska, 1970). Progestogen alone did not alter the platelet electrophoretic mobility pattern (Bolton *et al.*, 1968). In a further study platelet aggregability, measured by the screen filtration pressure technique, was found to be increased in eight out of 26 women during the two-month period after starting a combined contraceptive preparation (Bierenbaum *et al.*, 1979). A week's course of either natural or synthetic estrogens was reported to increase the platelet response to ADP in men as assessed by both elctrophoretic mobility and aggregability (Elkeles *et al.*, 1968).

Influence on the Coagulation System

There is general agreement that the plasma level of a number of the coagulation factors is raised in women receiving combined estrogen/progestogen preparations although changes may not be detected when measurements are made over a single cycle (Beller and Porges, 1967). The most consistently observed increases are in prothrombin and factors VII and X, assessed by either specific assays or the P & P test of Owren and Aas (1951). Such studies include those of Rutherford *et al.* (1964), Hougie *et al.* (1965), Donayre and Pincus (1965), Thomson and Poller (1965), Ambrus *et al.* (1969), Ygge *et al.* (1969), Meade *et al.* (1976) and Tsakok *et al.* (1979). The increase in these clotting factors has been shown to be persistent (Poller *et al.*, 1971). The changes appear to be the result of the estrogen component of the contraceptive pill since women on a parenteral progestogen preparation (medroxyprogesterone acetate) had no changes in these coagulation factors (Tsakok *et al.*, 1979).

Discrepant conclusions have emerged on the effect of oral contraceptives on the contact system. Rutherford and colleagues (1964) reported an increased level of factor XII, but no change in factor XI: increased factor XII activity, determined as prekallikrein-activating capacity, was found by Laake *et al.* (1974) in women receiving estrogen-containing preparations. Consistent with these findings was the report of Gordon and colleagues (1980) of an elevation of factor XII, both procoagulant activity and antigen titre, and unchanged factor XI, prekallikrein and HMW kininogen in combined oral contraceptive users. In contrast, Carvalho and associates (1977) obtained normal factor XII levels and a moderate elevation of prekallikrein in women taking estrogen/progestogen pills.

Agreement on the effect of estrogen/progestogen drugs on factor VIII levels is also lacking. Egeberg and Owren (1963) noted increased factor VIII coagulant activity in women taking ethinylestradiol and norethynodrel for 20 days: increased factor VIII levels were also reported by Rutherford and co-workers (1964). Others have found no change in the factor VIII level of women receiving combined oral contraceptives at normal dosage (Thomson and Poller, 1965; Nilsson and Kullander, 1967; Meade *et al.*, 1976). Ethinylestradiol given to post-menopausal women produced a marked rise in the level of factor VIII whereas estriol succinate decreased the factor VIII concentration (Davies *et al.*, 1976). In experiments on human umbilical vein endothelial cell cultures Harrison and McKee (1981) observed that 17β-estradiol stimulated the cells to increase their rate of production of von Willebrand factor.

Factor IX has been found by some investigators to be elevated in combined oral contraceptive users (Rutherford *et al.*, 1964; Davies *et al.*, 1976). In addition, the concentration of this factor is increased when high doses of estrogen in the form of diethyl stilbestrol are taken to suppress lactation (Daniel *et al.*, 1968; Hakim *et al.*, 1969).

Most studies have shown no change in factor V levels in women receiving oral contraceptives (Brakman *et al.*, 1967; Alkjaersig *et al.*, 1975; Meade *et al.*, 1976): the report of Rutherford and co-workers is an exception.

There has also been disagreement on the effect of estrogen/progestogen preparations on the plasma fibrinogen concentration. Among those who have found no change in level are Egeberg and Owren (1963), Powell *et al.* (1965), Amris and Starup (1967), Beller and Porges (1967) and Brakman *et al.* (1970). A large number of investigators, however, have described rises in the fibrinogen level in women taking oral contraceptives (Amundsen and Pilgeram, 1963; Brakman and Astrup, 1964; Donayre and Pincus, 1965; Miller *et al.*, 1965; Ambrus *et al.*, 1969; Howie *et al.*, 1970; Alkjaersig *et al.*, 1975; Meade *et al.*, 1976; Ball and McKee, 1977; Tsakok *et al.*, 1979). A slight increase in fibrinogen was noted in women given a parenteral progestogen contraceptive for two or more years (Tsakok *et al.*, 1979).

Factor XIII levels appear to be unchanged by oral contraceptives (Coopland *et al.*, 1969; Ball and McKee, 1977), although Alkjaersig and co-workers (1975) noted a wide distribution of values with a bimodal frequency distribution in the women on the estrogen/progestogen preparations.

Influence on Protease Inhibitors

A number of investigators have described decreased antithrombin activity and decreased levels of antithrombin III in the plasma of women taking either combined estrogen/progestogen preparations or estrogen alone (Fagerhol *et al.*, 1970; Howie *et al.*, 1970; Alkjaersig *et al.*, 1975; Davies *et al.*, 1976; Tsakok *et al.*, 1979). Estrogens given to post-menopausal women also cause a reduction in antithrombin III levels (von Kaulla *et al.*, 1975). The estrogen-induced fall in antithrombin III appears to be dose-dependent with a much higher proportion of the women taking a high estrogen content pill showed a decreased level (Conard *et al.*, 1972). A dose effect was also noted by Ball and McKee (1977). In most studies no change in the antithrombin III level has been seen in women receiving progestogen only (Howie *et al.*, 1970; Larsson-Cohn *et al.*, 1972; Tsakok *et al.*, 1979). In the report of Bounameaux *et al.* (1978), however, there was an equivalent reduction in antithrombin III, measured by a variety of methods, in the progestogen only and combined preparation groups of women: it was proposed that estrogen-like metabolites of the 19-nortestosterone progestogen used could have been responsible for the reduction in antithrombin III.

Women on estrogen/progestogen preparations have been found to have increased plasma α_2-macroglobulin levels by the majority (Laurell *et al.*, 1968; Horne *et al.*, 1970; Alkjaersig *et al.*, 1975; Teger-Nilsson, 1979): others have noted a tendency for an increase which did not reach statistical significance (Dawson *et al.*, 1976; Meade *et al.*, 1976).

The level of plasma α_1-antitrypsin is substantially increased in women receiving combined oral contraceptive pills (Laurell *et al.*, 1968; Alkjaersig *et al.*, 1975; Dawson *et al.*, 1976).

Estrogen/progestogen preparations have been reported to decrease the plasma concentration of C̄1 inactivator in association with the increase in the level of factor XII (Gordon *et al.*, 1980).

Influence on the Fibrinolytic System

In contrast to the marked depression of plasma fibrinolytic activity seen in pregnancy there is fairly general agreement that women taking estrogen/progestogen preparations have increased fibrinolytic activity when assessed by such techniques as the euglobulin clot lysis time or the dilute whole blood clot lysis time (Donayre and Pincus, 1965; Brakman *et al.*, 1970; Farbiszewski and Kurowska, 1970; Hedlin, 1975; Meade *et al.*, 1976; Ball and McKee, 1977; Tsakok *et al.*, 1979). This effect appears to be due to the estrogen component of the contraceptive pill since women on progestogen alone have been found to have either an unchanged euglobulin clot lysis time (Tsakok *et al.*, 1979) or a lengthened time (Masure *et al.*, 1979). A parenteral progestogen preparation decreased plasma euglobulin fibrinolytic activity measured on fibrin plates (Brakman *et al.*, 1970). The mechanism of the increased plasma fibrinolytic activity during the use of estrogens is uncertain. One possibility is increased synthesis or release of activator from the vessel wall. Using exceptionally high doses of ethinylestradiol (250 μg/day for 10 days) there was no suppression of the vein wall fibrinolytic activity (Nilsson and Astedt, 1971). Hedner *et al.* (1975), however, found the fibrinolytic activity in vein walls to be increased by the use of ethinylestrenol, while Hedlin and co-workers (1978) obtained an increase in the fibrinolytic response to exercise in oral contraceptive users. Women on the progestogen drug chlormadinone acetate were found to have a normal fibrinolytic response to venous occlusion and unchanged activator activity in the vein wall (Nilsson and Astedt, 1971).

There is general agreement that the plasma plasminogen concentration is raised in response to combined estrogen/progestogen preparations (Pilgeram *et al.*, 1964; Ygge *et al.*, 1969; Brakman *et al.*, 1970; Hedlin, 1975; Dawson *et al.*, 1976; Tsakok *et al.*, 1979).

Antiplasmin activity has been found to be increased in the plasma of women taking combined estrogen/progestogen drugs or estrogen alone, but was unchanged in those receiving progestogen alone (Howie *et al.*, 1970). Little information is presently available on the influence of estrogen or progestogen on α_2-antiplasmin: no change was seen in the plasma level in women taking either combination pills or low dosage progestogen (Teger-Nilsson *et al.*, 1978; Gordon *et al.*, 1980). Low estrogen content preparations were found to decrease the plasma level of histidine-rich glycoprotein (Jespersen and Kluft, 1982): since the plasminogen is increased it would appear that relatively more plasminogen would be available for binding to fibrinogen in women taking estrogen-containing drugs.

In contrast to the situation in pregnancy where urokinase inhibitory activity in plasma is prominent, no increase in such inhibition was detected in women receiving a combined estrogen/progestogen preparation (Brakman *et al.*, 1967) or progestogen alone (Brakman *et al.*, 1970).

Ovarian Plasminogen Activator and Ovulation

The mature mammalian ovarian follicle comprises a fluid-filled antrum surrounded by granulosa cells to which the ovum is attached. During ovulation the follicle wall must be degraded to allow release of the ovum into the peritoneal cavity.

It has been shown that bovine follicular fluid contains plasminogen, plasminogen activator and protease inhibitors while the follicular wall also contains plasminogen activator (Beers, 1975): it was suggested tha plasmin, its production being enhanced at the time of ovulation, might be involved in the degradation of the follicle wall during ovulation. in a further report by Beers and co-workers (1975) it was shown that ovarian granulosa cells in culture contain plasminogen activator, the appearance of the activator activity being correlated with ovulation. The cultured cells could be stimulated to release activator by exposure to luteinising hormone.

Androgens

The most important androgenic steroids produced by the testes are testosterone and androstenedione. Androgens are also secreted from the adrenal cortex, in particular, dehydroepiandrosterone.

Influence of Testosterone on Hemostatic Components

Intramuscular testosterone proprionate in a dose of 25 to 100 mg daily produced a shortening of the dilute whole blood clot lysis time in patients with ischemic vascular disease and in the single healthy subject tested (Fearnley and Chakrabarti, 1962): on withdrawing the hormone after eight days there was a gradual rise in the lysis time to the pre-treatment level. The enhancement of blood fibrinolytic activity in patients receiving parenteral testosterone was confirmed by Winther (1967).

Influence of Anabolic Steroids on Hemostatic Components

In addition to their action on the male reproductive tract, androgens have an anabolic action, stimulating protein synthesis. A number of steroids have been synthesised with prominent anabolic properties, but with minimal androgenic effects.

234

Figure 6.1: Structure of Testosterone, Ethylestrenol and Stanozolol

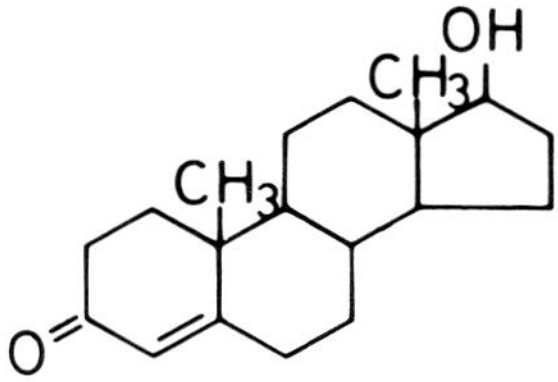

Testosterone

(Androst-4-ene-17β-ol-3-one)

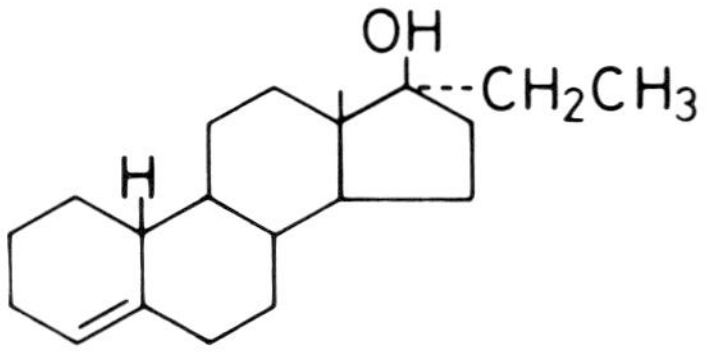

Ethyloestrenol

(17 α-ethyl-19-norandrost-4-en-17β-ol)

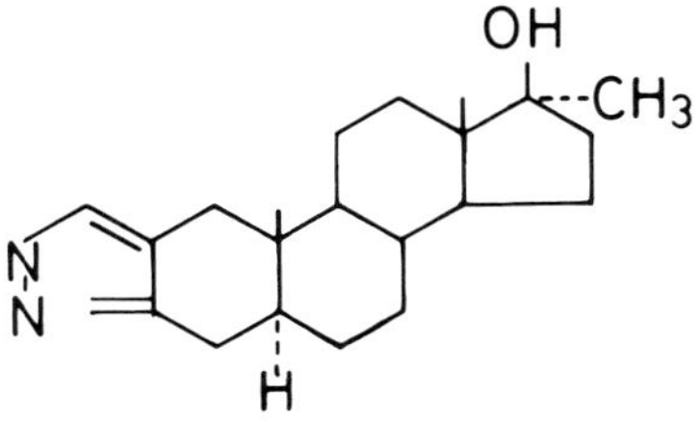

Stanozolol

(17-β-hydroxy-17 α-methyl androstano-(3.2,- c) pyrazole.

A number of synthetic anabolic steroids with a 17α-alkyl group have been found to enhance plasma plasminogen activator levels. The initial steroid of this nature used was ethylestrenol (Figure 6.1), but its effect was found to be unsustained. Stanozolol has been shown to induce a sustained increase in plasma fibrinolytic activity in man (Davidson *et al.*, 1972), the increase being evident within days of starting the drug (Preston *et al.*, 1981). The 17α-alkyl steroids also produce changes in other components of the hemostatic mechanism: plasma levels of fibrinogen and α_2-macroglobulin are reduced and that of plasminogen increased (Davidson *et al.*, 1975; Preston *et al.*, 1981). Antithrombin III levels are also increased (Davidson *et al.*, 1975) and there has been a recent report of increases in factor XII and $\mathrm{C\bar{I}}$ inactivator and falls in the plasma level of prekallikrein and histidine-rich glycoprotein (Kluft *et al.*, 1982). However, 17α-alkylated steroids do not appear to occur naturally in man while the role of androgens in the physiology of hemostasis is doubtful.

Pancreatic Hormones

Influence of Glucagon on Hemostatic Components

Glucagon is a small peptide synthesised by the α-cells of the pancreatic islets of Langerhans. Through the formation of cyclic AMP glucagon activates hepatic phosphorylase resulting in the liberation of glucose from glycogen in the liver: this is its principal physiological action. When administered in relatively large doses glucagon stimulates the secretion of insulin and can also stimulate the release of catecholamines, pituitary hormones and calcitonin.

Glucagon, examined as an activator of platelet adenylate cyclase, was found to have no effect on platelet aggregation in response to ADP or collagen when administered in a continuous infusion of 70 μg/kg over 90 minutes (Mannucci and Pareti, 1974): in these experiments the cyclic AMP levels were not affected. In a later study, using the higher dose of 1 mg glucagon intravenously, there was increased sensitivity of platelets to ADP in healthy subjects pre-treated with aspirin together with an increase in the plasma cyclic AMP level (Proctor *et al.*, 1980).

Platelet function may be influenced by glucagon indirectly since there is some evidence that plasma lipid may be transferred to platelets through the action of glucagon (Caren and Corbo, 1970).

Islets of Langerhans and Plasminogen Activator

Rat islets of Langerhans have been shown to secrete plasminogen activator, the secretion being modulated by glucose (Virji *et al.*, 1980). In further studies it was found that cyclic AMP, which augments glucose-induced insulin release, did not influence plasminogen activator production in the rat islets, and it was suggested that activator production may be related to insulin synthesis, possibly in the conversion of the proinsulin into insulin (Virji, 1982).

Insulin, secreted from the β-cells of the islets of Langerhans, induces an increase in blood fibrinolytic activity after injection into man. It has been found that the β-blocker propranolol prevents the insulin-induced increase in fibrinolytic activity in the rabbit suggesting that its effect is mediated by catecholamine release (Kleniewski *et al.*, 1974). Insulin-induced hypoglycemia increases factor VIII activity in normal subjects, but not in patients who have undergone pre-ganglionic sympathectomy (Corrall *et al.*, 1980): the increase in factor VIII was also suppressed by propranolol, but not by selective β_1-blockade and it was again concluded that the increase in factor VIII is mediated by epinephrine via a β_2-receptor.

Hypothalamic and Pituitary Hormones

Influence of Vasopressin on Hemostatic Components

Vasopressin (antidiuretic hormone: ADH) is an octapeptide which is synthesised in the supraoptic nucleus of the hypothalamus and secreted from the neurohypophysis in response to decreases in the blood volume and to increased plasma osmolarity.

Early observations indicated that intravenous vasopressin increased the fibrinolytic activity of blood (Schneck and von Kaulla, 1961). It was shown subsequently that intravenous lysine-vasopressin in man increases the level of circulating plasminogen activator and factor VIII, both coagulant activity and antigen (Mannucci *et al.*, 1972; Gader *et al.*, 1973b; Mannucci, 1974; Mannucci *et al.*, 1975). This effect was attributed initially to its potent vasoconstrictor activity, but it was found later that the analogue 1-desamino-8-D-arginine vasopressin (DDAVP) which is devoid of vasoactive properties retains its ability to increase plasma plasminogen activator and factor VIII levels (Gader *et al.*, 1974b; Cash *et al.*, 1978; Åberg *et al.*, 1979). The mode of action of DDAVP is not established. It does not affect the release of activator from the isolated pig ear (Markwardt and Klocking, 1976) and does not produce any change in activator level in the draining venous blood after intra-arterial perfusion (Cash *et al.*, 1978). A prolonged release of activator follows intranasal administration (Mannucci and Rota, 1980).

Cash (1978) has postulated the existence of a vasopressin-like octapeptide, possibly also synthesised in the hypothalamus and released from the neurohypophysis, which has the property of inducing the release of plasminogen activator from vascular endothelium. It was further suggested that such a hypothetical peptide might be the major mediator of activator release during normal daily activities, while the catecholamines are involved in activator release in response to severe stresses. In recent studies hypothalamic stimulation was found to cause the peripheral release of plasminogen activator in both normal and vasopressin-deficient rats (Prowse *et al.*, 1982): pituitary extracts could also induce the release of activator while the effect of hypothalamic stimulation was abolished after sectioning the hypophyseal stalk (Boulton, 1982). It was concluded that a hypothalamic agent stimulates the pituitary to release a factor which causes the peripheral release of plasminogen activator.

A direct action of vasopressin on activator synthesis is suggested by experiments in which agents which increase cyclic AMP levels in cultured porcine renal tubular cells increased the release of activator: salmon calcitonin was particularly potent, but vasopressin also produced some increase in activator production (Dayer *et al.*, 1981).

Arginine and lysine-vasopressin have been shown to be capable of inducing platelet aggregation and release (Haslam and Rosson, 1971).

Influence of Somatostatin on Hemostatic Components

Somatostatin is a peptide secreted by the hypothalamus and also found in other parts of the brain, the pancreas and the gastrointestinal tract. Its principal actions are the inhibition of release of growth hormone from the adenohypophysis, the inhibition of thyroid stimulating hormone synthesis and release, inhibition of the secretion of insulin and glucagon, and it has a number of effects on the gastrointestinal tract.

The effect of somatostatin on platelet function has been studied in the rat, rabbit, baboon and in man. An intravenous infusion of somatostatin (0.8 μg/kg/minute) for two hours in baboons was reported to produce a transient decrease in the platelet count, a reduction in platelet retention in glass bead columns and inhibition of aggregation induced by ADP, collagen and epinephrine (Koerker *et al.*, 1975), but the peptide did not directly influence platelet aggregation. Using a higher dose of somatostatin Chiang and co-workers (1975) obtained similar results in rabbits. Rats given a two-hour infusion of somatostatin (5 μg/kg/minute) also showed a fall in the platelet count and a reduction in ADP and collagen-induced aggregation whereas chronic administration resulted in enhanced aggregation with a return of the platelet count to normal (Long *et al.*, 1979).

Conclusions on the influence to somatostatin infusions on platelet function in man have been contradictory. Mielke and colleagues (1975) reported that synthetic linear or cyclic somatostatin (0.12 μg/ kg/minute for up to 18 hours) did not alter the platelet count, platelet adhesiveness or platelet aggregability in response to ADP, collagen or epinephrine in normal or diabetic subjects. In contrast, Besser and associates (1975) found that platelet aggregation in response to ADP and collagen was impaired in healthy men during and after an infusion of synthetic somatostatin at a dosage of 0.03 to 0.09 μg/kg/minute. In an attempt to clarify the previous discrepant results Bensoussan and co-workers (1977) infused ten healthy men with synthetic linear and cyclic somatostatin at a rate of 1500 μg over three hours after an initial dose of 250 μg: neither the platelet count nor ADP-induced aggregation were altered.

Somatostatin infusions did not alter the prothrombin time, the partial thromboplastin time or the fibrinogen concentration (Besser *et al.*, 1975; Mielke *et al.*, 1975), and there was no effect on factor VIII coagulant activity or factor VIII-related antigen levels in the study of Bensoussan *et al.* (1977).

The discordant conclusions on the effect of somatostatin on platelet function are not readily explained, but a physiological role for this peptide in the control of hemostatic function is improbable.

Influence of Growth Hormone on Hemostatic Components

The administration of growth hormone (1 to 2 I.U.) to patients with panhypopituitarism resulted in a rise in von Willebrand factor activity (Sarji *et al.*, 1977). In addition, there was a significant correlation between the level of growth hormone and von Willebrand factor activity in fasting normal subjects, and it was suggested that growth hormone might have a physiological role in the regulation of von Willebrand factor.

Production of Plasminogen Activator by the Pituitary

Cultures of the anterior lobe of the rat pituitary have been found to produce plasminogen activator of both urokinase and tissue activator type (Granelli-Piperno *et al.*, 1982). Its synthesis and secretion appeared to be under hormonal control with stimulation of production by dibutyryl cyclic AMP and depression by hydrocortisone.

References

Åberg, M., Nilsson, I.M. and Vilhardt, H. (1979) 'The release of fibrinolytic activator and factor VIII after injection of DDAVP', in J.F. Davison, V. Cepelak, M.M. Samama and P.C. Desnoyers (eds.) *Progress in Chemical Fibrinolysis and Thrombolysis*, Vol. 4, Raven Press, New York, pp. 92-7

Alkjaersig, N., Fletcher, A.P. and Burstein, R. (1975) 'Association between oral contraceptive use and thromboembolism: A new approach to its investigation based on plasma fibrinogen chromatography', *American Journal of Obstetrics and Gynecology, 122,* 199-211

Ambrus, J.L., Nieswander, K.R., Courey, N.G., Wamsteker, E.F. and Mink, I.B. (1969) 'Progestational agents and blood coagulation. II', *American Journal of Obstetrics and Gynecology, 103,* 994-1001

Amris, C.J. and Starup, J. (1967) 'The coagulation mechanism in oral contraception', *Acta Obstetrica et Gynecologica Scandinavia, 46,* 78-91

Amundson, B.A. and Pilgeram, L.O. (1964) 'Observations on a relationship between steroid metabolism and the concentration of plasma fibrinogen', *Thrombosis et Diathesis Haemorrhagica, 10,* 400-5

Ardlie, N.G., Glew, G. and Schwartz, C.J. (1966) 'Influence of catecholamines on nucleotide-induced platelet aggregation', *Nature, 212,* 415-7

Ball, A.P. and McKee, P.A. (1977) 'Fibrin formation and dissolution in women receiving oral contraceptive drugs', *Journal of Laboratory and Clinical Medicine, 80,* 751-62

Barrow, E.M., Dejanov, I.I., Ingram, G.I.C. and Knights, S.F. (1969) 'The rise in factor VIII after administration of adrenaline', *Coagulation, 2,* 203-11

Bechgaard, P. (1946) 'Tendency to hemorrhage in thyroioxicosis', *Acta Medica Scandinavica, 124,* 79-91

Beers, W.H. (1975) 'Follicular plasminogen and plasminogen activator and the effect of plasminogen ovarian follicle wall', *Cell, 6,* 379-86

Beers, W.H., Strickland, S. and Reich, E. (1975) 'Ovarian plasminogen activator: relationship to ovulation and hormonal regulation', *Cell, 6,* 387-94

Beller, F.K. and Porges, R.F. (1967) 'Blood coagulation and fibrinolytic enzyme studies during cyclic and continuous application of progestational agents', *American Journal of Obstetrics and Gynecology, 97,* 448-59

Bennett, J.S. and Vilaire, G. (1979) 'Exposure of platelet fibrinogen receptors by ADP and epinephrine', *Journal of Clinical Investigation, 64,* 1393-401

Bennett, N.B., Bennett, P.N., Fullerton, H.W., Ogston, C.M. and Ogston, D. (1966) 'Effect of oestriol on platelet adhesiveness and fibrinolysis in men', *Lancet, ii,* 881-2

Bennett, N.B., Ogston, C.M. and McAndrew, G.M. (1967) 'The thyroid and fibrinolysis', *British Medical Journal, 4,* 147-8

Bensoussan, D., Levy-Toledano, S., Passa Ph., Caen, J. and Canivet, J. (1977) 'Effect of somatostatin on platelet aggregation and plasma factor VIII level in normal man', *Diabète et Metabolisme (Paris), 3*, 223-7

Besser, G.M., Paxton, A.M., Johnson, S.A.N., Moody, E.J., Mortimer, C.H., Hall, R., Gomez-Pan, A., Schally, A.V., Kastin, A.J. and Coy, D.H. (1975) 'Impairment of platelet function by growth-hormone release-inhibiting hormone', *Lancet, i*, 1166-8

Bierenbaum, M.L., Fleischman, A.I., Stier, A., Watson, P., Somol, H., Naso, A.M. and Binder, M. (1979) 'Increased platelet aggregation and decreased high-density lipoprotein cholesterol in women on oral contraceptives', *American Journal of Obstetrics and Gynecology, 134*, 638-41

Biggs, R., Macfarlane, R.G. and Pilling, J. (1947) 'Observations on fibrinolysis. Experimental activity produced by exercise or adrenaline', *Lancet, i*, 403-5

Bolton, C.H., Hampton, J.R. and Mitchell, J.R.A. (1968) 'Effect of oral contraceptive agents on platelets and plasma phospholipids', *Lancet, i*, 1336-41

Boulton, F. (1982) Personal communication

Bounameaux, H., Duckert, F., Walter, M. and Bounameaux, Y. (1978) 'The determination of antithrombon III. Comparison of six methods. Effect of oral contraceptives', *Thrombosis and Haemostasis, 39*, 607-15

Brakman, P. and Astrup, T. (1964) 'Effects of female hormones, used as oral contraceptives, on the fibrinolytic system in blood', *Lancet, ii*, 10-2

Brakman, P., Albrechtsen, O.K. and Astrup, T. (1967) 'Blood coagulation, fibrinolysis and contraceptive hormones', *Journal of the American Medical Association, 199*, 69-74

Brakman, P., Sobrero, A.J. and Astrup, T. (1970) 'Effects of different systemic contaceptives on blood fibrinolysis', *American Journal of Obstetrics and Gynecology, 106*, 187-92

Branehög, I., Weinfield, A. and Roos, B. (1973) 'The exchangeable splenic platelet pool studied with epinephrine infusion in idiopathic thrombocytopenic purpura and in patients with splenomegaly', *British Journal of Haematology, 25*, 239-48

Britton, B.J., Wood, W.G., Peele, M., Hawkey, C. and Irving, M.H. (1975) 'Role of sympathoadrenal stimulation in the release of plasminogen activator', in J.F. Davidson, M.M. Damama and P.C. Desnoyers (eds.) *Progress in Chemical Fibrinolysis and Thrombolysis*, Vol. 1, Raven Press, New York, pp. 111-8

Bull, B.S. and Zucker, M.B. (1965) 'Changes in platelet volume produced by temperature, metabolic inhibitors, and aggregating agents', *Proceedings of the Society for Experimental Biology and Medicine, 120*, 296-301

Bygdeman, S. and Johnsen, O. (1969) 'Studies on the effect of adrenergic blocking drugs on catecholamine-induced platelet aggregation and uptake of noradrenaline and 5-hydroxytryptamine', *Acta Physiologica Scandinavica, 75*, 129-38

Cannon, W.B. and Gray, H. (1914) 'Factors affecting the coagulation time of blood. II. The hastening or retarding of coagulation by adrenalin injections', *American Journal of Physiology, 34*, 232-50

Caren, R. and Gorbo, L. (1970) 'Transfer of plasma lipid to platelets by action of glucagon', *Metabolism, 19*, 598-607

Carvalho, A.C.A., Vaillancourt, R.A., Cabral, R.B., Lees, R.S. and Colman, R.W. (1977) 'Coagulation abnormalities in women taking oral contaceptives', *Journal of the American Medical Association, 237*, 875-8

Cash, J.D. (1978) 'Control mechanism of activator release', in J.F. Davidson, R.M. Rowan, M.M. Samama and P.C. Desnoyers (eds.) *Progress in Chemical Fibrinolysis and Thrombolysis*, Vol. 3, Raven Press, New York, pp. 65-75

Cash, J.D. and Allan, C.E, (1967) 'The fibrinolytic response to moderate exercise and intravenous adrenaline in the same subjects', *British Journal of*

Haematology, 13, 376-83

Cash, J.D., Woodfield, D.G. and Allan, A.G.E. (1970) 'Adrenergic mechanisms in the systemic plasminogen activator response to adrenaline in man', *British Journal of Haematology, 18*, 487-94

Cash, J.D., Gader, A.M.A., Mulder, J.L. and Cort, J.H. (1978) 'Structure-activity relations of the fibrinolytic response to vasopressins in man', *Clinical Science and Molecular Medicine, 54*, 403-9

Cazenave, J.-P., Sutter, A., Hemmendinger, S., Wiszel, M.-L., Lanza, F. and Daver, J. (1981) 'Adrenaline activates human platelets but does not cause primary aggregation if thrombin generation is inhibited by hirudin', *Thrombosis and Haemostasis, 46*, 95

Chakrabarti, R., Fearnley, G.R. and Hocking, E.D. (1964) 'Effect of corticosteroid therapy on fibrinolysis in patients with inflammatory and non-inflammatory conditions', *British Medical Journal, 1*, 534-7

Chiang, T.M., Duckworth, W.C., Beachey, E.H. and Kang, A.H. (1975) 'The effect of somatostatin on platelet aggregation', *Endocrinology, 97*, 753-5

Clayton, S. and Cross, M.J. (1963) 'The aggregation of blood platelets by catecholamines and by thrombin', *Journal of Physiology, 163*, 82-3P

Cohen, R.J., Epstein, S.E., Cohen, L.S. and Dennis, L.H. (1968) 'Alterations of fibrinolysis and blood coagulation induced by exercise, and the role of beta-adrenergic-receptor stimulation', *Lancet, ii*, 1264-6

Conard, J., Samama, M. and Salomon, Y. (1972) 'Antithrombin III and the oestrogen content of combined oestro-progestogen contraceptives', *Lancet, ii*, 1148-9

Coopland, A., Alkjaersig, N. and Fletcher, A.P. (1969) 'Reduction in plasma factor XIII (fibrin stabilizing factor) concentration during pregnancy', *Journal of Laboratory and Clinical Medicine, 73*, 144-53

Corrall, R.J.M., Webber, R.G. and Frier, B.M. (1980) 'Increase in coagulation factor VIII activity in man following acute hypoglycaemia: mediation via an adrenergic mechanism', *British Journal of Haematology, 44*, 301-5

Crispell, K.R., Parson, W. and Hollifield, J. (1956) 'Study of rate of protein synthesis before and during administration of 1-triiodothyronine to patients with myxedema and healthy volunteers using N-15 glycine', *Journal of Clinical Investigation, 35*, 164-9

Crutchley, D.J., Conanan, L.B. and Maynard, J.R. (1981) 'Human fibroblasts produce inhibitor directed against plasminogen activator when treated with glucocorticoids', *Annals of the New York Academy of Sciences, 370*, 609-16

Daniel, D.G., Bloom, A.L., Giddings, J.C., Campbell, H. and Turnbull, A.C. (1968) 'Increased factor IX levels in puerperium during administration of diethylstilboestrol', *British Medical Journal, 1*, 801-3

Davidson, J.F., Lochhead, M., McDonald, G.A. and McNicol, G.P. (1972) 'Fibrinolytic enhancement by stanozolol: a double blind trial', *British Journal of Haematology, 22*, 543-59

Davidson, J.F., Walker, I.D. and McCallum, H.I. (1975) 'Study of the mechanism of action of anabolic steroids on fibrinolysis', in J.F. Davidson, M.M. Samama and P.C. Desnoyers (eds.) *Progress in Chemical Fibrinolysis and Thrombolysis*, Vol. 1, Raven Press, New York, pp. 311-9

Davies, T., Fieldhouse, G. and McNicol, G.P. (1976) 'The effects of therapy with oestriol succinate and ethinyl oestradiol on the haemostatic mechanism in post-menopausal women', *Thrombosis and Haemostasis, 35*, 403-14

Dawson, A.A., Allardyce, M., Allan, T.M., Ogston, D., Kerridge, D.F. and Lewis, H.B.M. (1976) 'Inter-relations between ABO blood group, plasminogen, alpha$_1$-antitrypsin, alpha$_2$-macroglobulin and the platelet count in blood donors', *Acta Haematologica, 56*, 19-26

Dayer, J.-M., Vassalli, J.-D., Bobbitt, J.L., Hull, R.N., Reich, E. and Krane, S.M. (1981) 'Calcitonin stimulates plasminogen activator in porcine renal tubular cells: LLC-PK$_1$', *Journal of Cell Biology, 91*, 195-200

Donayre, J. and Pincus, G. (1965) 'Effects of Enovid on blood clotting factors', *Metabolism, 14*, 418-28

Edson, J.R., Fecher, D.R. and Doe, R.P. (1975) 'Low platelet adhesiveness and other hemostatic abnormalities in hypothyroidism', *Annals of Internal Medicine, 82*, 342-6

Egeberg, O. (1963) 'Changes in the activity of antihemophilic A factor (f. VIII) and in the bleeding time associated with muscular exercise and adrenaline infusion', *Scandinavian Journal of Clinical and Laboratory Investigation, 15*, 539-49

Egeberg, O. and Owren, P.A. (1963) 'Oral contraception and blood coagulability', *British Medical Journal, 1*, 220-1

Elkeles, R.S., Hampton, J.R. and Mitchell, J.R.A. (1968) 'Effect of oestrogens on human platelet behaviour', *Lancet, ii*, 315-8

Fagerhol, M.K., Abildgaard, U., Bergsjo, P. and Jacobsen, J.H. (1970) 'Oral contraceptives and low antithrombin III concentration', *Lancet, i*, 1175

Farbiszewski, R. and Kurowska, T. (1970) 'Platelet adhesiveness, aggregation and fibrinolysis after the oral administration of the contraceptive "Lyndiol"', *Thrombosis et Diathesis Haemorrhagica, 24*, 304-7

Farid, N.R., Griffiths, B.L., Collins, J.R., Marshall, W.H. and Ingram, D.W. (1976) 'Blood coagulation and fibrinolysis in thyroid disease', *Thrombosis and Haemostasis, 35*, 415-22

Fearnley, G.R. and Chakrabarti, R. (1962) 'Increase of blood fibrinolytic activity by testosterone', *Lancet, ii*, 128-32

Fredén, K., Lundborg, P., Vilén, L. and Kutti, J. (1978) 'The peripheral platelet count in response to adrenergic alpha- and beta-1-receptor stimulation', *Scandinavian Journal of Haematology, 21*, 427-32

Gader, A.M.A., Clarkson, A.R. and Cash, J.D. (1973a) 'The plasminogen activator and coagulation factor VIII response to adrenaline, noradrenaline, isoprenaline and salbutamol in man', *Thrombosis Research, 2*, 9-10

Gader, A.M.A., Da Costa, J. and Cash, J.D. (1973b) 'A new vasopressin analogue and fibrinolysis', *Lancet, ii*, 1417-8

Gader, A.M.A., Da Costa, J. and Cash, J.D. (1974a) 'The effect of propranolol, alprenolol and practolol on the fibrinolytic and factor VIII responses to adrenaline and salbutamol in man', *Thrombosis Research, 4*, 25-33

Gader, A.M.A., Da Costa, J. and Cash, J.D. (1974b) '1-desamino-8-D-arginine vasopressin and fibrinolysis in man', *Scottish Medical Journal, 19*, 56-7

Genton, E., Kern, F. and von Kaulla, K. (1961) 'Fibrinolysis induced by pressor amines', *American Journal of Medicine, 31*, 564-71

Gordon, E.M., Ratnoff, O.D., Saito, H., Donaldson, V.H., Pensky, J. and Jones, P.K. (1980) 'Rapid fibrinolysis, augmented Hageman factor (factor XII) titres, and decreased Cl esterase inhibitor titres in women taking oral contraceptives', *Journal of Laboratory and Clinical Medicine, 96*, 762-9

Grabfield, G.P. (1916) 'Factors affecting the coagulation time of blood IX. The effect of adrenalin on the factors of coagulation', *American Journal of Physiology, 42*, 46-55

Granelli-Piperno, A., Belin, D., Schleuning, W.-D. and Reich, E. (1982) 'Pituitary plasminogen activators: immunological characterization and hormonal modulation', *Haemostasis, 11*, Suppl. 1, 63

Haft, J.I., Kranz, P.D., Albert, F.J. and Fani, K. (1972) 'Intravascular platelet aggregation in the heart induced by norepinephrine', *Circulation, 46*, 698-708

Hakim, C.A., Elder, M.G. and Hawkins, D.F. (1969) 'Plasma factor IX levels in patients given hexoestrol or stilboestrol to suppress lactation', *British Medical Journal, 4*, 82-4

Harrison, R.L. and McKee, P.A. (1981) 'The effects of estrogen on von Willebrand factor production and endothelial cell number in culture', *Thrombosis and Haemostasis, 46*, 20

Haslam, R.J. and Rosson, G.M. (1971) 'Effect of vasopressin on human blood platelets', *Journal of Physiology, 219*, 36-8P

Hawkey, C.M., Britton, B.J., Wood, W.G., Peele, M. and Irving, M.H. (1975) 'Changes in blood catecholamine levels and blood coagulation and fibrinolytic activity in response to graded exercise in man', *British Journal of Haematology, 29*, 377-84

Hedlin, A.M. (1975) 'The effect of oral contraceptive estrogen in blood coagulation and fibrinolysis', *Thrombosis et Diathesis Haemorrhagica, 33*, 370-8

Hedlin, A.M., Milojevic, S. and Korey, A. (1978) 'Plasminogen activator levels in plasma and urine during exercise and oral contaceptive use', *Thrombosis and Haemostasis, 39*, 743-50

Hedner, U., Nilsson, I.M. and Isacson, S. (1975) 'Effect of ethylestrenol on fibrinolysis in the vessel wall', *Thrombosis et Diathesis Haemorrhagica, 34*, 609

Hoffman, B.R., De Lean, A., Wood, C.L., Schocken, D.D. and Lefkowitz, R.J. (1979) 'Alpha-adrenergic receptor subtypes: quantitative assessment by ligand binding', *Life Sciences, 24*, 1739-46

Horne, C.H.W., Howie, P.W., Weir, R.J. and Goudie, R.B. (1970) 'Effect of combined oestrogen-progestogen oral contraceptives on serum-levels of α_2-macroglobulin, transferrin, albumin and IgG', *Lancet, i,* 49-50

Hougie, C., Rutherford, R.N., Banks, A.L. and Coburn, W.A. (1965) 'Effect of a progestin-estrogen oral contraceptive on blood clotting factors', *Metabolism, 14*, 411-7

Howie, P.W., Mallinson, A.C., Prentice, C.R.M., Horne, C.H.W. and McNicol, G.P. (1970) 'Effect of combined oestrogen-progestogen oral contraceptives, oestrogen, and progestogen on antiplasmin and antithrombin activity', *Lancet, ii*, 1329-32

Hume, R. (1965) 'Fibrinolytic activity and thyroid function', *British Medical Journal, 1*, 686-8

Hutton, R.A., Mikhailidis, D.P., Georgiadis, E., Hyden, A. and Ginsburg, J. (1980) 'The effect of tetracosactrin and corticosteroids on platelet aggregation: an *in vivo* and *in vitro* study', *Thrombosis Research, 17*, 5-11

Ikkala, E., Eisalo, A. and Heinivaara, O. (1962) 'Plasma coagulation factors in thyrotoxicosis — a useful test of thyroid hypofunction', *Acta Endocrinologica, 40*, 307-10

Ingram, G.I.C. (1961) 'Increase in antihaemophilic globulin activity following infusion of adrenaline', *Journal of Physiology, 156*, 217-24

Ingram, G.I.C. and Jones, R.V. (1966) 'The rise in clotting factor VIII induced in man by adrenaline: effect of α- and β-blocker', *Journal of Physiology, 187*, 447-54

Ingram, G.I.C., Jones, R.V., Hershgold, E.J., Denson, K.W.E. and Perkins, J.R. (1977) 'Factor-VIII activity and antigen, platelet count and biochemical changes after adrenoceptor stimulation', *British Journal of Haematology, 35*, 81-100

Isacson, S. (1970) 'Effect of prednisone on the coagulation and fibrinolytic systems', *Scandinavian Journal of Haematology, 7*, 212-6

Jespersen, J. and Kluft, C. (1982) 'Histidine-rich glycoprotein and plasminogen levels in plasma during the normal menstrual cycle and in women on oral contraceptives low in estrogen', *Haemostasis, 11*, Suppl. 1, 49

Jørgensen, K.A., Sørensen, P. and Freund, L. (1982) 'Effect of glucocorticosteroids on oral coagulation tests', *Acta Haematologica, 68*, 39-42

Käser-Glanzmann, R., Jákabová, M., George, J.N. and Lüscher, E.F. (1977) 'Stimulation of calcium uptake in platelet membrane vesicles by adenosine 3′,5′-cyclic monophosphate and protein kinase', *Biochimica et Biophysica Acta, 466*, 429-40

von Kaulla, E., Droegemueller, W. and von Kaulla, K.N. (1975) 'Conjugated estrogens and hypercoagulability', *American Journal of Obstetrics and Gynecolgy, 122*, 688-92

Kleniewski, J., Gladecki, K. and Cybulska, J. (1974) 'The influence of insulin on plasma fibrinolytic system', *Thrombosis Research, 4*, 137-45

Kluft, C., Preston, F.E., Malia, R.G. and Wijngaards, G. (1982) 'Characterization of the effect of stanozolol on fibrinolytic parameters in healthy volunteers', *Haemostasis, 11*, Suppl. 1, 56

Koerker, D.J., Harker, L.A. and Goodner, C.J. (1975) 'Effects of somatostatin on hemostasis in baboons', *New England Journal of Medicine, 293*, 476-9

Kwaan, H.C. and McFadzean, A.J.S. (1956) 'The inhibition of clot lysis by corticotrophin', *Lancet, i*, 136-7

Kwaan, H.C., McFadzean, A.J.S. and Cook, J. (1957) 'On plasma fibrinolytic activity in cryptogenic splenomegaly', *Scottish Medical Journal, 2*, 137-50

Laake, K., Venneröd, A.M., Haugen, G. and Gjönnaess, H. (1974) 'Cold-promoted activation of factor VII in human plasma: studies on the associated acyl-arginine esterase activity', *Thrombosis Research, 4*, 769-85

Lamberg, B.-A. and Gordin, R. (1954) 'Liver function in thyrotoxicosis. Studies on the cholesterol and prothrombin level in the blood during the treatment of thyrotoxicosis', *Acta Endocrinologica, 15*, 82-96

Larsson-Cohn, U., Fagerhol, M.K. and Abildgaard, U. (1972) 'Concentration of antithrombin III during combined and progestogen-only oral contraceptive treatment', *Acta Obstetrica et Gynaecologica Scandinavica, 51*, 315-7

Laurell, C.-B., Kullander, S. and Thorell, J. (1968) 'Effect of administration of a combined estrogen-progestin contraceptive on the level of individual plasma proteins', *Scandinavian Journal of Clinical and Laboratory Investigations, 21*, 337-43

Long, M.W., Micossi, P., Pontiroli, A.E., Dunbar, J.C., Foà, P.P. and Henry, R.L. (1979) 'Somatostatin-induced changes in the haemostasis of rats', *Thrombosis and Haemostasis, 41*, 601-7

Lord, J.W. and Andrus, W. DeW. (1941) 'Changes in the liver associated with hyperthyroidism', *Archives of Surgery, 42*, 643-60

Lynch, C.J. and Steer, M.L. (1981) 'Evidence for high and low affinity alpha$_2$-receptors. Comparison of [3H]norepinephrine and [3H]phentolamine binding in human platelet membranes', *Journal of Biological Chemistry, 256*, 3298-303

Mannucci, P.M. (1974) 'Enhancement of plasminogen activator by vasopressin and adrenaline: a role of cyclic AMP?', *Thrombosis Research, 4*, 539-49

Mannucci, P.M. and Pareti, F.I. (1974) 'Platelet functions after intravenous administration in man of cyclic-AMP and related drugs', *Journal of Laboratory and Clinical Medicine, 84*, 828-38

Mannucci, P.M. and Rota, L. (1980) 'Plasminogen activator response after DDAVP: A clinico-pharmacological study', *Thrombosis Research, 20*, 69-76

Mannucci, P.M., Ruggeri, Z.M. and Gagnatelli, G. (1971) 'Nervous regulation of factor-VIII levels in man', *British Journal of Haematology, 20*, 195-207

Mannucci, P.M., Gagnatelli, G. and D'Alonzo, R. (1972) 'Stress and blood coagulation', in K. Brinkhouse (ed.) *Thrombosis: Risk Factors and Diagnostic Approaches*, Schattauer, Stuttgart, pp. 105-13

Mannucci, P.M., Åberg, M., Nilsson, I.M. and Robertson, B. (1975) 'Mechanism of plasminogen activator and factor VIII increase after vasoactive drugs', *British Journal of Haematology, 30,* 81-93

Margulis, R.R., Ambrus, J.L., Mink, I.B. and Stryker, J.C. (1965) 'Progestational agents and blood coagulation', *American Journal of Obstetrics and Gynecology, 93,* 161-8

Markwardt, F. and Klöcking, H.-P. (1976) 'Studies on the nature of plasminogen activator', *Thrombosis Research, 8,* 217-23

Masure, R., Hurlet, A., Maes, E., Gillard, F., Col-DeBeys, C., Moriau, M., Meyer, C., Dubois-Pelerin, F. and Vekemans, M. (1979) 'Effects of hormonal contraception of haemostatic and lipid profiles', *Thrombosis and Haemostasis, 42,* 483

McClure, P.D., Ingram, G.I.C. and Jones R.V. (1965) 'Platelet changes after adrenaline infusions with and without adrenaline blockers', *Thrombosis et Diathesis Haemorrhagica, 13,* 136-9

McKenna, R., Bachmann, F., Pichairut, O. and Whittaker, B. (1981) 'Effect of prednisone on platelet function in patients with bleeding disorders', *Thrombosis and Haemostasis, 46,* 430

McMillan, R., Bakich, M.J. and Yelenosky, R.J. (1979) 'The adrenalin binding site on human platelets', *British Journal of Haematology, 41,* 597-604

Meade, T.W., Brozovic, M., Chakrabarti, R., Howarth, D.J., North, W.R.S. and Stirling, Y. (1976) 'An epidemiological study of the haemostatic and other effects of oral contraceptives', *British Journal of Haematology, 34,* 353-64

Menczel, J. and Dreyfuss, F. (1960) 'Effect of prednisone on blood coagulation time in patients on dicoumarol therapy', *Journal of Laboratory and Clinical Medicine, 56,* 14-20

Mielke, C.H., Gerich, J.E., Lorenzi, M., Tsalirian, E., Rodvien, R. and Forsham, P.H. (1975) 'The effect of somatostatin on coagulation and platelet function in man', *New England Journal of Medicine, 293,* 480-3

Miller, S.P., Lee, S.L. and Ritz, N. (1965) 'Progestin-estrogen (SC11, 800) therapy and the hemostatic mechanism — a controlled study', *Metabolism, 14,* 398-410

Mills, D.C.B. and Roberts, G.C.K. (1967) 'Effects of adrenaline on human blood platelets', *Journal of Physiology, 193,* 443-53

Mitchell, J.R.A. and Sharp, A.A. (1964) 'Platelet clumping *in vitro*', *British Journal of Haematology, 10,* 78-93

Nelson, W.R. and Taylor, G.A. (1975) '*In vitro* inhibition of endotoxin induced platelet aggregation with hydrocortisone sodium succinate (Solu-Cortef[R])', *Scandinavian Journal of Haematology, 15,* 35-44

Nilsson, I.M. and Astedt, B. (1971) 'Blood clotting and contraception', *British Medical Journal, 2,* 589-90

Nilsson, I.M. and Kullander, S. (1967) 'Coagulation and fibrinolytic studies during use of gestagens', *Acta Obstetrica et Gynaecologica Scandinavica, 46,* 286-303

O'Brien, J.R. (1963) 'Some effects of adrenaline and anti-adrenaline compounds on platelets *in vitro* and *in vivo*', *Nature, 200,* 763-4

Olsson, L.-B., Kutti, J., Lundborg, P. and Fredén, K. (1976) 'The peripheral platelet count in response to intravenous infusion of isoprenaline', *Scandinavian Journal of Haematology, 17,* 213-6

Owren, P.A. and Aas, K. (1951) 'Control of dicoumarol therapy and quantitative determination of prothrombin and proconvertin', *Scandinavian Journal of Clinical and Laboratory Investigation, 3,* 201-8

Ozge, A.H., Mustard, J.F., Hegardt, B., Rowsell, H.C. and Downie, H.G. (1963) 'The effect of adrenaline on blood coagulation, platelet economy and thrombus formation', *Canadian Medicial Association Journal, 88,* 265

Ozsoylu, S., Strauss, H.S. and Diamond, L.K. (1962) 'Effects of corticosteroids on coagulation of the blood', *Nature, 195*, 1214-5

Pilgeram, L.O., Amundson, B.A., Lofgren, P.E. (1964) 'Evidence for steroid control of the metabolism of profibrinolysin', *Thrombosis et Diathesis Haemorrhagica, 11*, 94-8

Pina-Cabral, J.M. and Rodriques, C. (1974) 'Blood catecholamine levels, factor VIII and fibrinolysis after therapeutic electroshock', *British Journal of Haematology, 28*, 371-80

Poller, L., Priest, C.M. and Thomson, J.M. (1969) 'Platelet aggregation during oral contraception', *British Medical Journal, 4*, 273-4

Poller, L., Thomson, J.M. and Thomas, W. (1971) 'Oestrogen/progestogen oral contaceptive and blood clotting: a long-term follow-up', *British Medical Journal, 4*, 648-50

Ponari, O., Civardi, E., Megha, A., Pini, M., Poti, R. and Dettori, A.G. (1973) 'Effect of alpha- and beta-blocking drugs on the clotting and fibrinolytic response to venous stasis in man', *British Journal of Haematology, 24*, 463-70

Powell, L.C., Guest, M.M. and Bond, T.P. (1965) 'Coagulation and fibrinolytic studies in women receiving an anovulatory drug (medroxyprogesterone acetate with estradiol)', *American Journal of Obstetrics and Gynecology, 93*, 167-72

Preston, F.E., Burakowski, B.K., Porter, N.R. and Malia, R.G. (1981) 'The fibrinolytic response to stanozolol in normal subjects', *Thrombosis Research, 22*, 543-51

Preston, F.E., Malia, R.G., Greaves, M., Wijngaards, G. and Kluft, C. (1982) 'The effect of stanozolol on thrombosis risk factors in healthy individuals', *Haemostasis, 11*, Suppl. 1, 56

Proctor, S.J., Davies, T.F., Oxley, A., Jones, P. and Heath, A. (1980) 'Increased platelet aggregation induced by glucagon administration', *Acta Haematologica, 64*, 6-11

Prowse, C., Dow, R. and Boulton, F. (1982) 'The plasminogen activator response to direct hypothalamic stimulation in normal and congenitally vasopressin-deficient rats', *Haemostasis, 11*, Suppl. 1, 36

Rennie, J.A.N., Bewsher, P.D., Murchison, L.E. and Ogston, D. (1978) 'Coagulation and fibrinolysis in thyroid disease', *Acta Haematologica, 59*, 171-7

Rifkin, D.B. (1978) 'Plasminogen activator synthesis by cultured human embryonic lung cells: characterization of the suppressive effect of corticosteroids', *Journal of Cellular Physiology, 97*, 421-8

Rutherford, R.N., Hougie, C., Banks, A.L. and Coburn, W.A. (1964) 'The effect of sex steroids and pregnancy on blood coagulability', *Obstetrics and Gynecology, 24*, 886-92

Sano, I., Kakimoto, Y., Taniguchi, K. and Takesdada, M. (1959) 'Active transport of epinephrine into blood platelets', *American Journal of Physiology, 197*, 81-4

Sarji, K.E., Levine, J.H., Nair, R.M.G., Sagel, J. and Corwell, J.A. (1977) 'Relation between growth hormone levels and von Willebrand factor activity', *Journal of Clinical Endocrinology and Metabolism, 45*, 853-6

Sawyer, W.D., Fletcher, A.P., Alkjaersig, N. and Sherry, S. (1960) 'Studies on the thrombolytic activity of human plasma', *Journal of Clinical Investigation, 39*, 426-34

Schneck, S.A. and von Kaulla, K.N. (1961) 'Fibrinolysis and the nervous system', *Neurology, 11*, 959-69

Sherry, S., Lindemeyer, R.I., Fletcher, A.P. and Alkjaersig, N. (1959) 'Studies on enhanced fibrinolytic activity in man', *Journal of Clinical Investigation, 38*, 810-22

Sjöberg, H.E., Blombäck, M. and Granberg, P.O. (1976) 'Thromboembolic complications, heparin treatment and increase in coagulation factors in

Cushing's syndrome', *Acta Medica Scandinavica, 199*, 95-8

Simone, J.V., Abildgaard, C.F. and Schulman, I. (1965) 'Blood coagulation in thyroid dysfunction', *New England Journal of Medicine, 273*, 1057-61

Tanser, A.R. and Smellie, H. (1964) 'Observations on adrenaline-induced fibrinolysis', *Clinical Science, 26*, 375-80

Teger-Nilsson, A.-C. (1979) 'Use of chromogenic substrates for screening of inhibitors of coagulation and fibrinolysis in patients', in M.F. Scully and V.V. Kakkar (eds.) *Chromogenic Peptide Substrates. Chemistry and Clinical Usage*, Churchill Livingstone, Edinburgh, pp. 269-76

Teger-Nilsson, A.-C., Gyzander, E., Myrwold, H., Noppa, H., Olsson, R. and Wallmo, L. (1978) 'Determination of fast-acting plasmin inhibitor (α_2-antiplasmin) in plasma from patients with tendency to thrombosis and increased fibrinolysis', *Haemostasis, 7*, 155-7

Thomas, D.P. (1967) 'Effect of catecholamines on platelet aggregation caused by thrombin', *Nature, 215*, 298-9

Thomson, J.M. and Poller, L. (1965) 'Oral contraceptive hormones and blood coagulability', *British Medical Journal, 2*, 270-3

Truelove, S.C. (1951) 'Fibrinolysis and the eosinophil count', *Clinical Science, 10*, 229-40

Truelove, S.C. (1952) 'Fibrinolysis in relation to ACTH and cortisone', *Clinical Science, 11*, 101-6

Taskok, F.H.M., Koh, S., Chua, L.M. and Ratnam, S.S. (1979) 'Coagulation changes in Asian women in long-term steroid contraception', *Thrombosis and Haemostasis, 42*, 391

Van Oosterom, A.T., Kerkhoven, P. and Veltkamp, J.J. (1979) 'Metabolism of the coagulation factors of the prothrombin complex in hypothyroidism in man', *Thrombosis and Haemostasis, 41*, 273-85

Virji, M.A.G. (1982) 'Plasminogen activator in the rat islets of Langerhans: response to cyclic AMP', *Haemostasis, 11*, Suppl. 1, 42

Virji, M.A.G., Vassalli, J.-D., Estensen, R.D. and Reich, E. (1980) 'Plasminogen activator of islets of Langerhans: modulation by glucose and correlation with insulin production', *Proceedings of the National Academy of Sciences, 77*, 875-9

Vosburgh, C.H. and Richards, A.N. (1903) 'An experimental study of the sugar content and extravascular coagulation of the blood after administration of adrenalin', *American Journal of Physiology, 9*, 35-51

Walker, I.D., Davidson, J.F., Young, P. and Conkie, J.A. (1975) 'Effect of anabolic steroids on plasma antithrombin III, α_2-macroglobulin and α_1-antitrypsin levels', *Thrombosis et Diathesis Haemorrhagica, 34*, 106-14

Wiggers, C.J. (1909) 'Studies in inaccessible internal haemorrhages. I. The effect of adrenaline on intestinal haemorrhage', *Archives of Internal Medicine, 3*, 139-59

Winther, O. (1967) 'Testosterone and fibrinolytic activity', *Scandinavian Journal of Clinical and Laboratory Investigation*, Suppl. 93, 207-10

Ygge, J., Brody, S., Korsan-Bengtsen, K. and Nilsson, L. (1969) 'Changes in blood coagulation and fibrinolysis in women receiving oral contaceptives', *American Journal of Obstetrics and Gynecology, 104*, 87-99

Age, Sex, Circadian and Seasonal Variations in the Hemostatic Mechanism

Influence of Age on Hemostatic Components

Fetal Life

Platelet Numbers and Function

Platelets have been found in embryos of eleven weeks' gestational age (Bleyer *et al.*, 1971) while clumps of platelets, suggesting their ability to aggregate, were seen by 12 to 15 weeks. Platelets aggregate in response to ADP by the 19th week, but less strongly than adult platelets (Pandolfi *et al.*, 1972): no response to epinephrine or collagen was seen. The number of platelets in the blood at 30 weeks' gestation was reported to be close to adult levels (Kalpaktsoglou and Emery, 1965).

Coagulation Factors

The blood of fetuses of under ten to eleven weeks' gestational age has been found not to clot, but after this stage of development rapid whole blood clotting times were found (Zilliacus *et al.*, 1966).

A number of studies have examined the time of appearance and level of individual coagulation factors in the fetus. Synthesis of fibrinogen was observed at 5.5 weeks' gestation (Gitlin and Biasucci, 1969). The level of clotting factors in a group of fetuses of 12 to 24 weeks' gestation was reported by Holmberg and associates (1974): the concentration of all the clotting factors measured was lower than adult values, in particular, the factor II-VII-X complex and factor IX. Factor V activity was noted to be close to adult levels by 12 to 15 weeks' gestation (Heikimheimo, 1964). The factor VIII:C antigen, measured by an immunoradiometric assay, has been found to range from 10 to 25 per cent in fetuses of 16 to 21 weeks (Peake *et al.*, 1979). Factor XIII levels of some 50 per cent of term values were noted in fetuses of 17 to 24 weeks' gestation (Henriksson *et al.*, 1974).

Protease Inhibitors

The synthesis of α_1-antitrypsin, α_2-macroglobulin and Cl inactivator has been detected in embryos of 29 days' gestation (Gitlin and Biasucci, 1969). The serum α_1-antitrypsin level was low in 6.5 week embryos, but had risen to over half the normal adult level by 9.5 weeks. Alpha$_2$-macroglobulin increased in concentration with gestation period to attain levels at term which were greater than that of maternal blood. The serum level of Cl inactivator at 6.5 weeks was 20 per cent of the normal adult level, and close to adult levels by 28 weeks' gestation.

Fibrinolytic System

Fibrinolytic activity appears to be present in the blood early in fetal life (Zilliacus *et al.*, 1966), and short euglobulin clot lysis times have been found from about 15 weeks' gestation (Ekelund *et al.*, 1970a). Plasminogen activator activity has been shown to be present in fetal tissues (Lieberman, 1959). Plasminogen has been detected in the blood of 10.5 week embryos, levels rising through fetal life (Gitlin and Biasucci, 1969): a plasminogen concentration of some 20 per cent of the adult level has been found at 13 weeks (Ekelund *et al.*, 1970a). The sole discordant result is that of Ambrus *et al.* (1965) who could not detect plasminogen in fetuses of up to eight months' gestation.

Infancy

Platelet Number and Function

The platelet count in both the pre-term and full-term infant has been reported to be close to, or a little lower than, those of the adult (Ablin *et al.*, 1961; Aballi *et al.*, 1968; Bleyer and Breckenridge, 1970; Appleyard and Brinton, 1971).

Platelet function appears to be impaired in the newborn. Reduced aggregability in response to ADP, collagen and thrombin in the newborn was reported by Mull and Hathaway (1970), while Corby and Schulman (1971) found that aggregation in response to collagen and epinephrine was reduced, although the response to ADP was similar to that of adults. Reduced aggregability of platelets from the neonate

250

has been noted by others (Hrodek, 1969; Ts'ao *et al.*, 1976; Foley *et al.*, 1977), and there is evidence that this results from an abnormality in the release reaction (Whaun, 1973; Ts'ao *et al.*, 1976; Corby and Zuck, 1976).

Further evidence of altered platelet function in the newborn has been provided by electrophoretic mobility studies (Kosztolanyi *et al.*, 1980): the mobility of washed platelets and platelets suspended in diluted plasma, related to the negative surface charge excess, was found to be similar for platelets obtained from adults and the newborn. The mobility changes induced by ADP, however, were different; platelets from the newborn failed to respond to low concentrations of ADP by an increase in electrophoretic mobility. This failure could be corrected by adult plasma, suggesting the presence in adult plasma of a factor responsible for the reaction to ADP.

Abnormalities in platelet function do not appear to be due to a defect in prostaglandin synthesis. There was a normal response of neonatal platelets to arachidonic acid (Corby *et al.*, 1977), and the platelet prostaglandin synthetic pathway has been found to be essentially normal in the neonatal period (Stuart, 1978; Del Principe *et al.*, 1979).

Coagulation Factors

The prothrombin time and the partial thromboplastin time have been found to be significantly longer in the newborn infant than in the mother (Aballi and De Lamerens, 1962; Cade *et al.*, 1969; Bleyer and Breckenridge, 1970). The prothrombin time reaches normal adult values a few days after birth (Aballi and De Lamerens, 1962; Cade *et al.*, 1969). The prolonged prothrombin time reflects the reduced concentration of the vitamin K-dependent coagulation factors found by a number of investigators in the newborn (Biland and Duckert, 1973; Barnard *et al.*, 1979). The adult level of prothrombin itself has been reported to be achieved by the age of four to seven months (Schettini *et al.*, 1976a).

An explanation for the prolonged partial thromboplastin time is provided by the low levels of the contact factors in the plasma of the newborn: a mean factor XI concentration of 38 per cent was found in the newborn infant by Hilgartner and Smith (1965), and others have confirmed the low factor XI levels at birth (Nossel *et al.*, 1966; Saito and Goldsmith, 1977; Barnard *et al.*, 1979). Adult levels of factor XI are reached one to two months after birth (Hilgartner and Smith, 1965). The concentration of factor XII in infancy has also been shown to be low by several investigators (Kurkcuoglu and McElfresh, 1960; Bleyer and Breckenridge, 1970; Saito *et al.*, 1976).

Factor V levels in the newborn are similar to those of adults (Douglas and Davies, 1955; Barnard *et al.*, 1979). Factor VIII coagulant activity in infants has also been found to be close to that of adults by most investigators (Sell and Corrigan, 1973; Fukui *et al.*, 1979), but some have reported increased levels (Johnson *et al.*, 1981). In one study factor VIII:C antigen levels were found to range between 25 and 65 per cent in normal term cord blood (Peake *et al.*, 1979). The level of factor VIIIR:Ag in full-term infants has been shown to lie close to adult levels (Barnard *et al.*, 1979; Henriksson *et al.*, 1979), or to be slightly lower (Fukui *et al.*, 1979). In contrast, elevated levels of factor VIIIR:Ag at birth have been reported by others with a return to adult levels by the third post-natal day (Maak *et al.*, 1978; Johnson *et al.*, 1981). Factor VIII ristocetin cofactor activity has also been noted to be increased in infancy (Ts'ao *et al.*, 1976; Johnson *et al.*, 1981). Crossed immunoelectrophoretic studies have suggested a qualitative abnormality of the factor VIII molecule in the newborn, but this was observed by Johnson and associates (1981) to be present only in severely ill infants.

The fibrinogen concentration in the newborn is close to that of adults (Fisher *et al.*, 1968; Bleyer and Breckenridge, 1970). The existence of a fetal form of fibrinogen has been discussed in Chapter 3. Some investigators have found the level of factor XIII in the newborn to be close to adult levels (Fisher *et al.*, 1968; Biland and Duckert, 1973), while others have reported that the level lies between 50 and 75 per cent of adult values (Bouhasin and Altay, 1968; Henriksson *et al.*, 1974): the relative unreliability of clot solubility assay methods may underlie this discrepancy.

The reduction in the level of a number of the coagulation factors in the newborn has been attributed to an impairment of the synthetic capacity of the liver and a deficiency of vitamin K. Muller and his associates (1977), however, found no evidence for vitamin K deficiency, but reported that infants aged three days had a heparin-like inhibitor of coagulation in their plasma. They suggested that this inhibitor is responsible for the apparently low coagulation factor levels in the newborn: at the present time the functional significance of this inhibitor is uncertain.

Protease Inhibitors

The level of α_2-macroglobulin in the full-term infant is higher than in

252

the adult (Ganrot and Scherstén, 1967; Gitlin and Biasucci, 1969; Ekelund *et al.*, 1970b). In contrast, antithrombin III levels are lower in the newborn (Weissbach *et al.*, 1974; Teger-Nilsson, 1975; Foley *et al.*, 1977; Hathaway *et al.*, 1978). The adult level of antithrombin III is reached at about 12 months. In one study the concentration of α_1-antitrypsin was found to be substantially greater in the newborn than in the adult (Cederholm-Williams *et al.*, 1981).

Fibrinolytic System

There are conflicting data on level of blood fibrinolytic activity in the newborn infant. Most have found that such activity, measured by techniques reflecting the level of circulating plasminogen activator, exceeds that of adults, but falls rapidly in the first few days of life (Cope and Simmons, 1958; Phillips and Skrodelis, 1958; Markarian *et al.*, 1967; Fisher *et al.*, 1968; Ekelund *et al.*, 1970b; Foley *et al.*, 1977). Others have reported that only small amounts of activator are present in the blood of normal full-term infants (Ambrus *et al.*, 1963, 1965).

There is general agreement that the plasma plasminogen concentration is low in the newborn, usually around 50 per cent of adult values (Phillips and Skrodelis, 1958; Quie and Wannamaker, 1960; Fisher *et al.*, 1968; Ekelund *et al.*, 1970b; Foley *et al.*, 1977). The plasminogen level rises rapidly and has been reported to attain adult levels in around two weeks (Schettini *et al.*, 1976b).

Antiplasmin activity, measured by a chromogenic substrate and reflecting the α_2-antiplasmin level, has been found to be slightly higher in the neonatal period (Cederholm-Williams *et al.*, 1981).

Adult Life

Platelet Numbers and Function

Platelet numbers in the peripheral blood do not appear to change with advancing age (Hamilton *et al.*, 1974b). In a study on women Korsan-Bengtsen and associates (1973) found no change in platelet adhesiveness associated with aging. However, there has been a report of greater aggregability in response to ADP, collagen, arachidonic acid and epinephrine with increasing age (Johnson *et al.*, 1975). Plasma levels of β-thromboglobulin and platelet factor 4 were found to be unrelated to age (Duncan and Didisheim, 1981).

Coagulation Factors

Increasing age has been associated with some increase in the plasma concentration of the majority of coagulation factors studied, but it is unknown whether this reflects a higher rate of synthesis or a slower metabolic decay. Most have found an increase in Factor VIII coagulant activity with advancing age (Cooperberg and Teitelbaum, 1960; Pitney *et al.*, 1962; Jeremic *et al.*, 1976), although others have been unable to confirm this (Preston and Barr, 1964; Korsan-Bengtsen *et al.*, 1973). Factor IX levels may rise slightly with age (Simpson and Biggs, 1962). Factor VII also increases with advancing years (Hamilton *et al.*, 1974b; Meade and North, 1977). The findings on factor V are conflicting: two studies have shown an increase in concentration with age (Hamilton *et al.*, 1974b; Gedde-Dahl *et al.*, 1975) while Tracy and associates (1981), using a radioimmunoassay, found no age trend. One study has suggested that factor X also rises with age (Hamilton *et al.*, 1974b).

There is general agreement that the plasma fibrinogen concentration increases with age (Pilgeram, 1961; Moser and Hajjar, 1966; Ogston and Ogston, 1966; Weisert and Jeremic, 1974; Meade *et al.*, 1979). The mean level rises from around 270 mg/100 ml in the 20-30 year age-group to about 360 mg/100 ml in the 50-70 year age-group. The rise in fibrinogen does not appear to progress into the very elderly and may indeed fall in those over the age of 85 (Hamilton *et al.*, 1974a): it is possible that such individuals are pre-selected by their longevity, the lower fibrinogen levels contributing to their unusually long life-span.

Protease Inhibitors

The concentration of serum α_2-macroglobulin is considerably higher in children than in adults; Ganrot and Scherstén (1967) found that the adult level was not reached until the age of 25 to 30 years, thereafter there was no increase up to the age of 65. A further rise in the elderly has been noted (Hamilton *et al.*, 1974a; Tunstall *et al.*, 1975). The changes in the α_1-antitrypsin level with age appear to be, at most, small. While Cleve (1966) was unable to find a significant age-related trend, Hamilton and associates (1974a) observed a modest increase in concentration with age. Antithrombin III levels were found to decrease in healthy men with advancing years whereas in women there was a slight but significant increase in the level around the time of the menopause (Fagerhol and Abildgaard, 1970).

254

Fibrinolytic System

The findings from most studies have led to the conclusion that plasma fibrinolytic activity is unrelated to age (Sawyer *et al.*, 1960; Moser and Hajjar, 1966; Rosing *et al.*, 1973). In one investigation there was no change in plasminogen activator level as assessed by the euglobulin clot lysis time up to the age of 75, thereafter an increased level of fibrinolytic activity was seen (Hamilton *et al.*, 1974a). In a study on a large group of men between 15 and 64 years Meade and co-workers (1979) found a decrease in fibrinolytic activity from the 20-25 up to the 50-55 year age-group, followed by a rise in activity up to the 60-65 years group.

Age-related changes in the plasma plasminogen level are, at most, small. In studies of relatively small groups of subjects Jacobsen (1964) and Hedner and Nilsson (1965) found no significant age differences in the plasminogen concentration, while a small rise was noted by Ogston and Ogston (1966). In a later study (Hamilton *et al.*, 1974a) the plasminogen level did not change with age except for a fall in the very old.

Influence of Sex on Hemostatic Components

Platelet Numbers and Function

It has been generally accepted that there is no difference in the peripheral platelet count between the sexes (Sloan, 1951). In more recent years, however, a number of studies have been reported showing a significantly higher mean platelet count in women (Dawson *et al.*, 1976; Stevens and Alexander, 1977; Kemona *et al.*, 1978; Bain and Forster, 1980). For example, in the study by Stevens and Alexander the 461 men had a mean platelet count of 225 ($\times\ 10^9$/liter) while the women who were not receiving a contraceptive pill had a mean count of 290 ($\times\ 10^9$/liter). There was no difference in the distribution of platelet size between men and women (Agarwal *et al.*, 1976).

Platelets from women have been found to be more reactive to aggregating agents than those from men. Johnson and associates (1975) observed that primary reversible aggregation induced by ADP was significantly increased in young healthy women compared to men: primary and secondary (irreversible) aggregation induced by epinephrine was also significantly greater in women, but there was no significant difference in secondary aggregation induced by either collagen or arachidonic acid. The increased sensitivity of female platelets to threshold levels of aggregating agents has been observed by other investigators (Nordøy *et al.*, 1978; Roper *et al.*, 1979; Meade *et al.*, 1981). A further study, however, has provided evidence that this sex difference is an artifact due to the lower hematocrit in women, the larger volume of distribution of the anticoagulant resulting in higher ionised calcium concentrations (Kelton *et al.*, 1980).

No difference has been found in the plasma β-thromboglobulin or platelet factor 4 levels between men and women (Duncan and Didisheim, 1981).

Coagulation Factors

Few consistent differences in the level of individual coagulation factors between men and women not receiving hormone preparations have been reported. Using a radioimmunoassay neither factor XI nor factor XII were found to differ between the sexes (Saito *et al.*, 1976; Saito and Goldsmith, 1977) and Proud and associates (1980) detected no sex difference in the level of HMW kininogen. Factor VIII coagulant activity was noted in two studies to be slightly but significantly greater in men (Cooper and Teitelbaum, 1960; Preston and Barr, 1964), but no difference was found by Pitney *et al.*, (1962) or by Jeremic and associates (1976) in a large group of blood donors aged between 20 and 60. The mean factor VII levels in men and women in the age-range 19 to 64 were reported to be similar (Brozović *et al.*, 1974), but the increase which occurred with age was noted to be more marked in women. The plasma factor V level, measured by a radioimmunoassay did not differ between the sexes (Tracy *et al.*, 1981). No sex difference in the fibrinogen concentration was found in any of the age-groups studied (Weisert and Jeremic, 1974).

Protease Inhibitors

The concentration of serum α_2-macroglobulin has been found to be higher in women than in men (Ganrot and Scherstén, 1967; Dawson *et al.*, 1976). While no sex difference was detected in the α_1-antitrypsin level in one study (Cleve, 1966), a small increase was noted in women by Dawson and colleagues (1976).

Antithrombin III levels, assayed by immunodiffusion, were found to be lower in women under the age of 45 than in men of comparable age: an increase at the time of the menopause brought the concentration above that of men (Fagerhol and Abildgaard, 1970). There

256

was, however, no difference in antithrombin III levels between adolescent girls and boys aged 13 to 19 years (Sveger, 1979).

Fibrinolytic System

There is general agreement that there is no difference in plasma fibrinolytic activity between men and women (Sawyer *et al.*, 1960; Beller *et al.*, 1964; Brakman *et al.*, 1966; Mann, 1967). A greater fibrinolytic response to exercise in women was reported by Cash (1966), but there was no sex difference in the response to venous occlusion (Robertson *et al.*, 1971). The excretion of urokinase was also equal in men and women (Wilcox *et al.*, 1978).

The plasma plasminogen concentration is the same in men and women (Jacobsen, 1964; Hedner and Nilsson, 1965; Ogston and Ogston, 1966).

Teger-Nilsson (1979) found a small but significant sex difference in the level of α_2-antiplasmin, measured with a chromogenic substrate assay, women having a higher mean level. There was no sex difference in the concentration of histidine-rich glycoprotein (Lijnen *et al.*, 1981).

Circadian Variations in Hemostatic Components

Data on circadian rhythms in platelet numbers or functions is fragmentary. Measurements on blood donors revealed no significant differences in the platelet count between those attending morning or afternoon sessions (Stevens and Alexander, 1977). Spontaneous platelet aggregability was reported to be at a minimum in the early afternoon (Walter *et al.*, 1979).

The circadian rhythm in fibrinolytic activity has received more study. Fearnley and his colleagues (1957) were the first to present evidence for a diurnal variation in blood fibrinolytic activity, fibrinolysis being lowest from midnight to 8 a.m. and rising progressively over the day. A basic rhythm of fibrinolytic activity was suggested from measurements on day and night nurses, increased fibrinolytic activity being found during the day irrespective of whether the nurse was asleep or ambulant. The existence of a circadian rhythm in fibrinolytic activity has been confirmed repeatedly (Billimoria *et al.*, 1959; Buckell and Elliott, 1959; Mann, 1967; Rosing *et al.*, 1973). While the increase in fibrinolytic activity during the day is found in subjects maintained at bed rest (Rosing *et al.*, 1970), it is much greater in those undergoing normal activities (Billimoria *et al.*, 1959). The circadian variation in adrenal cortical secretion prompted a study to compare levels of fibrinolytic activity and plasma 11-hydroxycorticosteroids at 9 a.m. and 4 p.m. (Menon *et al.*, 1967); the steroid level was lower and the fibrinolytic activity higher in the forenoon, but there was not a close causal relationship. Rosing and associates (1970) noted that the fibrinolytic response to exercise was greater at 4 p.m. than at 8 a.m. A study of mono- and dizygotic twins has suggested the existence of a genetically determined component in the circadian rhythm of fibrinolytic activity (Čepelak *et al.*, 1978).

A lack of apparent circadian variation in urokinase excretion in the study by Wilcox *et al.* (1978) contrasts with the finding of Fornasari and colleagues (1979) that both urinary urokinase and plasma fibrinolytic activity were highest in the late morning and early afternoon.

There appears to be no difference in the plasma concentrations of plasminogen and fibrinogen between the morning and afternoon (Ogston and Ogston, 1966; Walter *et al.*, 1979). Variations in the level of a number of protease inhibitors have been reported by Fornasari *et al.*, (1979); in this study blood samples were obtained every four hours over a 24-hour period from healthy volunteers. Alpha-antitrypsin, α_2-macroglobulin, Cl inactivator and inter-α-trypsin inhibitor all showed peak values in the late morning with the lowest values during the night. The highest level of antithrombin III was seen during the night and early morning.

Seasonal Variations in Hemostatic Components

Little information is currently available on possible circannual rhythms in hemostatic function or components. Even with such data the problem of unravelling the influences of such seasonally related factors as diet, physical activity and environmental temperature would be formidable.

In a study on men receiving long-term anticoagulant therapy Eastham and Avis (1966) found that the platelet count did not vary significantly over the year; there was, however, a seasonal swing in platelet adhesiveness with a peak in the early spring and a trough in the autumn. Patients with coronary atherosclerosis were observed to have maximal fibrinolytic activity in the summer months with minimal levels in winter and spring (Semenko, 1973).

258

Tromp (1972) reported higher mean fibrinogen values in the plasma of Dutch blood donors in the July/August period than in January. When the data were analysed by individual days it was found that very low fibrinogen levels were associated with cold weather conditions while high values were associated with warm weather.

References

Aballi, A.J. and de Lamerens, S. (1962) 'Coagulation changes in the neonatal period and in early infancy', *Pediatric Clinics of North America, 9*, 785-817

Aballi, A.J., Puapondh, Y. and Desposito, F. (1968) 'Platelet counts in thriving premature infants', *Pediatrics, 42*, 685-9

Ablin, A.R., Kushner, J.H., Murphy, A. and Zippin, C. (1961) 'Platelet enumeration in the neonatal period', *Pediatrics, 28*, 822-4

Agarwal, S., Roper, P.R., Johnson, D. and Drewinko, B. (1976) 'Platelet size distribution profiles in normal individuals and in patients with acute leukemia', *American Journal of Clinical Pathology, 65*, 266

Ambrus, C.M., Weintraub, D.H., Dunphy, D.H., Dowd, D., Pickren, J.E., Niswander, K.R. and Ambrus, J.L. (1963) 'Studies on hyaline membrane disease. I. The fibrinolysin system in pathogenesis and treatment', *Pediatrics, 32*, 10-24

Ambrus, C.M., Weintraub, D.H., Niswander, K.R. and Ambrus, J.L. (1965) 'Studies on hyaline membrane disease. II. The ontogeny of the fibrinolysin system', *Pediatrics, 35*, 91-6

Ambrus, C.M., Ambrus, J.L., Niswander, K.R., Weintraub, D.H., Bross, I.D.J. and Lassman, H.B. (1970) 'Changes in fibrin-stabilizing factor levels in relation to maternal hemorrhage and neonatal disease', *Pediatric Research, 4*, 82-8

Appleyard, W.J. and Brinton, A. (1971) 'Venous platelet counts in low birth weight infants', *Biology of the Neonate, 17*, 30-4

Bain, B. and Forster, T. (1980) 'A sex difference in the bleeding time', *Thrombosis and Haemostasis, 43*, 131-2

Barnard, D.R., Simmons, M.A. and Hathaway, W.E. (1979) 'Coagulation studies in extremely premature infants', *Pediatric Research, 13*, 1330-5

Beller, F.K., Goebelsmann, U., Douglas, G.W. and Johnson, A. (1964) 'The fibrinolytic system during the menstrual cycle', *Obstetrics and Gynecology, 23*, 12-6

Biland, L. and Duckert, F. (1973) 'Coagulation factors of the newborn and his mother', *Thrombosis et Diathesis Haemorrhagica, 29*, 644-51

Billimoria, J.D., Drysdale, J., James, D.C.O. and Maclagan, N.F. (1959) 'Determination of fibrinolytic activity of whole blood. With special reference to the effects of exercise and fat feeding', *Lancet, ii*, 471-5

Bleyer, W.A. and Breckenridge, R.T. (1970) 'Studies on the detection of adverse drug reactions in the newborn', *Journal of the American Medical Association, 213*, 2049-53

Bleyer, W.A., Hakami, N. and Shepard, T.H. (1971) 'The development of hemostasis in the human fetus and newborn infant', *Journal of Pediatrics, 79*, 838-53

Bouhasin, J.D. and Altay, C. (1968) 'Factor XIII deficiency: concentrations in relatives of patients and in normal infants', *Journal of Pediatrics, 72*, 336-41

Brakman, P., Albrechtsen, O.K. and Astrup, T. (1966) 'A comparative study of coagulation and fibrinolysis in blood from normal men and women', *British*

Journal of Haematology, 12, 74-85

Brozović, M., Stirling, Y., Harricks, C., North, W.R.S. and Meade, T.W. (1974) 'Factor VII in an industrial population', *British Journal of Haematology, 28,* 381-91

Buckell, M. and Elliott, F.A. (1959) 'Diurnal fluctuation of plasma-fibrinolytic activity in normal males', *Lancet, ii,* 660-2

Cade, J.F., Hirsh, J. and Martin, M. (1969) 'Placental barrier to coagulation factors: Its relevance to the coagulation defect at birth and to haemorrhage in the newborn', *British Medical Journal, 2,* 281-3

Cash, J.D. (1966) 'Effects of moderate exercise on the fibrinolytic system in normal young men and women', *British Medical Journal, 2,* 502-6

Cederholm-Williams, S.A., Spencer, J.A.D. and Wilkinson, A.R. (1981) 'Plasma levels of selected haemostatic factors in newborn babies', *Thrombosis Research, 23,* 555-8

Čepelák, V., Barcal, R., Čepeláková, H. and Mayer, O. (1978) 'Circadian rhythms of fibrinolysis', in J.F. Davidson, R.M. Rowan, M.M. Samama and P.C. Desnoyers (eds.) *Progress in Chemical Fibrinolysis and Thrombolysis,* Raven Press, New York, pp. 571-8

Cleve, H. (1966) 'Quantitative immunologische Bestimmung von saurem α_1-Glycoprotein und α_1-Antitrypsin. Serumkonzentrationen bei gesunden Blutspendern', *Klinische Wochenschrift, 44,* 1256-60

Cooperberg, A.A. and Teitelbaum, J. (1960) 'The concentration of antihaemophilic globulin (AHG) related to age', *British Journal of Haematology, 6,* 281-7

Cope, I. and Simmons, E. (1958) 'The fibrinolytic activity of umbilical cord blood', *Journal of Obstetrics and Gynaecology of the British Empire, 65,* 414-8

Corby, D.G. and Schulman, I. (1971) 'The effects of antenatal drug administration on aggregation of platelets of newborn infants', *Journal of Pediatrics, 79,* 307-13

Corby, D.G. and Zuck, T.F. (1976) 'Newborn platelet dysfunction: a storage pool and release defect', *Thrombosis and Haemostasis, 36,* 200-7

Corby, D.G., Goad, W.C., Barber, J. and O'Barr, T.P. (1977) 'Evaluation of cyclo-oxygenase pathway in platelets of the newborn', *Thrombosis and Haemostasis, 38,* 35

Corrigan, J.L., Sell, E.J. and Pagel, C. (1977) 'Hageman factor and disseminated intravascular coagulation (DIC) in newborns and rabbits', *Pediatric Research, 11,* 916-20

Dawson, A., Allardyce, M., Allan, T.M., Ogston, D., Kerridge, D.F. and Lewis, H.B.M. (1976) 'Interrelations between ABO blood groups, plasminogen, α_1-antitrypsin, α_2-macroglobulin and the platelet count in blood donors', *Acta Haematologica, 56,* 19-26

Del Principe, D., Gabriotti, M., Mastracchio, F., Meninchelli, A. and Giardini, O. (1979) 'Malonyldialdehyde formation, oxygen consumption, fatty acid composition in newborn platelets stimulated by thrombin', *Biology of the Neonate, 36,* 49-54

Douglas, A.S. and Davies, P. (1955) 'Hypoprothrombinaemia in the newborn', *Archives of Diseases in Childhood, 30,* 509-12

Duncan, A. and Didisheim, P. (1981) 'The influence of physiological parameters on the plasma concentration of β-thromboglobulin (β-TG) and platelet factor 4 (PF-4) in normal healthy subjects', *Thrombosis and Haemostasis, 46,* 28

Eastham, R.D. and Avis, P.R.D. (1966) 'Seasonal fluctuation in adhesive platelets during long-term anticoagulant therapy', *British Journal of Haematology, 12,* 39-43

Ekelund, H., Hedner, U. and Astedt, B. (1970a) 'Fibrinolysis in human fetuses',

Acta Paediatrica Scandinavica, 59, 369-79

Ekelund, H., Hedner, U. and Nilsson, I.M. (1970b) 'Fibrinolysis in newborns', *Acta Paediatrica Scandinavia, 59*, 33-43

Fagerhol, M.K. and Abildgaard, U. (1970) 'Immunological studies on human antithrombin III. Influence of age, sex and use of oral contraceptives on serum concentration', *Scandinavian Journal of Haematology, 7*, 10-7

Fearnley, G.R., Balmforth, G. and Fearnley, E. (1957) 'Evidence for a diurnal fibrinolysis rhythm: with a simple method of measuring natural fibrinolysis', *Clinical Science, 16*, 645-50

Fisher, Sh., Schwartz, M., Gottlieb, A., Ben Yaseph, N. and Schapiro, S. (1968) 'Fibrinolysis, fibrinogen and factor XIII in newborn infants', *Thrombosis et Diathesis Haemorrhagica, 30*, 542-7

Foley, M.E., Clayton, J.K. and McNicol, G.P. (1977) 'Haemostatic mechanisms in maternal umbilical vein and umbilical artery blood at the time of delivery', *British Journal of Obstetrics and Gynaecology, 84*, 81-7

Fornasari, P.M., Gamba, G., Dolei, D., Gratton, L. and Ascari, E. (1979) 'Circadian rhythms in fibrinolysis', in G.G. Neri Serneri and C.R.M. Prentice (eds.) *Haemostasis and Thrombosis*, Academic Press, London, pp. 773-7

Fukui, H., Takase, T., Ikari, H., Murakami, Y., Okubo, Y. and Nakamura, K. (1979) 'Factor VIII procoagulant activity, factor VIII related antigen and von Willebrand factor in newborn cord blood', *British Journal of Haematology, 42*, 637-46

Ganrot, P.O. and Schersténn, B. (1967) 'Serum α_2-macroglobulin concentration and its variation with age and sex', *Clinica Chimica Acta, 15*, 113-20

Gedde-Dahl, T.W., Jeremic, M. and Weisert, O. (1975) 'Factor V (proaccelerin) concentration in 1016 blood donors. The effects of age, sex, and ABO blood groups', *Scandinavian Journal of Clinical and Laboratory Investigation, 35*, 25-30

Gitlin, D. and Biasucci, A. (1969) 'Development of γG, γA, γM, β_{1C}/β_{1A}, C'1 esterase inhibitor, ceruloplasmin, transferrin, hemopexin, haptoglobin, fibrinogen, plasminogen, α_1-antitrypsin, orosomucoid, β-lipoprotein, α_2-macroglobulin, and prealbumin in the human conceptus', *Journal of Clinical Investigation, 48*, 1433-46

Hamilton, P.J., Dawson, A.A., Ogston, D. and Douglas, A.S. (1974a) 'The effect of age on the fibrinolytic enzyme system', *Journal of Clinical Pathology, 27*, 326-9

Hamilton, P.J., Allardyce, M., Ogston, D., Dawson, A.A. and Douglas, A.S. (1974b) 'The effect of age upon the coagulation system', *Journal of Clinical Pathology, 27*, 980-2

Hathaway, W.E., Neumann, L., Borden, C.A. and Jacobsen, L. (1978) 'Immunologic studies of antithrombin III-heparin cofactor in the newborn', *Thrombosis and Haemostasis, 36*, 624-30

Hedner, U. and Nilsson, I.M. (1965) 'Determination of plasminogen in human plasma by a casein method', *Thrombosis et Diathesis Haemorrhagica, 14*, 545-61

Heikinheimo, R. (1964) 'Coagulation studies with fetal blood', *Biologia Neonatorum, 7*, 319-27

Henriksson, P., Hedner, U., Nilsson, I.M., Boehm, J., Robertson, B. and Lorand, L. (1974) 'Fibrin-stabilizing factor (factor XIII) in the foetus and the newborn infant', *Pediatric Research, 8*, 789-91

Henriksson, P., Wesström, G. and Hedner, U. (1979) 'Umbilical artery catheterization in newborns. III. Thrombosis — A study of some predisposing factors', *Acta Paediatrica Scandinavica, 68*, 719-23

Hilgartner, M.W. and Smith, C.H. (1965) 'Plasma thromboplastin antecedent (factor XI) in the neonate', *Journal of Pediatrics, 66,* 747-52

Holmberg, L., Henriksson, P., Ekelund, H. and Astedt, B. (1974) 'Coagulation in the human fetus. Comparison with term newborn infants', *Journal of Pediatrics, 85,* 860-4

Hrodek, O. (1969) 'L'agrégation des plaquettes chez le nouveau-né', *Nouvelle Revue Française d'Hématologie, 9,* 569-73

Jacobsen, C.D. (1964) 'The caseinolytic method for determination of plasminogen-plasmin in human plasma and serum', *Scandinavian Journal of Clinical Investigation, 16,* 372-82

Jeremic, M., Weisert, O. and Gedde-Dahl, T.W. (1976) 'Factor VIII (AHG) levels in 1016 regular blood donors', *Scandinavian Journal of Clinical and Laboratory Investigation, 36,* 461-6

Johnson, M., Ramey, E. and Ramwell, P.W. (1975) 'Sex and age differences in human platelet aggregation', *Nature, 253,* 355-7

Johnson, S.S., Montgomery, R.R. and Hathaway, W.E. (1981) 'Newborn factor VIII complex: elevated activities in term infants and alterations in electrophoretic mobility related to illness and activated coagulation', *British Journal of Haematology, 47,* 597-606

Kalpaktsoglou, P.K. and Emery, J.L.(1965) 'Human bone marrow during the last three months of intrauterine life. A histological study', *Acta Haematologica, 34,* 228-38

Kelton, J.G., Powers, P., Julian, J., Boland, V., Carter, C.J., Gent, M. and Hirsh, J. (1980) 'Sex-related difference in platelet aggregation: influence of the haematocrit', *Blood, 56,* 38-41

Kemona, H., Prokopowicz, J. and Wotosowicz, N. (1978) 'The count of blood platelets and sex in humans', *Experientia, 34,* 257

Korsan-Bengsten, K., Bengtsson, C. and Tibblin, E. (1973) 'Blood coagulation, fibrinolysis and platelet function in women aged 38, 46, 50, 54 and 60', *Acta Medica Scandinavica, 193,* 543-6

Kosztolányi, G., Jobst, K., Kellermeyer, M. and Ludány, A. (1980) 'ADP induced surface charge changes of adult and newborn platelets', *British Journal of Haematology, 46,* 257-62

Kurkuoglu, N. and McElfresh, A.E. (1960) 'The Hageman factor: determination of its concentration during the neonatal period and presentation of a case of Hageman factor deficiency', *Journal of Pediatrics, 57,* 61-5

Lieberman, J. (1959) 'Clinical syndromes associated with deficient lung fibrinolytic activity', *New England Journal of Medicine, 260,* 619-26

Lijnen, H.R., Jacobs, G. and Collen, D. (1981) 'Histidine-rich glycoprotein in a normal and a clinical population', *Thrombosis Research, 22,* 519-23

Maak, B., Scheidt, B. and Frenzel, J. (1978) 'Factor VIII activity and factor VIII related antigen in newborns', *European Journal of Pediatrics, 128,* 283-9

Mann, R.D. (1967) 'Effect of age, sex, diurnal variation on the human fibrinolytic system', *Journal of Clinical Pathology, 20,* 223-6

Markarian, M., Githens, J.H., Jackson, J.J., Bannon, A.E., Lindley, A., Rosenblut, E., Martorell, R. and Lubchenco, L.O. (1967) 'Fibrinolytic activity in premature infants', *American Journal of Diseases of Children, 113,* 312-21

Meade, T.W. and North, W.R.S. (1977) 'Population-based distributions of haemostatic variables', *British Medical Bulletin, 33,* 283-8

Meade, T.W., Chakrabarti, R., Haines, A.P., North, W.R.S. and Stirling, Y. (1979) 'Characteristics affecting fibrinolytic activity and plasma fibrinogen concentrations', *British Medical Journal, 1,* 153-6

Meade, T.W., Thompson, S.G. and Vickers, M.V. (1981) 'The epidemiology of platelet aggregation', *Thrombosis and Haemostasis, 46,* 29

Menon, I.S., Smith, P.A., White, R.W.B. and Dewar, H.A. (1967) 'Diurnal

variations of fibrinolytic activity and plasma-ll-hydroxycorticosteroid levels',
Lancet, ii, 531-3

Moser, K.M. and Hajjar, G.C. (1966) 'Age and disease-related alterations in fibrinogen-euglobulin (fibrinolytic) behavior', *American Journal of Medical Sciences, 251,* 536-44

Mull, M.M. and Hathaway, W.E. (1970) 'Altered platelet function in newborns', *Pediatric Research, 4,* 229-37

Muller, A.D., van Doorm, J.M. and Hemker, H.C. (1977) 'Heparin-like inhibitor of blood coagulation in normal newborn', *Nature, 267,* 616-7

Nordøy, A., Svensson, B., Haycraft, D., Hoak, J.C. and Wiebe, D. (1978) 'The influence of age, sex, and the use of oral contraceptives on the inhibitory effects of endothelial cells and PGI_2 (prostacyclin) on platelet function', *Scandinavian Journal of Haematology, 21,* 179-87

Nossel, H.L., Lanzkowsky, P., Levy, S., Mibashan, R.S. and Hansen, J.D.L. (1966) 'A study of coagulation factor levels in women during labour and in their newborn infants', *Thrombosis et Diathesis Haemorrhagica, 16,* 185-97

Ogston, C.M. and Ogston, D. (1966) 'Plasma fibrinogen and plasminogen levels in health and in ischaemic heart disease', *Journal of Clinical Pathology, 19,* 352-6

Pandolfi, M., Astedt, B., Cronberg, L. and Nilssom, I.M. (1972) 'Failure of foetal platelets to aggregate in response to adrenaline and collagen', *Proceedings of the Society for Experimental Biology and Medicine, 141,* 1081-3

Peake, I.R., Bloom, A.L., Giddings, J.C. and Ludlam, C.A. (1979) 'An immunoradiometric assay for procoagulant factor VIII antigen: results in haemophilia, von Willebrand's disease and fetal plasma and serum', *British Journal of Haematology, 42,* 269-81

Phillips, L.L. and Skrodelis, V. (1958) 'A comparison of the fibrinolytic enzyme system in maternal and umbilical-cord blood', *Pediatrics, 22,* 715-26

Pilgeram, L.O. (1961) 'Relation of plasma fibrinogen concentration changes to human arteriosclerosis', *Journal of Applied Physiology, 16,* 660-4

Pitney, W.R., Kirk, R.L., Arnold, B.J. and Stenhouse, N.S. (1962) 'Plasma anti-haemophilic factor (factor VIII) concentrations in normal families', *British Journal of Haematology, 8,* 421-8

Preston, A.E. and Barr, A. (1964) 'The plasma concentration of factor VIII in the normal population', *British Journal of Haematology, 10,* 238-45

Proud, D., Pierce, J.V. and Pisano, J.J. (1980) 'Radioimmunoassay of human high molecular weight kininogen in normal and deficient plasmas', *Journal of Laboratory and Clinical Medicine, 95,* 563-74

Quie, P.G. and Wannamaker, L.W. (1960) 'The plasminogen-plasmin system of newborn infants', *American Journal of Diseases of Children, 100,* 836-43

Robertson, B.R., Pandolfi, M. and Nilsson, I.M. (1972) 'Response of local fibrinolytic activity to venous occlusion of arms and legs in healthy volunteers', *Acta Chirurgica Scandinavica, 138,* 437-40

Roper, P., Drewinko, B., Hasler, D., Johnston, D., Hester, J. and Freireich, E.J. (1979) 'Effects of time, platelet concentration, and sex on the human platelet aggregation response', *American Journal of Clinical Pathology, 71,* 263-5

Rosing, D.R., Brakman, P., Redwood, D.R., Goldstein, R.E., Beiser, G.D., Astrup, T. and Epstein, S.E. (1970) 'Blood fibrinolytic activity in man. Diurnal variation and the response to varying intensities of exercise', *Circulation Research, 27,* 171-84

Rosing, D.R., Redwood, D.R., Brakman, P., Astrup, T. and Epstein, S.E. (1973) 'Impairment of the diurnal fibrinolytic response in man. Effects of ageing, type IV hyperlipoproteinemia, and coronary artery disease', *Circulation Research, 32,* 752-8

Saito, H. and Goldsmith, G.H. (1977) 'Plasma thromboplastin antecedent (PTA,

factor XI): A specific and sensitive radioimmunoassay', *Blood, 50,* 377-85
Saito, H., Ratnoff, O.D. and Pensky, J. (1976) 'Radioimmunoassay of human
  Hageman factor (factor XII)', *Journal of Laboratory and Clinical Medicine, 88,*
  506-14
Sawyer, W.D., Fletcher, A.P., Alkjaersig, N. and Sherry, S. (1960) 'Studies on the
  thrombolytic activity of human plasma', *Journal of Clinical Investigation, 39,*
  426-34
Schettini, F., de Mattia, D., Mautone, A. and Altomare, M. (1976a) 'Post-natal
  development of factor II (pre-prothrombin and prothrombin) in man', *Biology
  of the Neonate, 29,* 82-8
Schettini, F., de Mattia, D., Montagna, O. and Altomore, M. (1976b) 'Sviluppo
  postnatale dell' antithrombina III, del plasminogeno e dell' α_2-macroglobulina
  nell'uomo', *Rivista Italiana Pediatrie, 2,* 315-22
Sell, E.J. and Corrigan, J.J. (1973) 'Platelet counts, fibrinogen concentration, and
  factor V and factor VIII levels in healthy infants according to gestational age',
  Journal of Pediatrics, 82, 1028-32
Semenko, J.F. (1973) 'Seasonal changes in the fibrinolytic system of the blood of
  patients with coronary atherosclerosis', *Klinicheskaia Meditsina (Moskva), 51,*
  57-63
Simpson, N.E. and Biggs, R. (1962) 'The Inheritance of Christmas factor', *British
  Journal of Haematology, 8,* 191-203
Sloan, A.W. (1951) 'The normal platelet count in man', *Journal of Clinical
  Pathology, 4,* 37-46
Stevens, R.F. and Alexander, M.K. (1977) 'A sex difference in the platelet count',
  British Journal of Haematology, 37, 295-300
Stuart, M.J. (1978) 'The neonatal platelet: evaluation of platelet malonyl
  dialdehyde formation as an indicator of prostaglandin synthesis', *British Journal
  of Haematology, 39,* 83-90
Sveger, T. (1979) 'Antithrombin III in adolescents', *Thrombosis Research, 15,*
  885-8
Teger-Nilsson, A.C. (1975) 'Antithrombin III in infancy and childhood', *Acta
  Paediatrica Scandinavica, 64,* 624-8
Teger-Nilsson, A.C. (1979) 'Use of chromogenic substrates for screening of
  inhibitors of coagulation and fibrinolysis in patients', in M.F. Scully and V.V.
  Kakkar (eds.) *Chromogenic Peptide Substrates: Chemistry and Clinical Usage,*
  Churchill Livingstone, Edinburgh, pp. 269-76
Tracy, P.B., Peterson, J.M., Nesheim, M.E., Katzmann, J.A. and Mann, K.G.
  (1981) 'Human plasma and platelet factor V levels as measured by
  radioimmunoassay', *Thrombosis and Haemostasis, 46,* 89
Tromp, S.W. (1972) 'Influence of weather and climate on the fibrinogen content of
  human blood', *International Journal of Biometeorology, 16,* 93-5
Ts'ao, C., Green, D. and Schultz, K. (1976) 'Function and ultrastructure of
  platelets of neonates: Enhanced ristocetin aggregation of neonatal platelets',
  British Journal of Haematology, 32, 225-32
Tunstall, A.M., Merriman, J.M.L., Milne, I. and James, K. (1975) 'Normal and
  pathological serum levels of α_2-macroglobulins in men and mice', *Journal of
  Clinical Pathology, 25,* 133-9
Walter, E., Birkholz, L.-B., Harenberg, J. and Weber, E. (1979) 'Circadian
  rhythms of platelet function, fibrinolytic activity of plasma, and the influence of
  reduced food uptake, fat loading and beer drinking on fibrinolytic activity', in
  G.G. Neri Serneri and C.R.M. Prentice (eds.) *Haemostasis and Thrombosis,*
  Academic Press, London, pp. 525-30
Weisert, O. and Jeremic, M. (1974) 'Plasma fibrinogen levels in 1,016 regular
  blood donors. I. The influence of age and sex on mean values and percentiles',

Vox Sanguinis, 27, 176-85
Weissbach, G., Domula, M., Lenk. H. and Schneider, P. (1974) 'The progressive antithrombin activity and its relations to other factors of the coagulation system in newborns', *Acta Paediatrica Scandinavica, 63*, 555-61
Whaun, J.M. (1973) 'The platelet of the newborn infant: 5-hydroxytryptamine uptake and release', *Thrombosis et Diathesis Haemorrhagica, 30*, 327-33
Wilcox, G.M., Hial, V., Horwitz, D., Pisano, J.J. and Keiser, H.R. (1978) 'Effect of orthostatic changes in urokinase excretion', *Thrombosis and Haemostasis, 39*, 69-73
Zilliacus, H., Ottelin, A.-M. and Mattsson, T. (1966) 'Blood clotting and fibrinolysis in human foetuses', *Biology of the Neonate, 10*, 108-12

8 Ethnic and Blood Group Variations in the Hemostatic Mechanism

Ethnic Influences on the Hemostatic Mechanism

The variations in incidence of thrombotic disorders between different ethnic groups have stimulated many studies of aspects of the hemostatic system in such groups. Firm conclusions on racially determined differences in platelet numbers and function, coagulation and fibrinolysis are lacking, principally because of the problems in eliminating the effect of factors which themselves influence hemostatic components and are inextricably interrelated with the ethnic group studied: these include diet, climate, body build, exercise, parasitic infestation and other diseases, and mental stress in volunteers unaccustomed to such procedures as venepuncture.

Platelet Numbers and Function

Information on possible variations in platelet numbers and function between different races is fragmentary. Peasant farmers in Nigeria were found to have significantly lower platelet counts than Nigerians of administrative and professional classes (Essien *et al.*, 1973): both groups had lower counts than Caucasians. Possible causes considered were the prevalence of malaria and the use of medicinal herbs. As a possible compensation for the lower platelet numbers the availability of platelet factor 3 was greater in the Nigerian farmers. Gill and associates (1979) studied 200 consecutive blood donors in Zambia who were well-nourished and in good health and found a low mean platelet count of $122,000/\mu l$ as measured by an electronic thrombocounter: concurrent measurements on a Caucasian population were not carried out. In a study on employees of a company in north-west London 141 black volunteers, 132 born in the West Indies and nine in Africa, were compared with age-matched whites in respect of a number of hemostatic variables (Meade *et al.*, 1978): the mean platelet count was higher in the white subjects, but there was no

266

difference in platelet adhesiveness. Shaper and his colleagues (1966) found no difference in platelet adhesiveness between indigenous Africans and Asians in Uganda. In a further study from East Africa Ferguson *et al.,* (1970) noted that the platelet count of young Africans and Asians in Kenya and that of young Europeans from Glasgow did not differ significantly, while platelet aggregation, as assessed in the Chandler tube, was inhibited by a saturated fat meal in the Europeans, but was unchanged in the Africans and Asians. Mackie and associates (1982) reported that platelet aggregation in response to collagen, ADP and ristocetin was significantly less in Negro subjects than in age and sex-matched Caucasians.

The importance of diet as an influence on hemostatic components in particular ethnic groups is well-illustrated by the Eskimos from Greenland who have been found to have longer bleeding times and reduced platelet aggregability in response to ADP and collagen (Dyerberg and Bang, 1979): this has been attributed to the high proportion of ω-3 polyunsaturated fatty acids such as 5,8,11,14,17-eicosapentaenoic acid in their platelets, consequent on the high content of such fatty acids in their diet. In comparison with age-matched Danish controls the Eskimos had similar platelet adhesiveness to glass, but had a significantly lower mean platelet count. The cause and ethnic relevance of this latter difference is uncertain.

The Coagulation System

A variety of tests of overall coagulability have been performed on groups of subjects of different ethnic origin. The whole blood clotting time in siliconised tubes, probably a crude reflection of the quantity of activated clotting factors in the circulating blood itself or generated at the time of blood withdrawal, did not differ significantly between Bantus and whites (Merskey *et al.,* 1960). Goldrick and White (1958), however, did find significantly faster whole blood clotting times in New Guinean natives than in Australians. Booth and MacGregor (1965) also reported shorter mean clotting times in Melanesians than in Caucasians, while Grace and colleagues (1970) observed shorter plasma recalcification times in the New Guinean natives. The general pattern of increased coagulability in the Melanesian was maintained by the finding of shorter 'Stypven' times in the fasting state than those of white Australians (Goldrick, 1961).

The one-stage prothrombin time, providing an index of the level of the components participating in the extrinsic coagulation pathway, has been found not to differ between Bantus and white South

Africans (Merskey *et al.*, 1960); to be slightly longer in Nigerians compared to Europeans (Howell, 1965); to be significantly longer in Melanesians (Booth and MacGregor, 1967) with a number of subjects with substantially prolonged times, possibly related to the prevalent liver disease in the population studied. Africans from Kenya also had significantly longer prothrombin times than Europeans (Barr *et al.*, 1973).

Some information is also available on the level of individual coagulation factors in different ethnic groups. In the study from South Africa by Merskey and co-workers (1960) the Bantus had lower levels of prothrombin and factor VII than the whites, but higher levels of factor VIII and the Bantus produced a greater amount of thromboplastin in the thromboplastin generation test. There was no difference between the races in the level of factor V. Australian aborigines have also been noted to have higher plasma factor VIII levels (Pitney and Elliott, 1960). In their study of workers in a London company Meade and his associates (1978) measured the plasma level of a number of clotting factors: the mean factor VII level was significantly higher in the 351 white men than in the 86 black men while the mean factor VIII level was higher in the black men. In further accord with the findings of Merskey and associates there was no difference in factor V levels. Black women also had significantly higher factor VIII levels than white women, but the level of factor VII did not differ between the two groups of women. Factor VIII:C levels have also been found to be high in a Nigerian population (Essien and Ayeni, 1978). In the study reported by Mackie *et al.* (1982) healthy Negro subjects were found to have higher factor VIII:C activity than age- and sex-matched whites whereas factor VIIIR:Ag and ristocetin cofactor levels were similar. In contrast to these studies Barr *et al.* (1973) found that factor VIII coagulant activity in Kenyan Africans were lower than in the Europeans; factor V levels were also higher in Europeans although the difference only reached statistical significance in women. Factor X assays were also performed, but no differences between the races were found. Prothrombin, assessed by a two-stage prothrombin time technique, was adjudged to be lower in Korean farmers on a low fat diet than in American soldiers (Lee *et al.*, 1966).

Plasma fibrinogen levels did not differ significantly between South African Bantus and whites (Merskey *et al.*, 1960), between Nigerians and British whites (Howell, 1963), between Korean farmers and American whites (Lee *et al.*, 1966) or between blacks and whites resident in London (Meade *et al.*, 1978), but Thai subjects had lower fibrinogen concentrations than American whites in Thailand (Visudhiphan *et al.*, 1982). Eskimos from Greenland were found to have higher plasma fibrinogen levels than a Danish group (Dyerberg and Bang, 1979).

Data on coagulation inhibitors in different ethnic groups is scanty: Visudhiphan and colleagues (1982) have reported that Thai men and women have higher levels of antithrombin III than Americans living in Thailand.

Fibrinolytic Enzyme System

Fibrinolytic Activity. Many studies have been reported on measurements of blood or plasma fibrinolytic activity using techniques which are primarily an assay of plasminogen activator in various ethnic groups. Practically all investigators have agreed that Bantu men possess greater fibrinolytic activity than South African whites (Gillman *et al.*, 1957; Merskey *et al.*, 1960; Walker, 1961; Franz *et al.*, 1961; Lackner and Sougin-Mibashan, 1964; Lackner and Johnson, 1967): the sole discordant finding was that of Antonis and Bersohn (1961) who were unable to obtain clear evidence for a racial difference between white and Bantu men on similar diets. Studies in East Africa have also shown increased fibrinolytic activity in the indigenous Africans compared to North Americans or Europeans (Shaper *et al.*, 1966; Ferguson *et al.*, 1970; Barr *et al.*, 1973). Yoruba tribesmen from Nigeria had greater fibrinolytic activity than white Britons in plasma tested on heated fibrin plates, but there was no difference in the dilute whole blood clot lysis time (Howell, 1963). Essien (1976) measured the euglobulin clot lysis time in a group of indigenous Nigerians and claimed that they had enhanced fibrinolysis when compared to the values reported from a Swedish laboratory. Black people of African origin tested in Britain appear to retain their increased fibrinolytic activity (Menon, 1968), even after residence and a similar diet to Britons for four years (Menon, 1967).

While there has been general agreement that indigenous African men have higher blood fibrinolytic activity than whites, some of the studies which have included women have found that the racial difference is not found in females. Walker (1961) reported that the fibrinolytic activity of Bantu women was lower than that of white women, and Essien (1976) noted that Nigerian women had longer lysis times than the men. There was no sex difference in Africans from Kenya (Barr *et al.*, 1973), and Meade and associates (1978) found no difference between black and white women living in London.

Some possible causes for the difference in fibrinolytic activity between Africans and Europeans have been examined. Gillman *et al.*, (1957) observed that fibrinolysis in Bantu men was particularly high in the presence of biopsy-established liver disease. Unrecognised liver disease may have contributed to the increased fibrinolytic activity in Bantus noted in other studies (Merskey *et al.*, 1960). Differences in physical activity and diet may also be a factor. For example, Walker (1961) found that fibrinolytic activity was greater in the rural Bantu consuming a diet low in animal protein and fat than in his urban compatriot eating a partially westernised diet. Overall, however, there seems reasonable grounds for concluding that the active fibrinolytic activity of the male African negro is a racial rather than solely an environmental or acquired characteristic. Indeed, it may be that it is the European's lesser fibrinolytic activity which is determined by dietary and other environmental factors. Such a conclusion is supported by observations on black and white adults in the United States who were comparable in age, weight, height and physical activity, but who differed in socio-economic status (Szczelik *et al.*, 1980); there was significantly greater plasma fibrinolytic activity in the black Americans, but fibrinolytic activity tended to decrease with rising social class.

There have been reports to indicate that Asians resident in Africa have blood fibrinolytic activity which does not deviate markedly from Europeans' values (Franz *et al.*, 1961) or is significantly less than that of indigenous Africans (Shaper *et al.*, 1966; Ferguson *et al.*, 1970). Indians and Pakistanis living in Britain also had significantly less fibrinolytic activity than West Africans, but similar lysis times to white Britons (Menon, 1968). In the Asians studied in Uganda lysis times were inversely related to skinfold thickness, a relationship not seen in the Africans (Shaper *et al.*, 1966). Information on Chinese people is very scanty, but Menon (1968) noted that four studied in Britain had much longer euglobulin clot lysis times than native Britons. Korean farmers had significantly shorter lysis times than white American soldiers stationed in Korea, but the ethnic relevance is again controversial in view of the very low dietary fat intake by the Koreans (Lee *et al.*, 1966). Thai men and women have also been found to have higher fibrinolytic activity than American whites resident in Thailand (Visudhiphan *et al.*, 1982), a difference attributed to the frequent ingestion of capsicum by Thai people (see page 313).

The fibrinolytic activity of the Melanesian people of New Guinea has also been the subject of investigation. Goldrick (1961) found them to have more rapid clot lysis times and observed that their fibrinolytic activity was positively related to body bulk reflecting muscle mass, but not, as in Europeans, inversely related to adiposity. In a later study by Grace and his colleagues (1970) the New Guineans had increased fibrinolytic activity as compared to young Australians when the dilute whole blood clot lysis time was used to assess fibrinolysis, but, inexplicably, not when the euglobulin clot lysis time was used. In this study there was no relationship between fibrinolytic activity and muscularity, but there was a significant inverse relationship between body fat and fibrinolysis: the authors suggest that differences in adiposity may account for much of the observed racial differences in fibrinolysis. Booth and MacGregor (1967) also reported that fibrinolytic activity, as assessed by the dilute whole blood clot lysis time, was higher in New Guineans. Liver disease is apparently widespread in Melanesians; presumably, as in the Bantu, this could be a contributory factor in the increased fibrinolysis.

Plasminogen. Measurement of other components of the fibrinolytic enzyme system in different racial groups has been reported much less frequently than overall blood or plasma fibrinolytic activity, but a number of observations on plasma plasminogen levels are available. Higher plasma concentrations of plasminogen were recorded in Bantus from South Africa (Lackner and Johnson, 1967) and in East Africans (Ferguson *et al.*, 1970; Barr *et al.*, 1973) in comparison to their European counterparts, but New Guinean Melanesians and Caucasians had similar levels (Booth and MacGregor, 1967). Rimon and Segal (1964) compared two ethnic groups living in Israel, the Ashkenazi Jews originating in Central and Eastern Europe and the Yemenite Jews originating in the Arabian Peninsula, in respect of the streptokinase-induced proteolytic activity in the euglobulin fraction of plasma. This technique effectively measures the plasma plasminogen concentration. The Yemenite Jews showed an increase in this proteolytic activity over the early years of life, but the adult levels were not markedly different from the values found in the Ashkenazim.

Inhibitors of Fibrinolysis. A number of the studies showing increased fibrinolytic activity in Bantus or Melanesians have also found reduced antiplasmin levels when compared to Caucasians, although some of the methods used do not clearly distinguish between plasmin and urokinase inhibition (Howell, 1965; Booth and MacGregor, 1967; Barr *et al.*, 1973). In contrast, others have been unable to detect any difference in inhibitor levels between Bantus and whites (Lackner and Johnson, 1967) or between Korean farmers and Americans (Lee *et al.*, 1966). In the study comparing black and white people in the United States overall antiplasmin levels were significantly less in the black group and they had lower serum concentrations of α_2-macroglobulin and α_1-antitrypsin (Szczeklick *et al.*, 1980). Further information on individual inhibitors of plasmin, in particular, α_2-antiplasmin, are not yet available. Yemenite Jews differed from Ashkenazim in showing a continued increase in antiplasmin levels after the age of 30 (Rimon and Segal, 1964).

Influence of Blood Group on Hemostatic Components

There is variation in the blood group frequencies between different ethnic groups while some reports have suggested an association between blood group and hemostatic component concentrations. Such an association may account, therefore, for part of the differences observed between races. For this reason this small section on blood groups and hemostatic factors is linked to the consideration of possible ethnic influences on hemostasis. Additionally, however, the association of particular blood groups with altered levels on hemostatic components could account for part of the observed variation in the levels of coagulation and other hemostatic variables within a single ethnic group.

The initial observation of an association between blood group and the level of a coagulation factor was that of Preston and Barr (1964) who reported lower levels of factor VIII in persons of blood group O than in those of blood group A. A further group of subjects from the same center gave a similar result (Kerr *et al.*, 1966). Korsan-Bengtsen and associates (1972) also found less factor VIII coagulant activity in men of group O than in men of groups A, B and AB, but there was no difference in the factor VIII level between women of different ABO blood group. Further confirmation of the lower factor VIII concentration in individuals of blood group O have come from the studies of Jeremic *et al.* (1976) and Colonia and Roisenberg (1979). Blood group O individuals have also been reported to have significantly lower factor VIIIR:Ag levels (Stormorken and Eriksson, 1977). Orstravik and Magnus (1981) have confirmed the lower level of factor VIIIR:Ag in those of blood group O and have found that within this blood group individuals with Lewis type Le(a—b—) have a much lower concentration of factor VIIR:Ag than those with types Le(a—b+) or Le(a+b—).

Other coagulation factor-blood group associations have been less conclusive. Ambrus and Mink (1970) reported that women of group A had slightly enhanced thromboplastin generation when compared with those of group O, but there was no difference between the groups in the factor II-VII-X complex, factor V, factor IX or fibrinogen. In the study of Colonia and Roisenberg (1979) both Caucasian and Negroid donors in Brazil were examined and both racial and blood group effects delineated. While there were no differences in the levels of prothrombin, the factor II-VII-X complex, factor V or fibrinogen between races or blood groups, there was significant lengthening of the plasma recalcification time, partial thromboplastin time, kaolin-activated partial thromboplastin time, and the *in vitro* heparin anticoagulation effect in group O individuals, although the changes in the recalcification time, kaolin-activated partial thromboplastin time and heparin anticoagulation effect were only significant in the Negroids. Higher levels of antithrombin III in subjects of group A than those of group O have been recorded (Fagerhol *et al.*, 1971).

Neither Ambrus and Mink (1970) nor Dawson and colleagues (1976) found any association between blood group and the platelet count: platelet adhesiveness also showed no linkage with blood group (Ambrus and Mink, 1970).

There has been concordance of view that the level of components of the fibrinolytic enzyme system are not associated with the blood group. The measurements have included the euglobulin clot lysis time (Ambrus and Mink, 1970; Essien, 1976; Colonia and Roisenberg, 1979), plasminogen (Ambrus and Mink, 1970; Dawson *et al.*, 1976), antiplasmin activity (Ambrus and Mink, 1970) and α_2-macroglobulin (Dawson *et al.*, 1976).

References

Ambrus, M.S. and Mink, I.B. (1970) 'ABO blood types and factors of the blood coagulation and fibrinolysis systems in women', *Journal of Medicine*, *1*, 236-41

Antonis, A. and Bersohn, I. (1961) 'Blood fibrinolytic activity in white and Bantu subjects', *Lancet, i*, 401-2

Barr, R.D., Ouna, N. and Kendall, A.G. (1973) 'The blood coagulation and fibrinolytic enzyme systems in healthy adult Africans and Europeans — a comparative study', *Scottish Medical Journal, 18*, 93-7

Booth, P.B. and MacGregor, A. (1965) 'Fibrinolytic activity and fibrinogen levels in Papuans and New Guineans', *Medical Journal of Australia, 2*, 405-9

Booth, P.B. and MacGregor, A. (1967) 'Comparisons of fibrinolysis and blood coagulation in Melanesians and Caucasians', *British Journal of Haematology, 13*, 779-89

Colonia, V.J. and Roisenberg, I. (1979) 'Investigation of associations between ABO blood groups and coagulation, fibrinolysis, total lipids, cholesterol and triglycerides', *Human Genetics, 48*, 221-30

Dawson, A.A., Allardyce, M., Allan, T.M., Ogston, D., Kerridge, D.F. and Lewis, H.B.M. (1976) 'Inter-relations between ABO blood group, plasminogen, $alpha_1$-antitrypsin, $alpha_2$-macroglobulin and the platelet count in blood donors', *Acta Haematolologica, 56*, 19-26

Dyerberg, J. and Bang, H.O. (1979) 'Haemostatic function and platelet polyunsaturated fatty acids in Eskimos', *Lancet, ii*, 433-5

Essien, E.M. (1976) 'Enhanced fibrinolysis in Nigerians — probable contributory factor to low prevalence of atherosclerosis in the Nigerian', *African Journal of Medicine and Medical Sciences, 5*, 287-90

Essien, E.M. and Ayeni, O. (1978) 'Factor VIII coagulant activity in an African population in relation to a recognized standard', *British Journal of Haematology, 29*, 225-31

Essien, E.M., Usanga, E.A. and Ayeni, O. (1973) 'The normal platelet count and platelet factor 3 availability in some Nigerian population groups', *Scandinavian Journal of Haematology, 10*, 378-83

Fagerhol, M.K., Abildgaard, U. and Kornstad, L. (1971) 'Antithrombin-III concentration and ABO blood groups', *Lancet, ii*, 664

Ferguson, J.C., Mackay, N. and McNicol, G.P. (1970) 'Effect of feeding fat on fibrinolysis, Stypven time, and platelet aggregation in Africans, Asians, and Europeans', *Journal of Clinical Pathology, 23*, 580-5

Franz, R.C., Kark, A.E. and Hathorn, M. (1961) 'Postoperative thrombosis and plasma fibrinolytic activity. A comparative study in Africans, Indians, and Whites', *Lancet, i*, 195-7

Gill, G.V., England, A. and Marshal, C. (1979) 'Low platelet counts in Zambians', *Transactions of the Royal Society of Tropical Medicine and Hygiene, 73*, 111-3

Gillman, T., Naidoo, S.S. and Hathorn, M. (1957) 'Fat, fibrinolysis, and atherosclerosis in Africans', *Lancet, ii*, 696-7

Goldrick, R.B. (1961) 'Fibrinolysis, blood clotting, serum lipids and body build in natives of New Guinea and Australians', *Australasian Annals of Medicine, 10*, 20-8

Goldrick, R.B. and Whyte, H.M. (1958) 'Serum lipoproteins and blood coagulation', *Nature, 182*, 1743-4

Grace, C.S., Sinnett, P.F. and Whyte, H.M. (1970) 'Blood fibrinolysis and coagulation in New Guineans and Australians', *Australian Annals of Medicine, 4*, 328-33

Howell, M. (1965) 'Comparison of fibrinolysis and coagulation in Nigerian and European men', *Journal of Atherosclerosis Research, 5*, 80-9

Jeremic, M., Weisert, O. and Gedde-Dahl, T.W. (1976) 'Factor VIII (AHG) levels in 1016 regular blood donors. The effects of age, sex, and ABO blood groups', *Scandinavian Journal of Clinical and Laboratory Investigation, 36*, 461-6

Kerr, C.B., Preston, A.E., Barr, A. and Biggs, R. (1966) 'Further studies on the inheritance of factor VIII', *British Journal of Haematology, 12*

Korsan-Bengtsen, K., Wilhelmsen, L., Nilsson, L.-A. and Tibblin, G. (1972)

'Blood coagulation and fibrinolysis in relation to ABO, Rh, MN, and Duffy blood groups in a random population sample of men aged 54 years', *Thrombosis Research, 1*, 549-58

Lackner, H. and Johnson, A.J. (1967) 'The fibrinolytic system in South African white and Bantu men', *Thrombosis et Diathesis Haemorrhagica, 18*, 456-61

Lackner, H. and Sougin-Mibashan, R. (1964) 'Fibrinolysis and alimentary lipaemia in whites and Bantus: their relationship and response to intravenous heparin', *Thrombosis et Diathesis Haemorrhagica, 11*, 108-18

Lee, K.T., Kim, D.N., Keokarn, Y. and Thomas, W.A. (1966) 'Geographic pathology of atherosclerosis and thrombosis. Coagulation and clot-lysis phenomena in Koreans on a low-fat diet', *Journal of Atherosclerosis Research, 6*, 203-13

Mackie, I., Bull, H., Kennedy, J. and Brozović, M. (1982) 'Ristocetin-induced platelet agglutination in Afro-Caribbean and Caucasian people', *British Journal of Haematology, 50*, 171-3

Meade, T.W., Brozović, M., Chakrabarti, R., Haines, A.P., North, W.R.S. and Stirling, Y. (1978) 'Ethnic group comparisons of variables associated with ischaemic heart disease', *British Heart Journal, 40*, 789-95

Menon, I.S. (1967) 'Fibrinolytic activity in the blood of Nigerian students after four years' residence in the United Kingdom', *Laboratory Practice, 16*, 574-7

Menon, I.S. (1968) 'Fibrinolytic activity in the blood of 99 healthy young British, African and Asian subjects', *Laboratory Practice, 17*, 194-5

Merskey, C., Gordon, H., Lackner, H., Schrire, V., Kaplan, B.J., Sougin-Mibashan, R., Nossel, H.L. and Moodie, A. (1960) 'Blood coagulation and fibrinolysis in relation to coronary artery disease. A comparative study of normal white men, white men with overt coronary heart disease, and normal Bantu men', *British Medical Journal, 1*, 219-27

Orstavik, K.H. and Magnus, P. (1981) 'Association between plasma concentration of factor VIII related antigen, ABO blood type and Lewis blood type', *Thrombosis and Haemostasis, 46*, 126

Pitney, W.R. and Elliott, M.H. (1960) 'Plasma antihaemophilic-factor concentrations in the Australian aborigine and in conditions associated with hyperglobulinaemia', *Nature, 185*, 397

Preston, A.E. and Barr, A. (1964) 'The plasma concentration of factor VIII in the normal population. II. The effects of age, sex and blood group', *British Journal of Haematology, 10*, 238-45

Rimon, A. and Segal, S. (1964) 'Streptokinase-induced proteolytic activity with respect to age in two ethnic groups in Israel', *Thrombosis et Diathesis Haemorrhagica, 11*, 119-26

Shaper, A.G., Jones, K.W., Kyobe, J. and Jones, M. (1966) 'Fibrinolysis in relation to body fatness, serum lipids and coronary heart disease in African and Asian men in Uganda', *Journal of Atherosclerosis Research, 6*, 313-27

Stormorken, H. and Erikssen, J. (1977) 'Plasma antithrombin III and factor VIII antigen in relation to angiographic findings, angina and blood groups in middle-aged men', *Thrombosis and Haemostasis, 38*, 874-80

Szczeklick, A., Dischinger, P., Kueppers, F., Tyroler, H.A., Hames, C.G., Cassel, J.C. and Creagan, S. (1980) 'Blood fibrinolytic activity, social class, and habitual physical activity. II. A study of black and white men in Southern Georgia', *Journal of Chronic Diseases, 33*, 291-9

Visudhiphan, S., Poolsuppasii, S., Piboonnukarintr, O. and Tumliang, S. (1982) 'The relationship between high fibrinolytic activity and daily capsicum ingestion in Thais', *American Journal of Clinical Nutrition, 35*, 1452-8

Walker, A.R.P. (1961) 'Fibrinolytic activity of whole blood from South African Bantu and white subjects', *American Journal of Clinical Nutrition, 9*, 461-72

9 Environmental Influences on the Hemostatic Mechanism

Man's quest for natural resources and his appetite for ever more arduous recreation have led to increasingly frequent exposure to extremes of pressure and temperature. In addition, the severe winters of many countries claim a high toll from accidental hypothermia, particularly in the elderly. While some of the patho-physiological effects of environmental hazards have received extensive study, their influence on the hemostatic mechanism have had relatively scant attention in spite of hints from clinical observations that hemostatic function can be altered in the presence of high ambient pressure, hypothermia and hyperthermia. This chapter reviews the available information on the influences of cold, heat, high altitude and compression-decompression on components of the hemostatic mechanism.

Effect of Environmental Cold on Hemostasis

Data on the influence of environmental cold on the hemostatic system of man are scanty although some studies on the influence of cold on platelet function in *in vitro* systems have been reported. Platelets in blood chilled to 0°C for ten minutes were shown to undergo shape change with transition from discoid to spherical shape (Zucker and Borrelli, 1954) and reversion to normal on rewarming to 37°C. This shape change on exposure to low temperatures was found by White and Krivit (1967) to be related to the depolymerisation and disappearance of the marginal bundle of microtubules: on rewarming the tubules and the discoid shape returned. Chilling platelets for 24 hours resulted in essentially irreversible morphological changes (Behnke, 1967). Platelets incubated at 0°C undergo spontaneous aggregation (Zucker and Borrelli, 1960; Anstall and Hawkey, 1962). Kattlove and Alexander (1971) found that platelets chilled to between 4 and 6°C aggregated spontaneously when warmed and stirred: this did not appear to be the result of ADP release from the chilled platelets. In a further study Kattlove and associates (1972) compared platelet-rich plasma chilled at 0°C or 6°C for four to 20 hours with that retained at 23°C and observed enhanced platelet aggregation in response to collagen, epinephrine and ADP after rewarming for one hour at 37°C: the rewarmed platelets also showed greater retention in glass bead columns. These observations on platelets chilled *in vitro* to these low temperatures could have no physiological significance except perhaps in localised circulations exposed to severe cold.

Few studies have been carried out on patients whose core temperature has fallen to hypothermic levels. In a report by Chadd and Gray (1972) babies with a rectal temperature of 34°C or less shortly after birth had a prolonged thrombin time, a higher level of fibrin degradation products and thrombocytopenia. Deep hypothermia (20 to 25°C) in experimental animals is accompanied by a decrease in the number of circulating platelets (Helmsworth *et al.*, 1955; Willson *et al.*, 1958; Wensel and Bigelow, 1959; Johansson and Nilsson, 1964): on rewarming the platelet count returns rapidly to normal levels (Villalobos *et al.*, 1955) indicating that intravascular consumption of platelets does not occur. Pina-Cabral and colleagues (1973/74), using ^{51}Cr-labelled platelets and external scanning, found that radioactivity increased over the hepatic area during cooling of dogs to 20°C and decreased during rewarming to 36°C: they concluded that the liver is the main organ in the dog responsible for platelet sequestration during hypothermia. Dogs subjected to cooling to 20°C were shown to develop increased fibrinolytic activity and decreases in factor VII, prothrombin and fibrinogen, changes attributed to consumption coagulopathy (Johansson and Nilsson, 1964). Dodman and colleagues (1973) found that volunteers exposed to an environment of 4°C for 15 minutes without a fall in the central body temperature or shivering had no change in their plasma fibrinolytic activity whereas localised cold increased fibrinolytic activity in the experimental arm. No other observations on the influence of environmental cold and such consequential bodily responses as shivering on platelet numbers and function, coagulation and fibrinolysis can be traced.

Effect of Environmental Heat on Hemostasis

The influence of environmental heat on hemostatic function and components has also received relatively little attention. Bedrak and co-workers (1964) showed that fibrinolytic activity increased in resting subjects exposed to an environment of 40 to 44°C for two hours: in heat-acclimatised subjects the increase in fibrinolysis was less and the effects of heat and physical activity were additive. Increase in the whole body temperature induced by immersion in a hot bath at 45°C for ten minutes increased plasma fibrinolytic activity (Dodman *et al.*, 1973): local heat also resulted in increased fibrinolysis in the blood withdrawn from the experimental arm. The effect of heat on fibrinolysis was attributed by Bedrak *et al.* (1964) to release of epinephrine, but in the study by Britton and associates (1974) the marked increase in fibrinolytic activity after exposure to heat in a sauna bath was not associated with an alteration in the plasma epinephrine level. It was proposed that the increased fibrinolysis resulted from vasodilatation and the consequential exposure of the vascular endothelium to increased blood flow. In the experiments of Britton and colleagues there were no changes in the factor VIII level.

In vitro studies on the effect of heat on platelets have been reported by White (1968). Platelets in plasma exposed to temperatures above 43°C for 15 minutes lost their discoid shape and failed to aggregate in response to ADP or thrombin as assessed by the 'snowstorm' effect. In other studies (Ogston and Taylor, 1981) platelet-rich plasma exposed to temperatures between 37 and 45°C for 20 minutes showed a subsequent progressive decrease in aggregability in response to ADP, collagen and 5-hydroxytryptamine with increasing temperature. The platelet count of platelet-rich plasma heated at 43°C for 20 minutes did not differ from that of plasma maintained at 20°C. Whether platelets exposed *in vivo* to increased temperatures display altered aggregability is unknown. Patients suffering from heat stroke with marked hyperthermia frequently develop hemorrhagic features with coagulation changes compatible with disseminated intravascular coagulation, including thrombocytopenia (Meikle and Graybill, 1967; Weber and Blakely, 1969; Stefanini and Spicer, 1971), but it is likely that these changes are secondary to endothelial damage rather than to a direct effect on platelets or megakaryocytes (Sohal *et al.*, 1968). A single patient with heat stroke showed falls in the levels of factors VII, IX, X, XII, XIII and prothrombin and an increase in both factor VIII:C and VIIIR:Ag (Beard and Hickton,

278

1982): hepatic dysfunction with failure of protein synthesis was considered to be the cause of the reduction in clotting factor levels.

Effect of Altitude on Hemostasis

Studies on the influence of altitude and the associated reduced atmospheric pressure have been made on permanent residents in mountainous regions while the acute effects have been examined in non-residents shortly after exposure to high altitude.

Platelet Numbers and Function

Investigators have reached varying conclusions on the effect of acute exposure to natural or simulated altitude on the platelet count and function. Gray and colleagues (1975) studied 16 men decompressed for two hours in an altitude chamber to a simulated height of 6096 meters while breathing oxygen: they found a ten per cent, statistically significant, drop in the circulating platelet count for three days following decompression. Most of the subjects studied had a reduction in the platelet half-life compared to control values. In further experiments men exposed to 2990 meters and then 5365 meters in a mountain environment also showed significant decreases in the platelet count. On the basis of an increase in lung radioactivity in rabbits previously infused with homologous ^{51}Cr-labelled platelets it was postulated that altitude decompression results in platelet sequestration in the pulmonary vascular bed. In contrast, Maker and associates (1976) reported that exposure to a simulated altitude of 4400 meters without oxygen for one, 24 and 48 hours did not alter the platelet count; platelet aggregation in response to ADP, epinephrine or collagen; platelet factor 3 availability; or circulating platelet aggregates.

Singh and Chohan (1972a, b, 1974) have reported on a series of studies examining the influence of acute and chronic exposure to high altitude on their subjects' platelets and the relationship of the changes to the development of high altitude pulmonary edema and pulmonary hypertension. Soldiers airlifted to 3692 meters who did not develop pulmonary edema had a significant increase in their platelet count on the first day at altitude; there was also an increase in platelet factor 3 at that time, but this increase did not attain statistical significance until seven days after arrival at altitude. The platelet count remained significantly elevated over the 14-day period of study at altitude. Soliders who had stayed for two years at an altitude between 3692 and 5538 meters without developing high altitude pulmonary hypertension had elevated platelet counts, but no significant alteration in platelet factor 3 or platelet adhesiveness. Lack of any significant effect on acute induction to altitude was also noted by Sharma and Hoon (1978), although they did find increased adhesiveness in those who developed symptoms of high altitude sickness.

The effect of altitude on the platelet count has also been studied by Sharma (1980a, 1980b). He found a gradual rise in the platelet count in subjects taking up residence at an altitude of 3658 meters with stabilisation between 181 and 360 days when the values were close to those of permanent residents at that altitude. The rise over the first few days at altitude was small. Consistent with these reports was the finding by Caen *et al.* (1973/74) of a 20 per cent increase in the platelet count on adaptation to altitude.

Coagulation Mechanism

The first observations recorded on the effect of altitude on the coagulation mechanism were made by Hurtado (1932) who noted that natives of the Andes living at 4575 meters had a slightly shortened coagulation time, but both the technique and the method of obtaining control values lacked precision. In the study by Maker and co-workers (1976) on eight young men assessed at sea level and after one, 24 and 48 hours at a simulated altitude of 4400 meters there was no change in the thrombin or prothrombin times, but the mean partial thromboplastin time was significantly shortened after one and 24 hours at altitude. Both the plasma fibrinogen and factor VIII levels showed a significant reduction at one hour, but had returned to the pre-exposure values by 24 hours. These reductions could not be attributed to plasma volume changes since the hematocrit was unchanged. Three of the men had a transient rise in fibrin(ogen) degradation products as detected as immunoassay, raising the possibility that the coagulation changes had resulted from mild disseminated intravascular coagulation.

A more comprehensive range of specific coagulation assays was carried out in the series of studies reported by Singh and Chohan. On the day of arrival at an altitude of 3692 meters the subjects were found to have significantly increased levels of factors X and XII and the thrombin time was increased. In constrast, the factor VIII level was significantly decreased and the mean fibrinogen level fell slightly (Singh and Chohan, 1972a). Following arrival at high altitude, however, factor VIII rose to levels significantly higher than the control sea-level values by the seventh day. Factor XII maintained its increased level to the 14th day, the end of the study period, while the factor X level had declined to control values at that time. Factor V was unchanged (Singh and Chohan, 1974). The only significant alteration found in soldiers who had resided at altitudes of between 3658 and 5486 meters for two years and remained healthy was an increase in the plasma fibrinogen (Singh and Chohan, 1972b): the soldiers who had developed pulmonary edema had also significant increases in factors V and VIII.

Fibrinolytic System

On arrival at altitude (3692 meters) blood fibrinolytic activity, measured by the dilute blood clot lysis time, was found to be increased (Singh and Chohan, 1972a). This increase was maintained over the next two weeks (Singh and Chohan, 1974) and was still present in those studied after two years at high altitude (Singh and Chohan, 1972b) and in permanent residents at altitudes between 3692 and 5780 meters (Chohan *et al.*, 1974). Fibrinolytic activity in response to local venous stasis was not increased in Bolivians living at altitude in the Andes (Caen *et al.*, 1973/74).

The influence of hypoxia *per se* on fibrinolytic activity was examined by Cunningham and colleagues (1971): volunteers exposed to an inspired oxygen concentration of 11.5 per cent for 15 minutes, inducing a mean alveolar oxygen partial pressure of 7.2 kPascals, did not show any change in plasma fibrinolytic activity. This finding suggests that factors other than hypoxia cause the increased fibrinolysis observed at altitude.

The mechanism of the reported changes in hemostatic components in subjects at altitude are not established. There is agreement that exposure to high altitude results in an increase in plasma and urinary catecholamine levels (Hoon *et al.*, 1976; Malhotra *et al.*, 1976), but whether these result from altitude *per se*, hypoxia, cold, physical activity or mental stress is uncertain. In any event it is likely that increased sympatho-adrenal activity has a role in some of the coagulation/platelet/fibrinolysis changes seen in subjects exposed to high altitudes.

Effect of Compression-Decompression on Hemostasis

Possible changes in hemostatic function following exposure to low pressures and subsequent decompression have received considerable attention because of man's endeavours to explore the depths of the sea and the possibility that activation of the hemostatic system might contribute to the pathogenesis of decompression sickness.

Platelet Numbers and Function

A number of studies have demonstrated that decompression from a hyperbaric environment in man is followed by a phase of thrombocytopenia, the fall in the platelet count being maximal 48 to 72 hours post-decompression (Bennett and Gray, 1971; Martin and Nichols, 1972; Valeri *et al.*, 1974). The mechanism of the reduction in the platelet count is uncertain. One suggestion is that it may result from the adherence of platelets to intravascular bubbles and their consequent trapping as thrombi in the microcirculation (Philp *et al.*, 1972a). In later studies Philp and his associates (1974) observed that the fall in the platelet count following decompression was reduced by drugs which inhibited platelet function. In contrast to these findings, Valeri and colleagues (1974) found no decrease in platelet survival after hyperbaric exposure, while Weatherley and co-workers (1977) reported no change in the platelet count following decompression, but observed that some subjects' platelets did not aggregate in response to ADP.

Coagulation Mechanism

Changes in components of the coagulation system after pressurisation to 300 feet were reported by Philp *et al.* (1972b); there was a significant lengthening of the one-stage prothrombin time and a decrease in the prothrombin consumption time. Antithrombin III activity was decreased, but only in the divers who developed decompression symptoms. Thromboelastography showed no significant changes and there were also no differences in the partial thromboplastin time or in the factor V and VIII assays. In contrast, using experimental exposure to similar pressures, Bonin and colleagues (1973) found only minor alterations in coagulation assays. There has been the suggestion from studies by Jacey *et al.* (1977) that repeated hyperbaric exposure produces additive effects in relation to changes in hemostatic factors.

Hallenbeck and co-workers (1973) investigated the possibility

282

that intravascular bubbles influence coagulation in addition to their possible role in causing thrombocytopenia: using indirect techniques they concluded that bubbles do activate a plasma factor with resultant acceleration of coagulation, but factor XII activation, although proposed, was not directly implicated. An effect of compression-decompression on the contact activation mechanism is supported by the finding of reductions in the level of factors XI and XII in volunteers exposed to a simulated depth of 100 feet for one hour (Weatherley *et al.*, 1977).

Fibrinolytic System

In the study by Philp and colleagues (1972b) increased plasma fibrinolytic activity was found after decompression from a hyperbaric environment. Small increases in fibrin degradation products as assayed by the tanned red cell hemagglutination test were observed, and there was a small decrease in plasma plasminogen. Bonin and associates (1973) found slight increases in blood fibrinolytic activity at the end of decompression in about half the 12 subjects of the study, but there was no increase in fibrin(ogen) degradation products and no change in the plasminogen level.

References

Anstall, H.B. and Hawkey, C.M. (1962) 'Observations on platelet clumping and related phenomena', *Transfusion, 2,* 44-51

Beard, M.E.J. and Hickton, C.M. (1982) 'Haemostasis in heat stroke', *British Journal of Haematology, 52,* 269-74

Bedrak, E., Beer, G. and Furman, K.I. (1964) 'Fibrinolytic activity and muscular exercise in heat', *Journal of Applied Physiology, 19,* 469-71

Behnke, O. (1967) 'Some possible practical implications of the lability of blood platelet microtubules', *Vox Sanguinis, 13,* 502-7

Bennett, P.B. and Gray, S.P. (1971) 'Changes in human urine and blood chemistry during a simulated oxygen-helium dive to 1,500 feet', *Aerospace Medicine, 42,* 868-74

Bonin, B., Straub, P.W., Schibli, R. and Buhlmann, A.A. (1973) 'Blood coagulation during critical decompression following diving experiments with oxygen/helium', *Aerospace Medicine, 44,* 508-12

Britton, B.J., Hawkey, C., Wood, W.G., Peele, M., Kaye, J. and Irving, M.H. (1974) 'Adrenergic, coagulation, and fibrinolytic responses to heat', *British Medical Journal, 4,* 139-41

Caen, J.P., Drouet, L., Bellanger, R., Michel, H. and Hénon, P. (1973/74) 'Thrombosis, platelet behaviour, fibrinolytic activity, and diet on the Andes plateau', *Haemostasis, 2,* 13-20

Chadd, M.A. and Gray, O.P. (1972) 'Hypothermia and coagulation defects in the newborn', *Archives of Diseases in Childhood, 47,* 819-21

Chohan, I.S., Singh, I. and Balakrihson, K. (1974) 'Fibrinolytic activity at high altitude and sodium acetate buffer', *Thrombosis et Diathesis Haemorrhagica, 32*, 65-70

Cunningham, G.M., Boyde, G., Windebank, J., Moran, F. and McNicol, G.P. (1971) 'The effect of oxygen on the fibrinolytic enzyme system, *in vivo*', *Journal of Clinical Pathology, 24*, 705-7

Dodman, B., Cunliffe, W.J., Roberts, B.E. and Buchan, C.W. (1973) 'Effects of changes in temperature (local and central) on plasma fibrinolytic activity', *Journal of Clinical Pathology, 26*, 248-9

Gray, G.W., Bryan, A.C., Freedman, M.H., Houston, C.S., Lewis, W.F., McFadden, D.M. and Newell, G. (1975) 'Effect of altitude exposure on platelets', *Journal of Applied Physiology, 39*, 648-51

Hallenbeck, J.M., Bove, A.A., Moquin, R.B. and Elliott, D.H. (1973) 'Accelerated coagulation of whole blood and cell-free plasma by bubbling *in vitro*', *Aerospace Medicine, 44*, 712-4

Helmsworth, J.A., Stiles, W.J. and Elstun, W. (1955) 'Changes in blood cellular elements in dogs during hypothermia', *Surgery, 38*, 843-6

Hoon, R.S., Sharma, S.C., Balasubramanian, V., Chadha, K.S. and Mathew, O.P. (1976) 'Urinary catecholamine excretion on acute induction to high altitude (3,658 m)', *Journal of Applied Physiology, 41*, 631-3

Hurtado, A. (1932) 'Studies at high altitude. Blood observations on the Indian natives of the Peruvian Andes', *American Journal of Physiology, 100*, 487-505

Jacey, M.J., Gonzales, A. and Tappan, D.V. (1977) 'Hematologic changes after two exposures to 6.7 ATA air at three-day intervals', *Journal of Applied Physiology, 42*, 838-44

Johansson, B.W. and Nilsson, I.M. (1964) 'The effect of heparin and ε-aminocaproic acid on the coagulation in hypothermic dogs', *Acta Physiologica Scandinavica, 60*, 267-77

Kattlove, H.E. and Alexander, B. (1974) 'The effect of cold on platelets. I. Cold-induced platelet aggregation', *Blood, 38*, 39-48

Kattlove, H.E., Alexander, B. and White, F. (1972) 'The effect of cold on platelets. II. Platelet function after short-term storage at room temperature', *Blood, 40*, 688-96

Maker, J.T., Levine, P.H. and Clymerman, A. (1976) 'Human coagulation abnormalities during acute exposure to hyperbaric hypoxia', *Journal of Applied Physiology, 41*, 702-7

Malhotra, M.S., Selvamurthy, W., Purkayastha, S.S., Mukherjee, A.K., Mathew, L. and Dua, G.L. (1976) 'Responses of the autonomic nervous system during acclimatization to high altitude in man', *Aviation, Space, and Environmental Medicine, 47*, 1076-9

Martin, K.J. and Nichols, G. (1972) 'Observations on platelet changes in man after simulated diving', *Aerospace Medicine, 43*, 827-30

Meikle, A.W. and Graybill, J.R. (1967) 'Fibrinolysis and hemorrhage in a fatal case of heat stroke', *New England Journal of Medicine, 276*, 911-3

Ogston, D. and Taylor, A.K.M. (1981) 'The effect of heat on platelet aggregation', *Journal of Physiology, 310*, 71-2P

Philp, R.B., Inwood, M.J. and Warren, B.A. (1972a) 'Interactions between gas bubbles and components in the blood: implictions in decompression sickness', *Aerospace Medicine, 43*, 946-53

Philp, R.B., Ackles, K.N., Inwood, M.J., Livingstone, S.D., Achimastos, A., Binns-Smith, M. and Radomski, M.W. (1972b) 'Changes in the hemostatic system and in blood and urine chemistry of human subjects following decompression from a hyperbaric environment', *Aerospace Medicine, 43*, 498-505

284

Philp, R.B., Inwood M.J., Ackles, K.N. and Radomski, M.W. (1974) 'Effects of decompression on platelets and haemostasis in men and the influence of antiplatelet drugs (RA 233 and VK 744)', *Aerospace Medicine, 45*, 231-40

Pina-Cabral, J.M., Amaral, I., Macedo Pinto, M. and Guerra e Paz, L.H. (1973/74) 'Hepatic and splenic platelet sequestration during deep hypothermia in the dog', *Haemostasis, 2*, 235-44

Sharma, S.C. (1980a) 'Platelet count in temporary residents of high altitude', *Journal of Applied Physiology, 49*, 1047-8

Sharma, S.C. (1980b) 'Platelet count on acute induction to high altitude', *Thrombosis and Haemostasis, 43*, 24

Sharma, S.C. and Hoon, R.S. (1978) 'Platelet adhesiveness on acute induction to high altitude', *Thrombosis Research, 13*, 725-32

Singh, I. and Chohan, I.S. (1972a) 'Abnormalities of blood coagulation at high altitude', *Journal of Biometeorology, 16*, 283-97

Singh, I. and Chohan, I.S. (1972b) 'Blood coagulation changes at high altitude predisposing to pulmonary hypertension', *British Heart Journal, 34*, 611-7

Singh, L. and Chohan, I.S. (1974) 'Adverse changes in fibrinolysis, blood coagulation and platelet function in high altitude pulmonary oedema and their role in its pathogenesis', *Journal of Biometeorology, 18*, 33-45

Sohal, R.S. Sun, S.C., Colcolough, H.L. and Burch, G.E. (1968) 'Heat stroke. An electron microscopic study of endothelial cell damage and disseminated intravascular coagulation', *Archives of Internal Medicine, 122*, 43-7

Stefanini, M. and Spicer, D.D. (1971) 'Hemostatic breakdown, fibrinolysis, and acquired hemolytic anemia in a patient with fatal heatstroke', *American Journal of Clinical Pathology, 55*, 180-6

Valeri, C.R., Feingold, H., Zaroulis, C.G., Sphar, R.L. and Adams, G.M. (1974) 'Effects of hyperbaric exposure on human platelets', *Aerospace Medicine, 45*, 610-6

Villalobos, T.J., Adeson, E. and Barila, T.G. (1955) 'Hematologic changes in hypothermic dogs', *Proceedings of the Society for Experimental Biology and Medicine, 89*, 192-6

Weatherley, C.R., Oxley, A., Jones, P. and Walder, D.N. (1977) 'Haematological studies following hyperbaric exposure', *Proceedings of the 6th International Congress on Hyperbaric Medicine*, Aberdeen University Press, pp. 140-5

Weber, M.B. and Blakely, J.A. (1969) 'The haemorrhagic diathesis of heatstroke', *Lancet, i*, 1190-2

Wensel, R.H. and Bigelow, W.G. (1959) 'The use of heparin to minimize thrombocytopenia and bleeding tendency during hypothermia', *Surgery, 45*, 223-8

White, J.G. (1968) 'Effects of heat on platelet structure and function', *Blood, 32*, 324-35

White, J.G. and Krivit, W. (1967) 'An ultrastructural basis for the shape changes induced in platelets by chilling', *Blood, 30*, 625-35

Willson, J.T., Miller, W.R. and Eliot, T.S. (1958) 'Blood studies in the hypothermic dog', *Surgery, 43*, 979-89

Zucker, M.B. and Borrelli, J. (1954) 'Reversible alteration in platelet morphology produced by anticoagulants and by cold', *Blood, 9*, 602-8

Zucker, M.B. and Borrelli, J. (1960) 'Viscous metamorphosis produced by chilling and clotting', *Thrombosis et Diathesis Haemorrhogica, 4*, 424-34

10 Nutritional Influences on the Hemostatic Mechanism

Nutritional influences on hemostasis have been a major source of interest to investigators for some three decades. The earlier studies concentrated on attempts to provide laboratory evidence to explain the epidemiological association between a high saturated fat diet and a high incidence of ischemic heart disease. These endeavours resulted in a multiplicity of reports on the effect of single high-fat meals on measures of coagulability or fibrinolysis. In more recent years, with increasing understanding of platelet biochemistry, have come studies on longer-term manipulation of the dietary fat pattern and some understanding of the mechanism of their influence on platelet function. Whether such studies will clarify our understanding of the causation of thrombosis or lead to rational measures for its prevention remains to be established. Additionally, there has been increasing recognition of other nutritional influences on hemostasis; undoubtedly many more remain to be identified and the mechanism of their effect elucidated. Even at our present incomplete state of knowledge, however, it is clear that strict control of diet is mandatory if studies on the effect of other physiological variables on hemostasis are to be valid.

Plasma Lipids

Dietary fat, principally triglyceride, is emulsified and broken down to fatty acids and glycerides in the small intestine. These enter the villi of the mucosal cells where the monoglycerides are re-esterified by activated fatty acids. The lipids in the mucosal cells pass out into the lymph in the form of chylomicrons composed mainly of triglycerides with some phospholipid, cholesterol and protein. The chylomicrons enter the blood via the thoracic duct and are carried to the tissues. The fatty acid pattern of the dietary fat and the thoracic duct lymph are

286

basically similar, but can differ because of admixture with fatty acid entering the gut mucosal cells from the blood.

Other lipid-protein complexes (lipoproteins) function to transport the blood lipids. The plasma lipoproteins have been classified according to ultracentrifugal or electrophoretic characteristics. The familial conditions in which the plasma lipids are abnormally high have been classified in different ways. The widely used World Health Organisation classification of the hyperlipoproteinemias (Table 10.1) is based on lipoprotein electrophoretic appearances (Beaumont *et al.*, 1970). Very low density lipoprotein (VLDL; pre-β-lipoprotein) is rich in endogenously synthesised triglyceride: it is metabolised through the stage of an intermediate density lipoprotein to low density lipoprotein (LDL), also termed β-lipoprotein, which is the major transport protein for cholesterol.

Table 10.1: World Health Organisation Classification of the Hyperlipoproteinemias (Beaumont *et al.*, 1970)

Lipoprotein type	Main lipoprotein abnormality	Serum lipids	
		Cholesterol	Triglyceride
I	increased chylomicrons	++	++++
IIa	increased LDL	++	N
IIb	increased LDL and VLDL	++	+
III	increased LDL	++	++
IV	increased VLDL	N or +	+++
V	increased chylomicrons and VLDL	++	+++

LDL = low density lipoprotein; VLDL = very low density lipoprotein

The metabolism of the chylomicrons and VLDL takes place at the endothelial surface to which the lipase is bound via glycosaminoglycans: through the action of the lipase the fatty acids are split from the triglycerides and are able to enter the tissues. Stored triglyceride can also be broken down, the liberated fatty acids entering the circulation to be oxidised as a source of energy by the tissues.

The nomenclature of a number of medium and long-chain fatty acids is detailed in Table 10.2. A further terminology used in the text indicates the location of the double bonds. For example, C20:1ω6 represents a fatty acid with 20 acyl carbons and one double bond located six carbons from the methyl terminal.

Table 10.2: Nomenclature of Some Medium and Long-chain Fatty Acids

	Common name	Systematic name
Saturated fatty acids		
C12:0	Lauric	Dodecanoic
C13:0	Myristic	Tetradecanoic
C16:0	Palmitic	Hexadecanoic
C18:0	Stearic	Octadecanoic
C20:0	Arachidic	Eicosadecanoic
Unsaturated fatty acids		
C16:1	Palmitoleic	9-hexadecenoic
C18:1	Oleic	9-octadecenoic
C18:2	Linoleic	9,12-octadecadienoic
C18:3	Linolenic	9,12,15-octadecatrienoic
C20:3	Dihomo-γ-linolenic	8,11,14-eicosatrienoic
C20:4	Arachidonic	5,8,11,14-eicosatetraenoic
C20:5	—	5,8,11,14,17-eicosapentaenoic

Effect of Lipids on Platelet Function

Alimentary Lipemia and Platelets

A number of studies have shown a significant decrease in platelet numbers after a high-fat meal (Philp and Wright, 1965; O'Brien *et al.*, 1976; Lowe *et al.*, 1979b), but others have been unable to confirm this (Dubber *et al.*, 1967).

As stressed elsewhere, the measurement of platelet adhesiveness is open to many sources of error, and the observations on the effect of fat feeding on platelet adhesiveness to glass must be viewed in this context. Both Philp and Wright (1965) and Coccheri and associates (1966) reported that adhesiveness was increased during alimentary lipemia, and Farbiszewski and co-workers (1968) postulated that coating of the platelet surface by β-lipoprotein is responsible for the increased adhesiveness.

The influence of a single high-fat meal on platelet aggregability has received relatively little study. Dubber and colleagues (1967) noted prolongation of the time for platelet aggregation in a Chandler's tube after a 95 g saturated fat meal. On the basis of the finding of increased circulating platelet aggregates Lowe and associates (1979) suggested that the ingestion of saturated fat causes activation of platelets. This concept would be in accord with the observation that platelet factor 4 activity, assessed by heparin-thrombin clotting time assays, is increased after a high saturated fat meal, but unchanged or reduced after a meal rich in unsaturated fat (Nordøy *et al.*, 1974; O'Brien *et al.*, 1976a). The addition of partially purified β-lipoprotein to plasma was reported to induce an acceleration of ADP and thrombin-induced platelet aggregation (Farbisjewski *et al.*, 1969).

Fatty Acids and Platelets

A number of studies have demonstrated that the addition of fatty acids to blood *in vitro* can cause platelet aggregation. The initial observations were made by Shore and Alpers (1963) and by Haslam (1964) who found that unbound long-chain saturated fatty acids produce rapid aggregation of washed human platelets, and it was proposed that fatty acids achieve this effect by releasing ADP from platelets with consequent aggregation. Kerr and colleagues (1965) reported that the sodium salts of free fatty acids, dissolved in lecithin, caused aggregation of platelets in platelet-rich plasma: linoleate and linolenate, however, did not induce aggregation and linoleate inhibited the aggregation of platelets by saturated fatty acids. In a further study Mahadevan *et al.* (1966) confirmed that long-chain saturated fatty acids and oleic acid cause platelet aggregation while linoleic and linolenic acids were ineffective. Hoak and associates (1967) were able to induce platelet aggregation by the sodium salts of stearic, oleic, linoleic and linolenic acids, but the long-chain saturated fatty acid was more potent than the unsaturated acids: it was also shown that the aggregation of platelets was diminished by the presence of albumin which binds fatty acids. In the studies of Prost-Dvojakovic and Samama (1972/73) erucic acid, a monounsaturated fatty acid (C22:1) was shown to be capable of inducing platelet aggregation. Further experiments by Hoak *et al.* (1970) showed that long-chain fatty acids bound to albumin have little direct platelet-aggregating action, but do enhance ADP-induced aggregation. Albumin-bound arachidonate, however, does induce platelet aggregation and secretion (Leonardi *et al.*, 1973) and potentiates aggregation induced by other agents (Silver *et al.*, 1973). Silver and colleagues (1973) also observed that platelet aggregation induced by arachidonic acid could be inhibited by certain unsaturated fatty acids such as 11, 14, 17-eicosatrienoic acid, 8,11,14-eicosatrienoic acid and 5,8.11,14,17-eicosapentaenoic acid.

Whole human platelets or isolated platelet membranes are capable of incorporating long-chain free fatty acids from plasma or solutions containing albumin-bound fatty acids (Deykin and Desser, 1968; Hoak *et al.*, 1970, 1972). Cohen *et al.* (1970) found that plasma-free fatty acids influence the fatty acid composition of platelet membranes, while Bills and associates (1976) provided evidence of ready incorporation of plasma arachidonic acid into the phospholipids of human platelets: its release from the platelet membrane is under the control of phospholipase and it can influence the platelet directly or through the synthesis of thromboxane A_2. It is probable that the effects of plasma fatty acids on platelets are not direct because of the binding by albumin, but are mediated through their incorporation into the phospholipids of the platelet membrane. It is in this way that long-term modification of the dietary fat influences platelet function.

Influence of Long-term Dietary Fat Modification on Platelet Function

The epidemiological evidence for a link between a high intake of saturated fat and atherosclerosis and ischemic heart disease, and the primary role of platelets in arterial thrombus formation, has stimulated a number of studies into the effect of long-term dietary lipid modification on platelet function. Mustard and Murphy (1962) studied seven men maintained successively on a low-fat diet, a saturated fat diet and a vegetable fat diet: platelet half-life was shortest and turnover greatest after the three weeks on the saturated fat diet. The platelet clumping time and platelet adhesiveness were also greatest on the saturated fat diet, but the differences did not reach statistical significance. Nordøy and Rodset (1971) compared two groups given isocaloric diets for 21 days: one group was given 40 per cent of the calories in the form of the unsaturated soybean oil (principal fatty acid linoleic) and the other group the same proportion of calories in the form of MCT oil with a saturated fatty acid composition (principally C8:0 and C10:0). After the soybean oil diet there was a significant increase in the linoleic acid content of the platelet phospholipids, but no change in platelet electrophoretic mobility or platelet aggregation. The platelet factor 3 activity in platelet-rich plasma decreased. The subjects given the MCT oil diet had no significant change in their platelet lipids, but had increased platelet factor 3 activity in platelet-rich plasma exposed to ADP. There was a tendency towards increased aggregability, but the changes were not statistically significant.

A number of subsequent investigators have shown by a variety of techniques that there is diminished platelet function when a saturated fat diet is replaced by a polyunsaturated fat diet. Hornstra and associates (1973) found that subjects maintained for some years on a diet rich in polyunsaturated fat had decreased platelet aggregability as assessed by the filter screen method compared to those on a normal Finnish diet containing a polyunsaturated to saturated fat ratio of 0.25. The two groups did not differ in platelet count or platelet adhesiveness measured by the rotating bulb technique. Fleischman and colleagues (1975) reported that replacement of a saturated fat diet with one rich in linoleic acid for two weeks resulted in a doubling of the platelet aggregation time, as determined by the filtragometer technique, while the disaggregation time was halved. Partial replacement of the fat in a normal diet by linoleic acid for five weeks was found by O'Brien *et al.* (1976b) to produce an increase in the heparin-thrombin clotting time (claimed to represent a reduction in platelet factor 4 release), a decrease in the platelet count, a reduction in platelet adhesiveness, and an increased aggregation rate in response to ADP and collagen: in contrast to the findings of Hornstra *et al.* (1973) and Fleischman *et al.* (1975), the polyunsaturated fat diet did not influence platelet aggregation as assessed by the filter screen method. Consistent with these earlier studies Jakubowski and Ardlie (1978) found that a period on a polyunsaturated fat diet resulted in a decrease in plasma heparin neutralising activity and a fall in the number of circulating platelet aggregates. There was an early reduction in the platelet count, but this returned to normal after a further period on the polyunsaturated fat.

Animal studies have also demonstrated that an increase in dietary unsaturated fat influences platelet function. For example, feeding increased amounts of linoleic acid to rats decreased the susceptibility of their platelets to thrombin-induced aggregation (McGregor and Renaud, 1978; McGregor *et al.*, 1980).

Renaud and colleagues (1978) have reported on studies of farmers from two regions in France, Var and Moselle. The intake of total calories, total lipids and saturated fats was higher in the Moselle farmers who had increased platelet aggregability in response to ADP and thrombin, but not to collagen or epinephrine. Platelet aggregability was significantly correlated with the saturated fat intake, mainly stearic acid. In a further study on British farmers platelet aggregability was again higher in the group with a dietary intake high in saturated fat and low in polyunsaturated fat (Renaud *et al.*, 1981): of the platelet phospholipid fatty acids C20:3ω9, C22:3ω9 and C20:4 were most closely related to the platelet function tests, while the fatty acid C20:3ω9 in the platelet phospholipid was significantly correlated with the intake of unsaturated fat.

Elucidation of the mechanisms by which modification of the dietary fat content can alter platelet function is in a phase of rapid acceleration. It is clear that the pattern of fatty acid in dietary lipid is reflected in the plasma and platelet lipids. It has been shown that linoleic acid is incorporated into the platelet phosphatidylcholine (Bills *et al.*, 1977), and that it interferes with the conversion of arachidonate into the cyclic endoperoxides and, consequently, with the production of thromboxane (Gerrard *et al.*, 1976). It is a reasonable hypothesis that it is this effect which explains the influence of dietary linoleic acid enrichment on platelet function. Gerrard and colleagues (1976) found that stearic acid, in contrast to linoleic acid, enhanced platelet aggregation: this appeared to be due to a direct action on the platelet without effect on cyclic endoperoxide production.

Eskimos from Greenland provide a further example of the influence of dietary fat on platelet function. They have high levels of ω-3 polyunsaturated fatty acids such as 5,8,11,14,17-eicosapentaenoic acid and low levels of arachidonic acid in their diet, serum lipids and platelet lipids (Dyerberg *et al.*, 1978; Dyerberg and Bang, 1979). Both eicosapentaenoate and arachidonate are readily acylated into platelet phospholipids and released on stimulation of phospholipase A_2 by thrombin. Eicosapentaenoate, however, has been shown to modify the normal pathways from arachidonate to thromboxane A_2 or prostacyclin (Dyerberg *et al.*, 1978; Needleman *et al.*, 1979). A comparison of the metabolic pathways of arachidonic and eicosapentaenoic acids is illustrated in Figure 10.1. From eicosapentaenoic acid the cyclic endoperoxide PGH_3 is formed by the action of cyclo-oxygenase; from PGH_3 there is generation of thromboxane A_3 in platelets and of PGI_3 in blood vessel walls. In contrast to PGH_2 and thromboxane A_2, PGH_3 and thromboxane A_3 lack platelet-aggregating properties whereas PGI_3 exerts similar effects as PGI_2 in causing inhibition of platelet aggregation and vasodilatation. Platelet cyclo-oxygenase is relatively inefficient at converting eicosapentaenoic acid, but the C20:5 fatty acid has a high binding affinity for the enzyme and effectively completes with arachidonate for the cyclo-oxygenase. In this way PGH_2 and thromboxane A_2 formation is suppressed (Needleman *et al.*, 1979). In a study on 24 Eskimos the bleeding time was found to be significantly longer than Danish

Figure 10.1: Metabolic Pathways of Arachidonic Acid and Eicosapentaenoic Acid

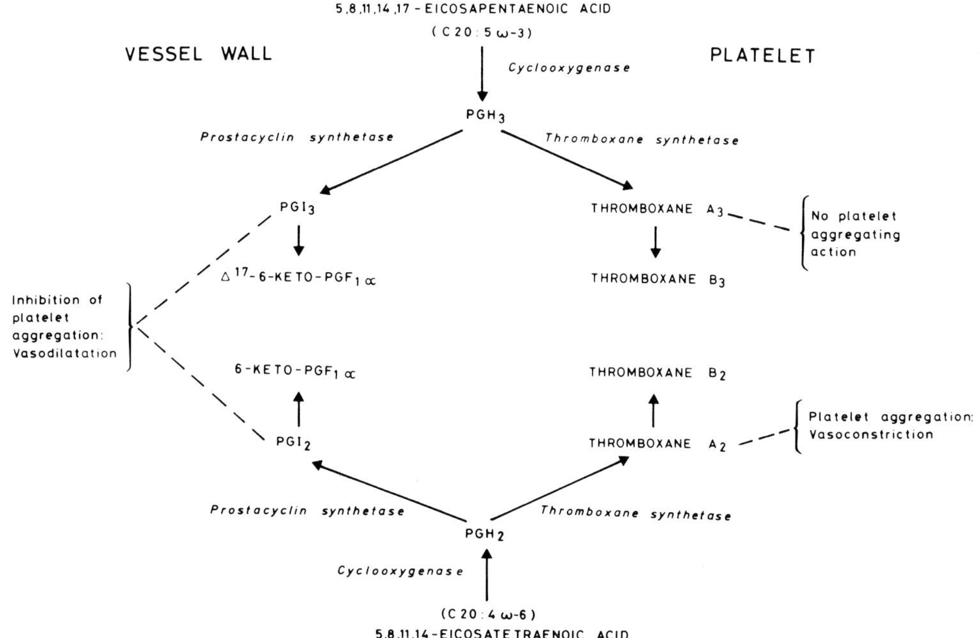

controls (Dyerberg and Bang, 1979), and there was a significantly higher threshold for ADP-induced secondary phase aggregation. In ten of the Eskimos no secondary phase aggregation could be induced. Platelet adhesion to glass bead columns, which is not dependent on prostaglandin synthesis, did not differ between the Eskimos and the Danes.

The effect of an increase in dietary eicosapentaenoic acid on platelet composition and reactivity has been tested directly by Siess and colleagues (1980). They fed a group of seven men a diet consisting of 500 to 800 g mackerel daily, a fish with a high content of C20:5 fatty acids, for one week. While on the diet the C20:5 level in the plasma and in the platelet membrane increased while the levels of C20:4 fell. Platelet aggregation and thromboxane A_2 synthesis after low-dose collagen stimulation in platelet-rich plasma were reduced after the week on the mackerel diet. Inexplicably platelet aggregability in response to epinephrine, ADP, and high concentrations of collagen was not significantly reduced: the investigators stress themselves that there are other possible explanations for the altered platelet function than the altered C20:5/C20:4 ratio. Results of an analogous study examining the effect of a four-week period on a diet containing 40 per cent fat derived from salmon oil comprising 18 per cent ω-3 fatty acids indicate that the C20:5ω3 content of the platelet phospholipids rose while both linoleic (C18:2ω6) and arachidonic acids fell (Goodnight *et al.*, 1981). The platelet count of the subjects decreased and platelet adhesion to glass beads was reduced while on the diet, but platelet aggregation and ^{14}C-serotonin release did not change significantly. Cod liver oil supplements for six weeks resulted in an increase in the ω-3 fatty acids of the plasma and platelet phospholipids in healthy men whereas the ω-6 fatty acids decreased: collagen-induced platelet aggregation and thromboxane B_2 formation were reduced (Nordøy *et al.*, 1981). In a further recent study a diet of fish oil providing 3.5 g eicosapentaenoic acid daily was given to 13 patients with ischemic heart disease for five weeks: this regime induced a ten per cent lengthening of the platelet survival time, a 15 per cent fall in the platelet count, a 75 per cent fall in plasma levels of platelet factor 4 and a 30 per cent fall in plasma β-thromboglobulin (Hay *et al.*, 1982).

While eicosapentaenoic acid is a precursor of prostaglandins of the '3' series, 8,11,14-eicosatrienoic acid (dihomo-γ-linolenic acid) is the precursor of prostaglandins of the '1' series (Willis *et al.*, 1974) Dihomo-γ-linolenic acid is only present in very low concentrations in

294

plasma (Dole *et al.*, 1959), and is absent from platelet phospholipids and neutral lipids (Cohen and Derkson, 1969; Marcus *et al.*, 1969). It has, however, been proposed that dietary supplements of this fatty acid might have a role in the prevention of thrombosis since thromboxane A_1 does not possess platelet aggregating properties. Oral administration of dihomo-γ-linolenic acid has been reported to produce some inhibition of ADP-induced platelet aggregation and a decrease in plasma heparin-neutralising activity (Kernoff *et al.*, 1977).

Platelet Function in Patients with Hyperlipoproteinemia

A different approach to the study of the effect of lipids on platelet function has been to examine the changes associated with the abnormalities in plasma fat found in the hereditary hyperlipoproteinemias. Patients with type II hyperlipidemia, characterised by increased low density lipoprotein and raised cholesterol, were found to have platelets which were abnormally sensitive to aggregation by ADP, collagen and epinephrine (Carvalho *et al.*, 1974): platelet adhesiveness was normal in these patients. The platelet membranes of patients with this form of hyperlipidemia were reported to have a high cholesterol:phospholipid ratio and this abnormality correlated with the increased sensitivity to aggregating agents (Shattil *et al.*, 1977). Shattil and associates (1975) also found that normal platelets incubated with cholesterol/dipalmitoyl lecithin liposomes of high cholesterol:phospholipid ratio acquired excess cholesterol without change in phospholipid: the cholesterol-loaded platelets showed a marked increase in sensitivity to ADP and epinephrine-induced aggregation while loss of platelet cholesterol resulting from incubation with cholesterol-poor liposomes was associated with a reduction in sensitivity to epinephrine. Other studies have been carried out on the effect of altering the platelet cholesterol content on platelet function. Stuart and colleagues (1980) found that cholesterol-enriched platelets, obtained by incubation with cholesterol-rich liposomes, released a greater quantity of arachidonic acid in response to thrombin compared to cholesterol-depleted platelets and conversion into thromboxane was also increased. Enhancement of the platelet cholesterol content by incubation with low density lipoprotein has also been found to increase platelet aggregation in response to thrombin (Aviram and Brook, 1981). Jamieson and associates (1980) have reported that cholesterol-enriched platelets have an increased ability to bind thrombin through an increase in the number of available high affinity sites. In keeping with this evidence that platelets alter their function on taking up cholesterol are the findings that there is a relationship between serum cholesterol concentration and platelet thromboxane B_2 production (Tremoli *et al.*, 1979) and that there is an increased production of thromboxane A_2 in patients with type II hyperlipoproteinemia (Bizios *et al.*, 1977).

Other findings relating hyperlipoproteinemia to platelet function have included a reduced platelet survival time in a patient with hypercholesterolemia (Corash *et al.*, 1976) and increased circulating platelet aggregates in patients with type II hyperlipoproteinemia (Lowe *et al.*, 1979a). Platelets from patients with type IV hyperlipoproteinemia had normal sensitivity to ADP and collagen (Carvalho *et al.*, 1974). In a study by Zahavi and associates (1981) plasma levels of platelet factor 4 and β-thromboglobulin, presumably indicating enhanced platelet activation *in vivo*, were found to be significantly increased in patients with hyperlipoproteinemias of types IIa, IIb and IV: malondialdehyde formation, representing prostaglandin synthesis, was also higher in the hyperlipidemic patients than in controls.

Effect of Lipids on Coagulation

Effect of Alimentary Lipemia on Coagulation

The earliest recorded observations on the effect of alimentary lipemia on blood coagulability were made by Waldron and associates (1951) who showed that the whole blood clotting time in non-wettable glassware shortened after the ingestion of 60 ml of 20 per cent cream. This finding was confirmed by some later investigators (Fullerton *et al.*, 1953; Mustard, 1957), but not by others (Tulloch *et al.*, 1953; Merskey and Nossel, 1957; Sheehy and Eichelberger, 1958).

A consistent finding has been that the recalcified plasma clotting time in the presence of Russell's viper venom ('Stypven' time) is shortened after the ingestion of fat (Fullerton *et al.*, 1953; O'Brien, 1955, 1956; Mustard, 1957; Merskey and Nossel, 1957; Nitzberg *et al.*, 1959; Dubber *et al.*, 1967). In a series of studies on rat plasma Poole and Robinson concluded that the activity of chylomicra in accelerating the 'Stypven' time resides in the phospholipid surface layer of the chylomicra, in particular, ethanolamine phosphatide (Poole and Robinson, 1956a, b; Robinson and Poole, 1956). The 'Stypven' time has been found to be shortened after meals of both animal and vegetable fat (O'Brien, 1956; Merskey and Nossel, 1957; Maclagan *et al.*, 1958), although Maclagan and colleagues (1958) noted that butter added to plasma *in vitro* was more active in accelerating clotting than margarine. It is a long-standing observation that Russell's viper venom has little effect on plasma depleted of platelets and fat (Macfarlane *et al.*, 1941). Since the studies on the effect of fat meals on the 'Stypven' time were carried out on essentially platelet-poor plasma it is likely that the effect of lipemia is to provide the phospholipid normally contributed by the platelet membrane.

The plasma recalcified clotting time in the absence of Russell's viper venom has also been found by some investigators to be shortened following a fat-enriched meal (O'Brien, 1955; Pilkington, 1957; McDonald and Fullerton, 1958), although this finding is not universal (Merskey and Nossel, 1957; Meuwissen *et al.*, 1968). An explanation for these differences in experimental results is not readily apparent. In some studies meals rich in unsaturated fat did not shorten the recalcified plasma clotting time while saturated fats did so (Pilkington, 1957). In other studies unsaturated fats produced an acceleration of the clotting time equal to that produced by saturated fats. Using the thromboelastogram as an index of coagulability Sheehy and Eichelberger (1958) found no acceleration during the alimentary lipemia induced by 100 g animal fat, 250 ml 38 per cent cream or 50 ml olive oil.

Information on the possible influence of lipemia on the plasma level of individual coagulation factors is fragmentary. Egeberg (1966) reported an increase in factor XI activity maximum some four to seven hours after a meal rich in margarine, but no increase after butter or cream-enriched meals. The increase in factor XI activity was less after a meal in which the margarine was manufactured from hydrogenated vegetable oils and was prevented by prior intake of linseed oil. There was no change in the levels of factors VII, VIII or IX. In other studies a high animal fat meal produced no change in fibrinogen, factor V or factor VII (Mustard, 1957; Nitzberg *et al.*, 1959), but an increase in factor IX has been claimed (Mustard, 1957).

Fatty Acids and Coagulation

The concentration of fatty acids in the blood increases after a fat meal (Grossman *et al.*, 1954). A possible explanation for the postulated influence of ingested fat on blood coagulability stems from the observation that the soaps of long-chain fatty acids accelerate coagulation.

Connor and Poole (1961) and Connor (1962) showed that the sodium salts of long-chain saturated fatty acids accelerate the re-calcified clotting time of whole blood in siliconised tubes or in rotating plastic loops. The activity of the saturated fatty acids on coagulation increases with the length of the carbon chain (Connor and Poole, 1961; Margolis, 1962).

The acceleration of clotting by fatty acids has been attributed to activation of factor XII (Connor, 1962; Margolis, 1962), although Botti and Ratnoff (1963) concluded from their experiments that the procoagulant effect of the sodium salts of long-chain saturated fatty acids is due to the activation of factor XI. It is believed that fatty acids exert their effects through their negative charges (Hubbard and Lucas, 1960; Nossel *et al.*, 1969).

Fatty acids are known to be bound by albumin, lipoproteins, calcium and other plasma constituents (Goodman, 1958; Tompkins and Dayton, 1965): in the bound state the saturated acids have no effect on coagulation (Connor and Poole, 1961). Any hypothesis linking fatty acids with the increased coagulability after a fat meal must be reconciled with these observations. Overall it is doubtful whether the effect of fatty acids on coagulation has any *in vivo* relevance although a high molar ratio of fatty acid to albumin might theoretically allow binding of fatty acids to other plasma proteins including coagulation factors with their consequent activation (Verstraete, 1972).

Influence of Long-term Dietary Fat Modification on Coagulation

Considering the lack of uniformity in the conclusions reached on the effect of single fat meals on the coagulation system it is not surprising that experiments on the long-term effect of fat diets on coagulability and coagulation factors have failed to provide definitive data. Mustard and Murphy (1962) studied a group of men maintained for a period on a low-fat, high saturated fat and high unsaturated fat diet and found that the greatest coagulant activity, as assessed by the whole blood clotting time, one-stage prothrombin time and partial thromboplastin time, was during the period on a diet rich in saturated fat. However, the correlations between serum cholesterol and phospholipid levels and the coagulation tests were poor. In contrast, Tompkins and associates (1964) found no difference in whole blood clotting times carried out with or without glass contact between two groups of elderly men maintained for three years on a diet of either saturated or unsaturated fat, in each the fat contributing 40 per cent

298

of the diet's calories: in this study no correlation was found between the whole blood clotting time and the serum cholesterol or saturated free fatty acid concentrations.

In a study of nine patients with primary hypercholesterolemia and one with idiopathic hyperlipidemia Peyman and associates (1960) observed that long-term treatment with essential fatty acids did not influence the level of factors V, VII and prothrombin or a variety of tests of overall coagulability. Partial replacement of the fat in a normal diet by linoleic acid for five weeks produced an increase in the plasma fibrinogen concentration, an increase in antithrombin III and a prolongation of the 'Stypven' time (O'Brien *et al.*, 1976b).

Effect of Lipids on the Fibrinolytic System

Alimentary Lipemia and Fibrinolytic Activity

The initial observation suggesting that alimentary lipemia inhibits circulating fibrinolytic activity was reported by Greig (1956): he found that the lysis of clots made from diluted plasma was diminished after a meal rich in animal fat compared with that found after a low fat meal. A succession of studies on the effect of meals rich in animal fat on blood or plasma fibrinolytic activity using a variety of assay techniques were published over the next few years, but the conclusions reached were contradictory. A number of these studies are summarised in Table 10.3. The results obtained from studies comparing the effects of animal and vegetable fat meals on fibrinolytic activity have also been discordant: some of the reported comparisons between the two types of fat meal are shown in Table 10.4. Experiments have been performed on the effect of adding chylomicra to plasma *in vitro*: inhibition of the plasma fibrinolytic activity which had been enhanced by prior administration of epinephrine was observed in the presence of chylomicra derived from the ingestion of both animal and vegetable fats (Farquhar *et al.*, 1961).

Blood Lipid Levels and Fibrinolysis

The results from studies on the relationship between blood lipid levels and components of the fibrinolytic system have not been illuminating. In healthy men Sweet and colleagues (1966) found a significant negative correlation between plasma triglyceride levels and euglobulin clot lysis times, but there was no correlation between serum cholesterol, triglyceride or phospholipid and the plasma level

Table 10.3: Effect of High Animal (Saturated) Fat Meals on Circulating Fibrinolytic Activity in Healthy Subjects

Study	Fat content of meal	Fibrinolytic activity assessment technique	Number of subjects	Time after meal (hours)	Conclusion
Greig (1956)	'high fat'	DPCL	25	3½-4	inhibitory
Buckell and Elliot (1959)	50 g butter	ECL	17	3	slightly inhibitory
		DPCL	10	3	inhibitory
Billimoria *et al.* (1959)	42 g butter	DBCL	20	2	inhibitory
				4	no effect
Nitzberg *et al.* (1959)	85 g animal fat	DBCL	10	4	no effect
Hougie and Ayers (1960)	'high fat'	DPCL	9	3½-4	no effect
Gajewski (1961)	250 ml 22% cream	ECL	12	3	no effect
Farquhar *et al.* (1961)	1.5 g fat/kg body weight	DPCL	8	4	inhibitory
Ogston and Fullerton (1962)	8.5 g animal fat	DPCL	9	3½	no effect
Dubber *et al.* (1967)	270 g cream (95 g fat)	ECL	10	3	inhibitory
Meuwissen *et al.* (1968)	57 g fat	ECL lysis of ^{131}I-fibrin clots	17	3	no effect
	118 g fat	ECL/lysis of ^{131}I-fibrin clots	6	3	no effect

DPCL = Dilute plasma clot lysis time; DBCL = Dilute blood clot lysis time; ECL = Euglobulin clot lysis time

of fibrinogen or plasminogen. In contrast, Spöttl and associates (1970) obtained no correlation between blood levels of triglyceride, cholesterol or phospholipid and plasminogen activator activity in either normal subjects or patients with hypertriglyceridemia.

A number of reports have appeared on fibrinolytic activity in patients with hyperlipoproteinemia. Although Nitzberg *et al.* (1959) found no alteration in the dilute whole blood clot lysis time in patients with either idiopathic hyperlipidemia or primary hypercholesterolemia, other studies have shown that patients with type IV hyperlipo-

300

Table 10.4: Comparison of Effect of High Animal and Vegetable Fat Meals on Circulating Fibrinolytic Activity

Study	Animal fat (saturated)	Vegetable fat (unsaturated)
Greig (1956)	inhibitory	no effect
Gajewski (1961)	no effect	no effect
Farquhar *et al.* (1961)	inhibitory	inhibitory
Ogston and Fullerton (1962)	no effect	no effect
Meuwissen *et al.* (1968)	no effect	no effect

proteinemia have decreased plasma fibrinolytic activity (Sweet *et al.*, 1966; Spöfttl *et al.*, 1970) and a reduced fibrinolytic response to venous occlusion (Spöttl *et al.*, 1969). McArdle and colleagues (1982) studied 51 patients with type II hyperlipoproteinemia and reported that plasma activator levels were normal whole α_2-antiplasmin and plasminogen levels were increased.

Fatty Acids and the Fibrinolytic System

Evidence has been presented that long-chain saturated fatty acids induce fibrinolytic activity through the factor XII-dependent pathway (Ogston *et al.*, 1973), although the physiological significance of this observation is limited by the binding of fatty acids *in vivo* to albumin and other plasma constituents. A number of long-chain unsaturated fatty acids cause lysis of fibrin plates (Surgala *et al.*, 1967), but this is probably a non-enzymatic action. In a purified system free fatty acids, particularly long-chain unsaturated fatty acids, were found to inhibit plasmin (Muraoka and Okuda, 1977), and this inhibition was only partially removed in the presence of albumin.

Low-density Lipoprotein and Plasmin

It has been reported that the addition of β-lipoprotein *in vitro* prolongs the euglobulin clot lysis time (Pappenhagen *et al.*, 1963). Experiments by Riding and Ellis (1964) and by Skrzydlewski *et al.* (1966) suggested that β-lipoprotein has antiplasmin activity, but the preparations of β-lipoprotein used in these experiments were crude and probably contaminated with α_2-macroglobulin and other protease inhibitors which could have accounted for the plasmin inhibition found. A more highly purified preparation of β-lipoprotein failed to inhibit the caseinolytic or fibrinolytic activity of plasmin (Herbert *et al.*, 1973). Steele (1979) has shown subsequently that plasmin can cleave the protein component of plasma low-density lipoprotein and suggested that low-density lipoprotein has the potential to act as a competing substrate for plasmin.

Effect of Carbohydrates on the Hemostatic Mechanism

In contrast to the plethora of observations on the influence of lipids on hemostatic processes, only scanty experimental data are available on the possible effects of carbohydrates. While there is some evidence pointing to an influence of dietary sugar on the hemostatic mechanism, it is probable that the mode of production of effect is indirect.

A high intake of sucrose for 14 days was claimed to increase platelet adhesiveness in six of the 19 subjects studied (Szanto and Yudkin, 1969): these six also differed from the other subjects in gaining weight and in showing a rise in the serum insulin level. In studies on men with peripheral vascular disease there was a significant positive correlation between sucrose intake and platelet adhesiveness, but this was not seen in healthy men (Yudkin *et al.*, 1969). In contrast, a high glucose diet for 23 days was reported to decrease platelet adhesiveness (Korsan-Bengtsen *et al.*, 1972a). There has also been a suggestion that platelet aggregation may be suppressed by oral glucose (Sagel *et al.*, 1975).

The isocaloric substitution of mono- and disaccharides for starches has been reported to produce a shortening of the 'Stypven' time and a reduction in the thrombus formation time (Antar, 1968); again the relationship is likely to be indirect since such a diet caused an increase in saturated and a decrease in polyunsaturated fatty acids. In another dietary experiment measurements were made before and after a 23-day period on a diet in which glucose contributed 93 per cent of the caloric intake (Korsan-Bengtsen *et al.*, 1972a): the recalcified plasma clotting time, the 'Stypven' time and the activated partial thromboplastin time were all shortened on this diet without change in factors V and VIII or the protease inhibitors antithrombin III, α_1-antitrypsin and α_2-macroglobulin. Fibrinogen and plasminogen concentrations fell and there was a decrease in the activity of the factor complex II-VII-X.

Glucose *per se* does not influence fibrinolysis. Whether an oral load of glucose does so is questionable. Tsapogas *et al.* (1962) reported an increase in the blood fibrinolytic activity of patients with vascular disease after 50 g glucose. Tanser (1966) found an increase in fibrinolytic activity only in those who had a low fasting level of fibrinolytic activity whereas those with high fasting activity showed a lengthening of the lysis time after glucose. The significance and mechanism of these results is obscure. In the report of Korsan-Bengtsen and co-workers (1972a) subjects given a high glucose diet for 23 days showed no significant change in plasma fibrinolytic activity.

Pyridoxal 5′-Phosphate and Hemostasis

Vitamin B_6 is a co-enzyme involved in a number of different enzymatic reactions, particularly in protein metabolism. Metabolic processes in which it is known to be concerned include transamination; decarboxylation to form amines, for example, 5-hydroxytryptamine from 5-hydroxytryptophan; formation of melanin; metabolism of tryptophan; and conversion of linoleic acid into its 20-carbon metabolites ($C20:3\omega6$ and $C20:4\omega6$). Pyridoxine, pyridoxal and pyridoxamine (Figure 10.2) are equally effective in vitamin B_6 activity in animal nutrition. The biocatalytically active form of pyridoxal is the phosphorylated derivative. Vitamin B_6 is widely distributed in food-stuffs; rich sources are meat, vegetables, whole grain cereals and egg yolk. The daily intake is in the range of 2-3 mg.

Pyridoxal 5′-phosphate has been found to influence both coagulation and platelet function. The secondary wave of ADP-induced aggregation and thrombin and collagen-induced aggregation were inhibited by pyridoxal 5′-phosphate at a concentration of 0.8 mM (Subbarao *et al.*, 1977): epinephrine-induced aggregation could also be inhibited (Subbarao *et al.*, 1978). Other studies have shown that pyridoxal 5′-phosphate can inhibit ADP-induced shape change as well as aggregation (Kornecki and Feinberg, 1980), and inhibit release of 5-hydroxytryptamine (Subbarao *et al.*, 1977; Pollard *et al.*, 1977; Kornecki and Feinberg, 1980). Pyridoxal 5′-phosphate was ten-fold more potent than pyridoxamine 5′-phosphate in inhibiting platelet aggregation (Subbarao *et al.*, 1977); pyridoxine also inhibited secondary platelet aggregation, but only when present at a concentration above 10 mM (Subbarao *et al.*, 1979). Ingestion of 100 mg of pyridoxine was followed by a phase of inhibition of ADP,

Figure 10.2: Structure of Compounds with Vitamin B$_6$ Activity

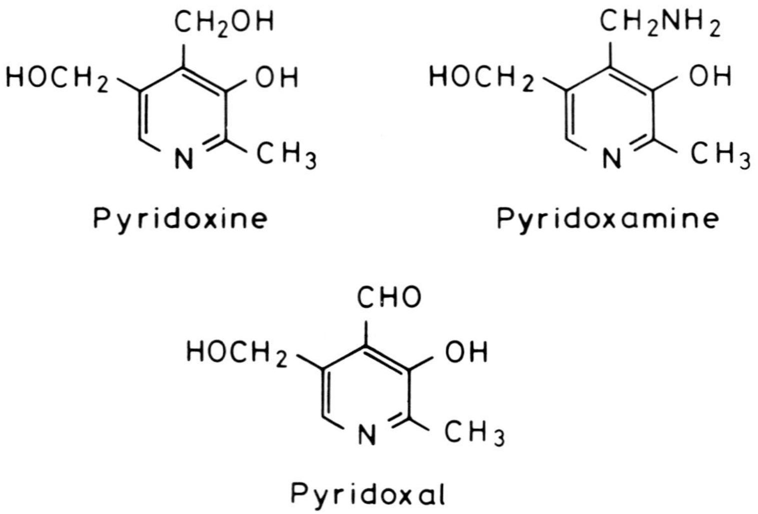

collagen and arachidonic acid-induced aggregation which persisted for 24 to 72 hours (Subbarao *et al.*, 1977).

Pyridoxal 5′-phosphate is also capable of inhibiting the esterase activity of thrombin on *p*-tosyl-L-arginine methyl ester (Kornecki and Feinberg, 1979) and on fibrinogen (Griffith, 1979): it achieves this through the formation of a Schiff base between its reactive aldehyde group and lysine residues of the enzyme. In a further study pyridoxal 5′-phosphate was shown to exert a concentration-dependent prolongation of the coagulation time of human plasma induced by thrombin (Subbarao *et al.*, 1979).

The mechanism of the inhibitory action of pyridoxal 5′-phosphate on platelet function is uncertain. Subbarao and colleagues (1978) reported that the vitamin interacts strongly with three specific platelet membrane components and suggested that its inhibitory effect results from this binding. Formation of a Schiff base with platelet surface amino groups was put forward as a possible explanation for the effect of pyridoxal phosphate on ADP-induced platelet aggregation

(Kornecki and Feinberg, 1980). The experiments of Lam and associates (1980) suggested that the inhibitory effect of pyridoxal phosphate was unlikely to be due to interference with nucleotide diphosphokinase, increase in platelet AMP levels or inhibition of fibrinogen binding.

Vitamin B$_{12}$ and Platelets

Thrombocytopenia is a common finding in patients with vitamin B$_{12}$ deficiency. Abnormal platelet aggregation has also been demonstrated in a proportion of such patients. Levine (1973) found impaired platelet aggregation in response to ADP, epinephrine and collagen while Ingeberg and Stoffersen (1979) reported reduced aggregation to ADP, epinephrine and ristocetin in some of the patients studied. The platelet aggregation abnormalities in both studies were corrected by administration of vitamin B$_{12}$.

Influence of Ascorbic Acid on the Hemostatic Mechanism

Some attention has been given to the possibility that ascorbic acid (vitamin C) might influence the hemostatic mechanism, but the experimental findings available do not permit any definitive conclusions.

The effect of vitamin C deficiency on platelet function has been examined both in experimental animals and in man. Scorbutic guinea-pigs were found to have a diminished velocity of ADP-induced platelet aggregation (Born and Wright, 1967), and this was confirmed in the same species by Purcell and Constantine (1972) for both ADP and collagen-induced aggregation. Patients with scurvy have also been studied: impaired platelet adhesiveness was reported in two patients by Wilson and colleagues (1967) with correction by administration of ascorbic acid. Platelet aggregation was found to be abnormal in elderly patients with scurvy in one study (Banerjee and Etherington, 1974) whereas the patients of Wallerstein and Wallerstein (1976) had normal platelet aggregability, but three of the five had decreased platelet adhesiveness. One patient with scurvy and five normal subjects rendered vitamin C-deficient by diet were studied by Johnson and colleagues (1981): there was no alteration in platelet aggregation in response to ADP, collagen, epinephrine, arachidonic acid or ristocetin, but platelet retention during bleeding decreased in the normal subjects when they became vitamin C-deficient. The retention of platelets in glass bead columns was normal.

The influence of ascorbic acid on platelet function was assessed by Sarji *et al.* (1979). The addition of ascorbic acid to platelet-rich plasma to a concentration of 1000 µg/ml inhibited platelet aggregation in response to ADP, epinephrine and collagen, but enhanced aggregation induced by arachidonic acid. Using washed platelets, however, ascorbic acid inhibited arachidonic acid-induced aggregation. In the same study the oral administration of ascorbic acid to volunteers (2 g daily for seven days) markedly inhibited platelet aggregation. Proposed mechanisms for the inhibition of platelet aggregation by ascorbic acid included inhibition of cyclo-oxygenase and a rise in the level of cyclic AMP. Similar findings have been reported by Cordova and associates (1982): inhibition of platelet aggregation and reduced production of malondialdehyde was seen after the addition of ascorbic acid to platelet-rich plasma and after the intravenous administration of 2 g ascorbic acid to healthy subjects. Crawford and colleagues (1975) reported that the ingestion of 1 g of ascorbic acid daily for three months did not significantly alter platelet adhesiveness as assessed by the rotating bulb technique.

The reported changes in fibrinolysis by ascorbic acid have been contradictory. While Shimizu and associates (1970) found that the blood and urinary fibrinolytic activity of patients with chronic nephritis was inhibited within three days of the start of ascorbic acid in a dosage of 0.5 to 1 g/day, Bordia *et al.* (1978) claimed that fibrinolysis was significantly increased three and six hours after a single 1 g dose of ascorbic acid: the fall in fibrinolytic activity seen after a high fat meal was also prevented by ascorbic acid. The long-term influence of ascorbic acid on a number of hemostatic parameters was studied by Crawford *et al.* (1975): the ingestion of 1 g ascorbic acid daily over a three-month period failed to influence plasma levels of plasminogen, fibrinogen, plasminogen activator as assessed by the euglobulin clot lysis time, fibrin(ogen) degradation products, α_1-antitrypsin, α_2-macroglobulin, or the partial thromboplastin time.

Influence of Vitamin E on the Hemostatic Mechanism

Vitamin E and Platelet Function

Evidence for a role for vitamin E in platelet function has been provided from a number of directions. Infants with low serum α-tocopherol levels were reported to have an increased platelet count with reduction to normal following administration of vitamin E (Melhorn and Gross, 1971). In a study on rats it was found that platelet aggregation in response to collagen was increased in vitamin E-deficient animals while a more prolonged deficiency produced, in addition, an elevation in the platelet count (Machlin *et al.*, 1975). A report of two children with vitamin E deficiency who had increased platelet aggregability in response to ADP, epinephrine and collagen provides further evidence for an influence of α-tocopherol on platelet function (Lake *et al.*, 1977): the changes were reversed after oral administration of vitamin E. Vitamin E is known to prevent the formation of peroxidases and it was suggested that vitamin E might prevent the peroxidation of arachidonate in platelets, perhaps at the step of phospholipase A activation of the conversion of arachidonic acid into cyclic endoperoxides by cyclo-oxygenase. Another possibility is that lipid peroxidases accumulated in the vessel wall might inhibit PGI_2 synthetase: this receives support from a study by Okuma and colleagues (1980) showing that the release of a prostacyclin-like substance from aortic rings was significantly reduced in vitamin E-deficient rats. This effect was reversed by feeding a vitamin E-supplemented diet.

Based on the finding that the platelet release reaction is associated with a sudden increase in lipid peroxidation and the ability of α-tocopherol to reduce the rate of lipid peroxidation in stored platelet suspensions Steiner and Anastasi (1976) investigated the influence of added α-tocopherol on platelet aggregation. Inhibition of the platelet release reaction in response to ADP, collagen and epinephrine in the presence of added α-tocopherol was shown by the abolition of the secondary wave of aggregation and by the decrease in aggregation-induced release of 5-hydroxy[^{14}C]tryptamine from

Figure 10.3: Structure of Alpha-Tocopherol

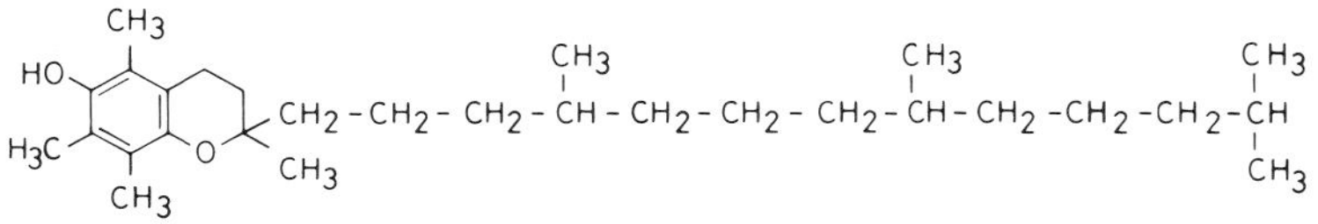

$$HO-\text{(chromanol ring: } CH_3, H_3C, CH_3, O)\text{-}CH_2-CH_2-CH_2-\underset{\underset{CH_3}{|}}{CH}-CH_2-CH_2-CH_2-\underset{\underset{CH_3}{|}}{CH}-CH_2-CH_2-CH_2-\underset{\underset{CH_3}{|}}{CH}-CH_3$$

prelabelled platelets. Oral vitamin E was shown in this study to increase the plasma and platelet α-tocopherol concentrations, but the effect on platelet aggregation was not reported. Inhibition of ADP-induced aggregation after the addition of α-tocopherol to platelet-rich plasma was observed by Creter and colleagues (1979): in contrast, increased aggregation in response to arachidonic acid was seen in the presence of added α-tocopherol. The oxidation product of vitamin E (vitamin E quinone) has been found to be a more potent inhibitor of platelet function than the parent compound (Rao *et al.*, 1981): it was shown to inhibit platelet aggregation and [^{14}C]-serotonin release induced by collagen, epinephrine, ADP, thrombin and arachidonate. It appeared to act by inhibiting phospholipase activity since it blocked the release of [^{14}C]-arachidonic acid from platelet membranes stirred with a calcium ionophore.

Differing from the effect of α-tocopherol added directly to plasma, the administration of 1000 I.U. tocopherol acetate daily for eight days did not influence platelet aggregation in response to ADP or epinephrine (Gomes *et al.*,1976). There appears, therefore, to be considerable evidence for an effect of α-tocopherol on platelet aggregation, but its physiological role in platelet function and hemostasis is unknown.

Platelet adhesiveness was reported to be unaltered in patients with ischemic heart disease given 300 mg/day of α-tocopherol over many weeks (Korsan-Bengtsen *et al.*, 1974).

Vitamin E and the Coagulation System

The results of some early studies suggested that α-tocopherol had antithrombin properties (Zierler *et al.*, 1948; Kay *et al.*, 1950), but subsequent investigation indicated that the antithrombin effect of tocopherol phosphate in *in vitro* systems was due to the phosphate ion rather than to tocopherol *per se* (Olson, 1973). Korsan-Bengtsen *et al.* (1974) found a highly significant prolongation of the plasma clotting time after 18 to 64 weeks of oral α-tocopherol (200 mg/day): they attributed this to a change in the platelet membrane and consequent reduction in platelet factor 3 availability.

Vitamin E and Fibrinolytic Activity

Moroz and Gilmore (1976) reported that vitamin E caused inhibition of plasmin-mediated fibrinolysis in a solid phase [^{125}I]-fibrin technique, but *in vivo* studies on the effect of oral vitamin E on fibrinolytic activity were not performed. In our own studies the inhibitory effect of α-tocopherol on plasmin activity was confirmed and, additionally, α-tocopherol was found to inhibit the activity of the plasminogen activator prepared from porcine heart and of urokinase in *in vitro* systems, both on fibrin plates and on amidolytic and esterolytic substrates (Ogston and West, 1981). The presence of whole plasma, plasma fractions or purified proteins including albumin, abolished the inhibitory effect of α-tocopherol on plasmin, urokinase and tissue activator. It is concluded that vitamin E is unlikely to influence plasmin or plasminogen activator activity under *in vivo* conditions. In further experiments it was observed that urokinase is not inhibited by phytol, menadione or hydroquinone, suggesting that both the isoprenoid chain and the ring component of α-tocopherol (Figure 10.3) are required for inhibitory activity (Ogston, 1982).

Effect of Alcoholic Beverages on the Hemostatic Mechanism

While alcohol-containing drinks may be considered as peripheral to physiological nutrition, their widespread use to supplement other foods merits some discussion on their influence on the hemostatic mechanism. The changes which arise as a complication of alcoholic hepatic cirrhosis are, however, outside the scope of this volume and will not be considered: for a review of these changes see Ratnoff (1977).

Alcohol and Platelets

Ethanol ingestion can induce thrombocytopenia: platelet production is impaired in spite of an increase in the number of megakaryocytes. The cause is unknown, but a direct toxic effect has been suggested (Sullivan and Herbert, 1964; Lindenbaum and Hargrove, 1968; Post and Desforges, 1968). In addition to decreased platelet production there is evidence for an increased rate of destruction of circulating platelets in the presence of sustained high blood levels of alcohol (Sullivan and Herbert, 1964; Cowan, 1973). Platelets from patients with alcohol-induced thrombocytopenia show a number of ultrastructural alterations (Cowan and Graham, 1975): these include absence or fragmentation of the circumferential band of microtubules, variation in the size and shape of granules, and the formation of vacuoles in the cytoplasm.

Platelet aggregability is affected by alcohol ingestion. Patients whose blood ethanol levels are elevated above 280 mg/dl for a week or more showed decreased platelet aggregability in response to ADP, epinephrine, thrombin and collagen, and reduced platelet factor 3 availability was found in some subjects (Haut and Cowan, 1974). The addition of ethanol to normal platelet-rich plasma to a concentration of 240 or 400 mg/dl resulted in impairment of secondary aggregation, reduction in platelet factor 3 availability and retardation in the rate of nucleotide release (Haut and Cowan, 1974). It was concluded that ethanol or a metabolite impairs platelet function through both extraneous factors and platelet injury. In the study by Davis and Phillips (1970) the addition of ethanol *in vitro* (final concentration 240 mg/dl) was also found to inhibit the aggregation of platelets in response to collagen and norepinephrine, but not to ADP.

The ingestion of alcohol is followed by a number of alterations in platelet metabolism, although the precise relationship of these to the alterations in platelet function is uncertain. Metabolic abnormalities include a decrease in storage pool nucleotides, a decrease in platelet cyclic AMP concentration and a reduction in platelet monoamine oxidase activity (Cowan, 1980). Thromboxane synthesis does not appear to be a factor since very high concentrations of ethanol are required to inhibit thromboxane formation from exogenous arachidonic acid (Pennington and Smith, 1979). Part or all of the platelet function abnormalities following alcohol ingestion may be mediated by the accompanying changes in plasma osmolarity (Cowan *et al.*, 1976), while some of the effects of alcohol may be due to metabolites of ethanol, for example, acetaldehyde (Cowan, 1980).

Influence of Beer, Cider and Wine on Fibrinolysis

The marked inhibition of blood fibrinolytic activity lasting some hours after drinking beer, cider or wine was first noted by Fearnley and colleagues (1960): gin, whisky and absolute ethanol did not affect fibrinolytic activity indicating that the inhibition does not result from the alcohol content. Subsequent studies confirmed that beer and wine decreased plasma fibrinolytic activity while increasing the capacity of serum to inhibit tissue activator (Nilsson *et al.*, 1961): *in vitro* experiments by these investigators demonstrated that wine and beer contain a heat-stable agent which can inhibit plasminogen activator. Analysis of this substance isolated from grape pulp showed it to be a pectin. In our own studies we have confirmed the inhibition of fibrinolysis after drinking cider, and have shown that cider inhibits tissue activator, urokinase and plasmin in *in vitro* systems (Anderson *et al* ., 1983).

310

Effect of Onion and Garlic on Hemostatic Function

Interest in a possible influence of onion (*Allium capa*) on the hemostatic mechanism originated in a report by Gupta and colleagues (1966) that ingestion of onions prevented the decrease in fibrinolytic activity which followed a fat-enriched breakfast. This effect was confirmed with raw, fried or boiled onions (Menon *et al.*, 1968; Menon, 1969). Menon and colleagues (1969) also noted that onions did not change the recalcified plasma clotting time, fibrinogen concentration or cholesterol level. In a later study Menon (1970) found that dried onions reconstituted with water and lyophilised onions also accelerated fibrinolytic activity. Bordia *et al.* (1975) reported that the freshly extracted juice or the essential oil fraction of both onions and garlic (*Allium sativum*) could prevent the reduced fibrinolytic activity produced by alimentary lipemia. The essential oil of garlic fed for 20 days to patients in the immediate post-infarction period was claimed to enhance blood fibrinolytic activity (Bordia *et al.*, 1977), while Chutani and Bordia (1981) noted that both raw and fried garlic increased fibrinolysis, but that frying removed the acrid smell of garlic.

In addition to its effect on fibrinolysis it has been found that onion and garlic influence platelet function. Bordia (1978) reported that the addition of essential oil of garlic to plasma inhibited platelet aggregation induced by ADP, epinephrine and collagen: the oral administration of garlic over a five-day period also produced inhibition of platelet aggregation. In another study the increased rate of platelet aggregation in response to ADP after a high saturated fat meal was reversed by the addition of 75 g fried onions to the meal (Baghurst *et al.*,1977), although the onions did not prevent the rise in plasma triglyceride after the fat meal.

The active principles of onion and garlic and the mechanism by which they influence fibrinolysis and platelet function is not fully elucidated. Augusti and associates (1975) observed that a steam-distilled and an ether extract of onion enhanced blood fibrinolytic activity as assessed by the euglobulin clot lysis time: components of the extract of onion were identified and the sulphur-containing amino acid cyclo-allicin found to have fibrinolysis-enhancing properties when fed to volunteers. Dipropyl disulfide was tested in a single individual and also shortened the euglobulin clot lysis time. The mechanism of fibrinolysis enhancement by these agents is unknown, and there are no data yet available on the influence of long-term intake of large quantities of onion or garlic on hemostatis.

Aqueous and alcoholic extracts of onion inhibited platelet aggregation in response to arachidonic acid, ADP and collagen, but not to thrombin (Phillips and Poyser, 1978): attempts to isolate the active constituent in onion led to the conclusion that several anti-aggregation substances were present. Makheja and colleagues (1979) isolated anti-platelet activity from onion and garlic which blocked thromboxane synthesis by inhibiting cyclo-oxygenase. In a later report inhibition of platelet arachidonic acid metabolism was claimed to be due principally to paraffinic polysulfides, particularly dimethyl trisulfide (Makheja *et al.*, 1981): adenosine and allicin, also found in the extracts of onion and garlic, inhibited platelet aggregation, but not arachidonic acid metabolism.

Miscellaneous Ingestants and Hemostatic Function

Black Tree Fungus (Mo-er)

A Chinese food termed Ma-po dou-fu (Szechwanese hot bean curd) which contains mo-er or black tree fungus (*Auricularia polytricha*) has been reported to inhibit ADP-induced platelet aggregation (Hammerschmidt, 1980). Saline extracts of mo-er added to platelet-rich plasma also inhibited aggregation in response to ADP and blocked serotonin release. Ingestion of 70 g of mo-er by itself produced marked inhibition of ADP- induced platelet aggregation. The component of mo-er responsible for this effect was water-soluble and heat-stable, and had a molecular weight of under 10,000 daltons, but was otherwise uncharacterised. Makheja and associates (1981) have confirmed the inhibition of ADP, arachidonic acid and collagen-induced platelet aggregation by extracts by Chinese black tree fungus and found that adenosine accounted for the anti-platelet activity.

*Ginger (*Zingiber officinale Roscoe*)*

The ingestion of a brand of marmalade containing ginger as a major ingredient was incriminated as a possible cause of failure of platelet aggregation in response to arachidonic acid (Dorso *et al.*, 1980). Extracts of ground ginger, the rootstock of the tropical plant *Zingiber officinale Roscoe*, were found to inhibit completely arachidonate-induced platelet aggregation. The tentative suggestion was made that the anti-platelet activity resides in gingerol, a component of the ginger plant with a structural similarity to aspirin.

Capsicum

There have been published hints of a possible relationship between the ingestion of capsicum (peppers) and the induction of increased blood fibrinolytic activity. Glatzel and colleagues (1965) reported increased fibrinolytic activity in healthy persons after eating chillies. Wasantapruer *et al.*, (1974) claimed subsequently that a meal of noodles to which two teaspoonfuls of ground fresh capsicum (*Capsicum frutescens*) had been added caused a significant shortening of the euglobulin clot lysis time immediately after the meal: fibrinolytic activity had returned to normal 30 minutes later, indicating that any effect was likely to have taken place before absorption. Further data on this phenomenon were presented by Visudhiphan and associates (1982), and it was claimed that the higher plasma fibrinolytic activity found in Thai subjects compared to American whites residing in Thailand was the result of the frequent use of capsicum for seasoning and as an appetiser by Thais.

Caffeine

Al Samarrae and Truswell (1977) reported that the dilute whole blood clot lysis time was reduced 30 and 60 minutes after drinking coffee containing about 150 mg of caffeine. The increase in fibrinolytic activity was largely abolished when decaffeinated coffee was taken, leading to the conclusion that caffeine was responsible for the stimulation of fibrinolytic activity.

Influence of Obesity on Hemostatic Components

Platelet Function in Obesity

Little data is available on platelet numbers or function alterations in obesity. Using the rotating bulb technique Warlow and co-workers (1972) found no difference in platelet adhesiveness between groups of subjects of varying observed/standard weight ratios.

Coagulation Mechanism in Obesity

Obesity does not appear to influence measurements of overall coagulability. Grace and associates (1970) found no significant relationship between adiposity and the recalcified plasma clotting time or the partial thromboplastin time while in the study by Warlow *et al.* (1972) obesity was not associated with any change in the whole blood clotting time in plastic tubes. Burt and colleagues (1962) was also unable to find any relationship between the recalcified plasma clotting time and body fat content in 54 men aged 18 to 33 years. The factor II-VII-X activity was found to be higher in the obese subjects than in age and sex-matched controls in the study by Korsan-Bengtsen and co-workers (1972b). A number of investigators have demonstrated an association between obesity and increased plasma fibrinogen concentration in both men and women (Ogston and McAndrew, 1964; Bennett *et al.*, 1966; Grace and Goldrick, 1968; Korsan-Bengtsen *et al.*, 1972b; Warlow *et al.*, 1972; Meade *et al.*, 1979), but the mechanism and significance of this rise in fibrinogen level is obscure.

Fibrinolytic System in Obesity

A significant relationship between obesity and low fibrinolytic activity was first noted in white Australians by Goldrick (1961). Using skinfold thickness as an index of adiposity, Shaw and MacNaughton (1963) confirmed this finding. In earlier investigations it had been noted that patients with a previous myocardial infarction could be divided into those who were overweight and had diminished plasma fibrinolytic activity and those who were of average body weight and had normal fibrinolytic activity (Ogston, 1962), while Fearnley and his colleagues (1963) found a significant relationship between obesity and low fibrinolytic activity in all groups of patients with diabetes mellitus and controls except the female controls. This association between obesity and reduced fibrinolytic activity has been confirmed repeatedly (Ogston and McAndrew, 1964; Grace and Goldrick, 1968; Warlow *et al.*, 1972; Meade *et al.*, 1979).

The explanation for the low plasma fibrinolytic activity in obese persons is not established. It is not an artifact resulting from the increased fibrinogen concentration in the euglobulin precipitate (Warlow *et al.*, 1972). No change in the plasma inhibition of the caseinolytic activity of plasmin or in serum inhibition of urokinase in a clot lysis system has been detected (Bennett *et al.*, 1966; Grace and Goldrick, 1968), but no data are available on the level of inhibitors of vascular activator or the individual plasmin inhibitors. The release of plasminogen activator in response to venous occlusion has been found to be normal (Grace, 1968; Korsan-Bengtsen *et al.*, 1972b), but obese subjects had a smaller response to a standard exercise stimulus than those of normal physique (Ogston and McAndrew, 1964; Korsan-Bengtsen *et al.*, 1972b). The level of activator in the tissues was reported to be the same in the obese and the non-obese (Grace and Goldrick, 1969), and the reduced plasma activator levels could not be attributed to changes in plasma triglyceride, free fatty acid, insulin or glucose (Grace and Goldrick, 1968).

Dietary restriction sufficient to induce rapid weight loss (140 to 600 kilocalories/day) increased plasma fibrinolytic activity in obese persons, but the increase was only maintained while weight loss continued (Ogston and McAndrew, 1964). It was concluded that the rise in plasminogen activator activity is associated with the metabolic changes of starvation.

Effect of Starvation on Hemostasis

Acute Effects

Egberg and associates (1977) examined the effect of an eleven-day fast in healthy subjects: during this period only calorie-free fluids and sodium bicarbonate and potassium chloride as indicated from the serum electrolyte measurements were taken. The mean level of factor VIII, both coagulant activity and antigen, were reduced by the tenth day of starvation, but only the changes in the factor VIII coagulant activity reached statistical significance. The reduction in factor VIII was attributed to decreased rate of synthesis. There was no change in the plasma levels of factor V, IX or fibrinogen, and no consistent changes in fibrinopeptide A or fibrin(ogen) degradation products were found.

Normal volunteers maintained on a 140 kilocalorie diet for three days showed a substantial increase in plasma fibrinolytic activity (Ogston and McAndrew, 1964).

Chronic Effects

Data on the effects of chronic starvation on the hemostatic mechanism are also scanty. Protein malnutrition (kwashiorkor) in children was noted to be associated with deficiency of factor VII while some of the children had thrombocytopenia (Merskey and Hansen, 1957). In a later study from Libya Hassanain and Tankowsky (1973) found significant reductions in factors VII, X and prothrombin in children with kwashiorkor; the factor V level was within normal. After administration of vitamin K the factor VII level increased, but the factor X and prothrombin returned to normal levels only after a period on a high protein diet. In contrast to the changes in kwashiorkor, children with calorie malnutrition (marasmus) had mean values for factors V, VII, X and prothrombin which did not differ from normal. It was suggested that this difference stems from an impairment of protein synthesis in kwashiorkor and its continuance in marasmus.

Bosch and colleagues (1979) examined platelet function in severely undernourished children in Venezuela. Platelet adhesiveness to glass beads, factor 3 availability and total platelet factor 3 and ATP were found to be significantly diminished. Most of the children had some alteration in platelet aggregability; half had absent or reduced collagen-induced aggregation, and some had a diminished rate of aggregation in response to ADP.

Effect of Zinc Deficiency on Hemostasis

Zinc-deficient rats have been reported to have a prolonged bleeding time which is corrected by the administration of zinc supplements (Gordon and O'Dell, 1980a). There were no significant differences between the zinc-deficient and control rats in respect of the whole blood clotting time, one-stage prothrombin time or partial thromboplastin time, but impairment of platelet aggregation was demonstrated (Gordon and O'Dell, 1980b). In further studies three men were given a low zinc diet and it was observed that platelet aggregation in response to ADP and arachidonate was impaired at times when the plasma zinc level was low (60 µg/dl or less). A normal aggregation response was restored by oral zinc supplements (Gordon *et al.*, 1982). The mechanism by which zinc deficiency influences platelet function is unknown; possibilities include interference with prostaglandin synthesis or an effect on the platelet membrane.

References

Al Samarrae, W. and Truswell, A.S. (1977) 'Short-term effect of coffee on blood fibrinolytic activity in healthy adults', *Atherosclerosis, 26*, 255-60
Anderson, J.A., Gow, L.A. and Ogston, D. (1983) 'Influence of cider on the fibrinolytic enzyme system', *Acta Haematologica*, in press
Antar, M.A. (1968) 'Studies of thrombus formation and fatty acid composition of serum lecithins and cephalins in man during high sugar and starch diets', *Journal of Atherosclerosis Research, 8*, 596-73
Augusti, K.T., Benaim, M.E., Dewar, H.A. and Virden, R. (1975) 'Partial identification of the fibrinolytic activators in onion', *Atherosclerosis,21*, 409-16

Aviram, M. and Brook, J.G. (1981) 'Lipoprotein platele,
   platelet cholesterol content and platelet function', *Thr*
   Haemostasis, 46, 195
Baghurst, K.I., Raj, M.J. and Truswell, A.S. (1977) 'Onions
   aggregation', *Lancet, i,* 101
Banerjee, A.K. and Etherington, M. (1974) 'Platelet function .
   Age and Ageing, 3, 97-105
Beaumont, J.L., Carlson, L.A., Cooper, G.R., Fejfar, Z., Fredric
   Strasser, T. (1970) 'Classification of hyperlipidaemias and
   hyperlipoproteinaemias', *Bulletin of the World Health Organisa*
   891-915
Bennett, N.B., Ogston, C.M., McAndrew, G.M. and Ogston, D. (1966) 'Studies on
   the fibrinolytic enzyme system in obesity', *Journal of Clinical Pathology, 19,*
   241-3
Billimoria, J.D., Drysdale, J., James, D.C.O. and Maclagan, N.F. (1959)
   'Determination of fibrinolytic activity of whole blood. With special reference to
   the effects of exercise and fat feeding', *Lancet, ii,* 472-5
Bills, T.K., Smith, J.B. and Silver, M.J. (1976) 'Metabolism of ^{14}C arachidonic acid
   by human platelets', *Biochimica et Biophysica Acta, 424,* 303-14
Bills, T.K., Smith, J.B. and Silver, M.J. (1977) 'Selective release of arachidonic
   acid from the phospholipids of human platelets in response to thrombin',
   Journal of Clinical Investigation, 60, 1-6
Bizios, R., Wong, L.K., Vaillancourt, R., Lees, R.S. and Carvalho, A.C. (1977)
   'Platelet prostacyclin endoperoxides formation in hyperlipidemias', *Thrombosis*
   and Haemostasis, 38, 228
Bordia, A. (1978) 'Effect of garlic on human platelet aggregation',
   Atherosclerosis, 30, 355-60
Bordia, A., Bansal, H.C., Arora, S.K. and Singh, S.V. (1975) 'Effect of the
   essential oils of garlic and onion on alimentary hyperlipemia', *Atherosclerosis,*
   21, 15-9
Bordia, A., Joshi, H.K., Sanadhya, Y.K. and Bhu, N. (1977) 'Effect of essential oil
   of garlic on serum fibrinolytic activity in patients with coronary artery disease',
   Atherosclerosis, 28, 155-9
Bordia, A., Paliwal, D.K., Jain, K. and Kothari, L.K. (1978) 'Acute effect of
   ascorbic acid on fibrinolytic activity', *Atherosclerosis, 30,* 351-4
Born, G.V.R. and Wright, H.P. (1967) 'Platelet adhesiveness in experimental
   scurvy', *Lancet, i,* 477-8
Bosch, N.B., Sacz, A., Bosch, H., Camarillo, H., Fonseca, G. and Arguello, A.
   (1979) 'Platelet functions in undernourished children', *Thrombosis and*
   Haemostasis, 42, 470
Botti, R.E. and Ratnoff, O.D. (1963) 'The clot-promoting effect of soaps on
   long-chain saturated fatty acids', *Journal of Clinical Investigation, 42,* 1569-77
Buckell, M. and Elliott, F.A. (1959) 'Effect of butter lipaemia on the rate of clot
   lysis in normal males', *Lancet, i,* 662-3
Burt, J.J., Blyth, C.S. and Rierson, H. (1962) 'Body fat, blood coagulation time,
   and the Harvard step test recovery index', *Research Quarterly, 33,* 339-42
Carvalho, A.C.A., Colman, R.W. and Lees, R.S. (1974) 'Platelet function in
   hyperproteinemia', *New England Journal of Medicine, 290,* 434-8
Chutani, S.K. and Bordia, A. (1981) 'The effect of fried versus raw garlic on
   fibrinolytic activity in man', *Atherosclerosis, 38,* 417-21
Coccheri, S., d'Antuono, G., Alessandri, M. and Garaja Condebru, J. (1966) 'The
   initiation of clotting in lipaemic blood: contact phase and platelet
   participation', *Hémostase, 6,* 339-49
Cohen, P. and Derksen, A. (1969) 'Comparison of phospholipid and fatty acid

...position of human erythrocytes and platelets', *British Journal of Haematology, 17*, 359-71

Cohen, P., Derksen, A. and van den Bosch, H. (1970) 'Pathways of fatty acid metabolism in human platelets', *Journal of Clinical Investigation, 49*, 128-39

Connor, W.E. (1962) 'The acceleration of thrombus formation by certain fatty acids', *Journal of Clinical Investigation, 41*, 1199-205

Connor, W.E. and Poole, J.C.F. (1961) 'The effect of fatty acids on the formation of thrombi', *Quarterly Journal of Experimental Physiology, 46*, 1-7

Corash, L., Schaefer, E., Poindexter, E. and Anderson, J. (1976) 'Platelet function and survival in familial hypercholesterolemia', *Circulation, 54*, Suppl. II, 117

Cordova, C., Musca, A., Violi, F., Perrone, A. and Alessandri, C. (1981) 'Influence of ascorbic acid on platelet aggregation *in vitro* and *in vivo*', *Atherosclerosis, 41*, 15-9

Cowan, D.H. (1973) 'Thrombokinetic studies in alcohol-related thrombocytopenia', *Journal of Laboratory and Clinical Medicine, 81*, 54-76

Cowan, D.H. (1980) 'Effect of alcoholism on hemostasis', *Seminars in Hematology, 17*, 137-47

Cowan, D.H. and Graham, R.C. (1975) 'Studies on the platelet defect in alcoholism', *Thrombosis et Diathesis Haemorrhagica, 33*, 310-27

Cowan, D.H., Graham, R.C. and Shook, P. (1976) 'Hyperosmolality: A factor in ethanol-related platelet dysfunction?' *Seminars in Hematology, 13*, 103-14

Crawford, G.P.M., Warlow, C.P., Bennett, B., Dawson, A.A., Douglas, A.S., Kerridge, D.F. and Ogston, D. (1975) 'The effect of vitamin C supplements on serum cholesterol, coagulation, fibrinolysis and platelet adhesiveness', *Atherosclerosis, 21*, 451-4

Creter, D., Pavlotzky, F. and Savir, H. (1979) 'Effect of vitamin E on platelet aggregation in diabetic retinopathy', *Acta Haematologica, 62*, 74-7

Davis, J.W. and Phillips, P.E. (1970) 'The effect of ethanol on human platelet aggregation *in vitro*', *Atherosclerosis, 11*, 473-7

Deykin, D. and Desser, R.K. (1968) 'The incorporation of acetate and palmitate into lipids by human platelets', *Journal of Clinical Investigation, 47*, 1590-602

Dole, V.R., James, A.T., Webb, J.P.W., Rizack, M.A. and Sturman, M.F. (1959) 'The fatty acid patterns of plasma lipids during alimentary lipemia', *Journal of Clinical Investigation, 38*, 1544-54

Dorso, C.R., Levin, R.I., Eldor, A., Jaffe, E.A. and Weksler, B.B. (1980) 'Letter', *New England Journal of Medicine, 303*, 756-7

Dubber, A.H.C., Rifkind, B., Gale, M., McNicol, G.P. and Douglas, A.S. (1967) 'The effect of fat feeding on fibrinolysis, 'Stypven' time and platelet aggregation', *Journal of Atherosclerosis Research, 7*, 225-35

Dyerberg, J. and Bang, H.O. (1979) 'Haemostatic function and platelet polyunsaturated fatty acids in Eskimos', *Lancet, ii*, 433-5

Dyerberg, J., Bang, H.O., Stoffersen, E., Moncada, S. and Vane, J.R. (1978) 'Eicosapentaenoic acid and prevention of thrombosis and atherosclerosis', *Lancet, ii*, 117-9

Egberg, N., Kockum, C. and Palmblad, J. (1977) 'Fasting (acute energy deprivation) in man: effect on blood coagulation and fibrinolysis', *American Journal of Clinical Nutrition, 30*, 1963-7

Egeberg, O. (1966) 'Blood factor XI after fat-rich meals', *Thrombosis et Diathesis Haemorrhagica, 15*, 390-402

Farbiszewski, R. and Worowski, K. (1968) 'Enhancement of platelet aggregation and adhesiveness by β-lipoprotein', *Journal of Atherosclerosis Research, 8*, 988-90

Farquhar, J.W., Merigan, T.C. and Sokolow, M. (1961) 'Plasma fibrinolysis in man: the effect of chylomicrons derived from different dietary fats', *Journal of*

Experimental Medicine, 113, 587-97

Fearnley, G.R., Ferguson, J., Chakrabarti, R. and Vincent, C.T. (1960) 'Effect of beer on blood fibrinolytic activity', *Lancet, i*, 184-6

Fearnley, G.R., Chakrabarti, R. and Avis, P.D.R. (1963) 'Blood fibrinolytic activity in diabetes mellitus and its bearing on ischaemic heart disease and obesity', *British Medical Journal, 1*, 921-3

Ferguson, J.C., Mackay, N. and McNicol, G.P. (1970) 'Effect of feeding fat on fibrinolysis, Stypven time, and platelet aggregation in Africans, Asians, and Europeans', *Journal of Clinical Pathology, 23*, 580-5

Fleischman, A.J., Justice, D., Bierenbaum, M.L., Stier, A. and Sullivan, A. (1975) 'Beneficial effect of increased dietary linoleate upon *in vivo* platelet function in man', *Journal of Nutrition, 105*, 1286-90

Fullerton, H.W., Davie, W.J.A. and Anastaspoulos, G. (1953) 'Relationship of alimentary lipaemia to blood coagulability', *British Medical Journal, 2*, 250-3

Gajewski, J. (1961) 'Effect of the ingestion of various fats on the fibrinolytic activity in normal subjects and patients with coronary heart disease', *Journal of Atherosclerosis Research, 1*, 222-8

Gerrard, J.M., White, J.G. and Krivit, W. (1976) 'Labile aggregation stimulating substance, free fatty acids, and platelet aggregation', *Journal of Laboratory and Clinical Medicine, 87*, 73-82

Glatzel, H. and Ruberg-Schweer, M. (1965) 'Kreislaufregulation und Ernahrung. 7 Aktivierung der Fibrinolyse durch Gewurze (Chillies)', *Zeitschrift fur Kreislaufforschung, 53*, 374-84

Goldrick, R.B. (1961) 'Fibrinolysis, blood clotting, serum lipids and body build of natives of New Guinea and Australians', *Australasian Annals of Medicine, 10*, 20-8

Gomes, J.A.C., Verkatachalapathy, D. and Haft, J. (1976) 'The effect of vitamin E on platelet aggregation', *American Heart Journal, 91*, 425-9

Goodman, D.S. (1958) 'The interaction of human serum albumin with long-chain fatty acid anions', *Journal of the American Chemical Society, 80*, 3892-8

Goodnight, S.H., Harris, W.S. and Connor, W.E. (1981) 'The effects of dietary ω3 fatty acids on platelet composition and function in man: a prospective controlled study', *Blood, 58*, 880-5

Gordon, P.R. and O'Dell, B.L. (1980a) 'Short-term zinc deficiency and haemostasis in the rat', *Proceedings of the Society for Experimental Biology and Medicine, 163*, 240-4

Gordon, P.R. and O'Dell, B.L. (1980b) 'Rat platelet aggregation impaired by short-term zinc deficiency', *Journal of Nutrition, 110*, 2125-9

Gordon, P.R., Woodruff, C.W., Anderson, H.L. and O'Dell, B.L. (1982) 'Effect of acute zinc deprivation on plasma zinc and platelet aggregation in adult males', *American Journal of Clinical Nutrition, 35*, 113-9

Grace, C.S. (1968) 'The fibrinolytic enzyme system in obesity: the effects of venous occlusion and *in vitro* activation by surface contact', *Clinical Science, 34*, 497-504

Grace, C.S. and Goldrick, R.B. (1968) 'Fibrinolysis and body build. Interrelationships between blood fibrinolysis, body composition and parameters of lipid and carbohydrate metabolism', *Journal of Atherosclerosis Research, 8*, 705-19

Grace, C.S. and Goldrick, R.B. (1969) 'Tissue fibrinolytic activity in obesity', *Australian Journal of Experimental Biology and Medical Science, 47*, 397-400

Grace, C.S., Sinnett, P.F. and Whyte, H.M. (1970) 'Blood fibrinolysis and coagulation in New Guineans and Australians', *Australasian Annals of Medicine, 4*, 328-33

Greig, H.B. (1956) 'Inhibition of fibrinolysis by alimentary lipaemia', *Lancet, ii*, 16-8

Griffith, M.J. (1979) 'Covalent modification of human α-thrombin with pyridoxal 5'-phosphate', *Journal of Biological Chemistry, 254*, 3401-6

Grossman, M.I., Palm, L., Becker, G.H. and Moeller, H.C. (1954) 'Effects of lipemia and heparin on free fatty acid content of rat plasma', *Proceedings of the Society for Experimental Biology and Medicine, 87*, 312-5

Gupta, N.N., Mehrotra, R.M.L. and Sircar, A.R. (1966) 'Effect of onion on serum cholesterol, blood coagulation factors and fibrinolytic activity in alimentary lipaemia', *Indian Journal of Medical Research, 54*, 48-53

Hammerschmidt, D.E. (1980) 'Szechwan purpura', *New England Journal of Medicine, 302*, 1191-3

Haslam, R.J. (1964) 'Role of adenosine diphosphate in the aggregation of human blood-platelets by thrombin and by fatty acids', *Nature, 202*, 765-8

Hassanein, E.A. and Tankovsky, I. (1973) 'Disturbances of coagulation mechanism in protein-calorie malnutrition', *Tropical and Geographical Medicine, 25*, 158-62

Haut, M.J. and Cowan, D.H. (1974) 'The effect of ethanol on hemostatic properties of human blood platelets', *American Journal of Medicine, 56*, 22-33

Hay, C.R.M., Durber, A.P. and Saynor, R. (1982) 'Effect of fish oil on platelet kinetics in patients with ischaemic heart disease', *Lancet, i*, 1269-72

Herbert, R.J., Bennett, B. and Ogston, D. (1973) 'Effect of low-density lipoprotein preparations on plasmin', *Clinical Science and Molecular Medicine, 45*, 129-32

Hoak, J.C., Warner, E.D. and Connor, W.E. (1967) 'Platelets, fatty acids and thrombosis', *Circulation Research, 20*, 11-7

Hoak, J.C., Spector, A.A., Fry, G.L. and Warner, E.D. (1970) 'Effect of free fatty acids on ADP-induced platelet aggregation', *Nature, 228*, 1330-2

Hoak, J.C., Spector, A.A., Fry, G.L. and Barnes, B.C. (1972) 'Localization of free fatty acids taken up by human platelets', *Blood, 40*, 16-22

Hornstra, G., Lewis, B., Chait, A., Turpeinen, O., Karvonen, M.J. and Vergroesen, A.J. (1973) 'Influence of dietary fat on platelet function in men', *Lancet, i*, 1155-7

Hougie, C. and Ayers, F. (1960) 'Lipaemia and fibrinolytic potentiality', *Lancet, i*, 186-8

Hubbard, D. and Lucas, G.L. (1960) 'Ionic charges of glass surfaces and other materials, and their possible role in the coagulation of blood', *Journal of Applied Physiology, 15*, 265-70

Hutton, R.A., Mikhailidis, D., Dormandy, K.M. and Ginsburg, J. (1979) 'Platelet aggregation studies during transient hypoglycaemia', *Journal of Clinical Pathology, 32*, 434-8

Ingeberg, S. and Stoffersen, E. (1979) 'Platelet dysfunction in patients with vitamin B_{12} deficiency', *Acta Haematologica, 61*, 75-9

Jakubowski, J.A. and Ardlie, N.G. (1978) 'Modification of human platelet function by a diet enriched in saturated or polyunsaturated fat', *Atherosclerosis, 31*, 335-44

Jamieson, G.A., Jung, S.M., Ordinas, A. and Marcum, J.M. (1980) 'The influence of receptor mobility and association on the high affinity binding of thrombin to platelets', in A. Rotman, F.A. Meyer, G. Gitler and A. Silverberg (eds.) *Platelets: Cellular Response Mechanisms and their Biological Significance*, Wiley, New York, pp. 151-6

Johnson, G.J., Holloway, D.E., Hutton, S.W. and Duane, W.C. (1981) 'Platelet function in scurvy and experimental human vitamin C deficiency', *Thrombosis Research, 24*, 85-93

Kay, J.H., Hutton, S.B., Weiss, G.N. and Ochsner, A. (1950) 'Studies on antithrombin. III. A plasma antithrombin test for the prediction of intravascular clotting', *Surgery, 28*, 24-8

Kernoff, P.B.A., Willis, A.L., Stone, K.J., Davies, J.A. and McNicol, G.P. (1977), 'Antithrombotic potential of dihomo-gamma-linolenic acid', *British Medical Journal, 4*, 1441-4

Kerr, J.W., Pirrie, R., MacAulay, I. and Bronte-Stewart, B. (1965) 'Platelet-aggregation by phospholipids and free fatty acids', *Lancet, i*, 1296-9

Kornecki, E. and Feinberg, H. (1979) 'Mechanism of inhibition of thrombin-induced platelet aggregation by pyridoxal phosphate', *Biochemical and Biophysical Research Communications, 90*, 963-8

Kornecki, E. and Feinberg, H. (1980) 'Pyridoxal phosphate inhibition of platelet function', *American Journal of Physiology, 238*, H54-60

Korsan-Bengtsen, K., Gustavsson, A., Sjöström, L. and Björntorp, P. (1972a) 'Effects of carbohydrate feeding on blood coagulation, fibrinolysis and platelet adhesiveness — relations to serum lipids and lipoproteins', *Thrombosis Research, 1*, 407-26

Korsan-Bengtsen, K., Stenberg, J., Sjöström, L., Björntorp, P. and Sullivan, L. (1972b) 'Blood coagulation, fibrinolysis and platelet function in obese subjects at rest and after maximal exercise', *Thrombosis Research, 1*, 389-406

Korsan-Bengtsen, K., Elmfeldt, D. and Holm, T. (1974) 'Prolonged plasma clotting time and decreased fibrinolysis after long term treatment with α-tocopherol', *Thrombosis et Diathesis Haemorrhagica, 31*, 505-12

Lackner, H. and Sougin-Mibashan, R. (1964) 'Fibrinolysis and alimentary lipaemia in whites and Bantus; their relationship and response to intravenous heparin', *Thrombosis et Diathesis Haemorrhagica, 11*, 108-18

Lake, A.M., Stuart, M.J. and Oski, F.A. (1977) 'Vitamin E deficiency and enhanced platelet function: Reversal following E supplementation', *Journal of Pediatrics, 90*, 722-5

Lam, S.C.-T., Harfenist, E.J., Packham, M.A. and Mustard, J.F. (1980) 'Investigation of possible mechanisms of pyridoxal 5'-phosphate inhibition of platelet reactions', *Thrombosis Research, 20*, 633-45

Leonardi, R.G., Alexander, B. and White, F. (1973) 'Prevention of aspirin inhibition of platelet release reaction by the fatty acid precursor of platelet prostaglandins', *Thrombosis Research, 3*, 327-38

Levine, P.H. (1973) 'A qualitative platelet defect in severe vitamin B_{12} deficiency', *Annals of Internal Medicine, 78*, 533-9

Lindenbaum, J. and Hargrove, R.L. (1968) 'Thrombocytopenia in alcoholics', *Annals of Internal Medicine, 68*, 526-32

Lowe, G.D.O., Drummond, M.M., Third, J.L.H.C., Bremner, W.F., Forbes, C.D., Prentice, C.R.M. and Lawrie, T.D.V. (1979a) 'Increased plasma fibrinogen and platelet-aggregates in type II hyperproteinaemia', *Thrombosis and Haemostasis, 42*, 1503-7

Lowe, G.D.O., Johnston, R.V., Drummond, M.M., Forbes, C.D. and Prentice, C.R.M. (1979b) 'Induction of circulating platelet-aggregates in healthy subjects by a saturated fat meal', *Thrombosis Research, 16*, 565-8

Macfarlane, R.G., Trevan, J.W. and Attwood, A.M.P. (1941) 'Participation of a fat soluble substance in the coagulation of blood', *Journal of Physiology, 99*, 7P

Machlin, L.J., Filipski, R., Willis, A.L., Kuhn, D.C. and Brin, M. (1975) 'Influence of vitamin E on platelet aggregation and thrombocythemia', *Proceedings of the Society for Experimental Biology and Medicine, 149*, 275-7

Maclagan, N.F., Billimoria, J.D. and Curtis, C. (1958) 'Lipaemia and blood coagulation with special reference to the Stypven technique', *Lancet, ii*, 865-9

Mahadevan, V., Singh, H. and Lundberg, W.O. (1966) 'Effects of saturated and unsaturated fatty acids in blood platelet aggregation *in vitro*', *Proceedings of the*

Society for Experimental Biology and Medicine, 121, 82-5

Makheja, A.N., Vanderhoek, J.Y. and Bailey, J.M. (1979) 'Inhibition of platelet aggregation and thromboxane synthesis by onion and garlic', *Lancet, i*, 781

Makheja, A.N., Low, C.E. and Bailey, J.M. (1981) 'Biological nature of platelet inhibitors from *Allium cepa, Allium sativum* and *Auricularia polytricha*', *Thombosis and Haemostasis, 46*, 148

Marcus, A.J., Ullman, H.L. and Safier, L.B. (1969) 'Lipid composition of subcellular particles of human blood platelets', *Journal of Lipid Research, 10*, 108-11

Margolis, J. (1962) 'Activation of Hageman factor by saturated fatty acids', *Australian Journal of Experimental Biology, 40*, 505-14

McArdle, B.M., Lowe, G.D.O., Stromberg, P., Lorimer, A.R., Forbes, C.D. and Prentice, C.R.M. (1982) 'Fibrinolysis in type II hyperlipoproteinaemia', *Haemostasis 11*, Suppl. 1, 93

McDonald, G.A. and Fullerton, H.W. (1958) 'Comparison of animal and vegetable fats in increasing blood coagulability', *Lancet, ii*, 598-600

McGregor, L. and Renaud, S. (1978) 'Effect of dietary linoleic acid deficiency on platelet aggregation and phospholipid fatty acids of rats', *Thrombosis Research, 12*, 921-7

McGregor, L., Morazain, R. and Renaud, S. (1980) 'Effect of dietary linoleic acid on platelet function in the rat', *Thrombosis Research, 20*, 499-507

Meade, T.W., Chakrabarti, R., Haines, A.P., North, W.R.S. and Stirling, Y. (1979) 'Characteristics affecting fibrinolytic activity and plasma fibrinogen concentrations', *British Medical Journal, 1*, 153-6

Melhorn, D.K. and Gross, S. (1971) 'Vitamin E-dependent anemia in the premature infant. II. Relationships between gestational age and absorption of vitamin E', *Journal of Pediatrics, 79*, 581-8

Menon, I.S. (1969) 'Fresh onions and blood fibrinolysis', *British Medical Journal, 1*, 845

Menon, I.S. (1970) 'Onions and blood fibrinolysis', *British Medical Journal, 2*, 421

Menon, I.S., Kendal, R.Y., Dewar, H.A. and Newell, D.J. (1968) 'Effect of onions on blood fibrinolytic activity', *British Medical Journal, 3*, 351-2

Merskey, C. and Hansen, J.D.L. (1957) 'Blood coagulation defects in kwashiorkor and infantile gastroenteritis', *British Journal of Haemotology, 3*, 39-49

Merskey, C. and Nossel, H.L. (1957) 'Blood coagulation after the ingestion of saturated and unsaturated fats', *Lancet, i*, 806-11

Meuwissen, O.J.A.Th., Hart, H.Ch. and van Hemel-Rupert, M.S.E. (1968) 'The influence of fats on blood coagulation and fibrinolysis', *Thrombosis et Diathesis Haemorrhagica, 19*, 267-78

Moroz, L.A. and Gilmore, N.J. (1976) 'Inhibition of plasmin-mediated fibrinolysis by vitamin E', *Nature, 259*, 235-7

Muraoka, T. and Okuda, H. (1977) 'Effects of free fatty acids on fibrinolytic activity', *Journal of Biochemistry, 82*, 529-33

Mustard, J.F. (1957) 'Increased activity of the coagulation mechanism during alimentary lipaemia: its significance with regard to thrombosis and atherosclerosis', *Canadian Medical Association Journal, 77*, 308-14

Mustard, J.F. and Murphy, E.A. (1962) 'Effect of different dietary fats on blood coagulation, platelet economy, and blood lipids', *British Medical Journal, 1*, 1651-5

Needleman, P., Raz, A., Minkes, M.S., Ferrendelli, J.A. and Sprecher, H. (1979) 'Triene prostaglandins: prostacyclin and thromboxane biosynthesis and unique biological properties', *Proceedings of the National Academy of Sciences, 76*, 944-8

Nilsson, I.M., Björkman, S.E., von Studnitz, W. and Hallén, A. (1961)

'Antifibrinolytic activity of certain pectins', *Thrombosis et Diathesis Haemorrhagica, 6,* 177-87

Nitzberg, S.I., Peyman, A., Goldstein, R. and Proger, S. (1959) 'Studies of blood coagulation and fibrinolysis in patients with idiopathic hyperlipemia and primary hypercholesteremia before and after a fatty meal', *Circulation, 19,* 676-90

Nordøy, A. and Rodset, M. (1971) 'The influence of dietary fats on platelets in man', *Acta Medica Scandinavica, 190,* 27-34

Nordøy, A., Strøm, E. and Gjesdal, K. (1974) 'The effect of alimentary lipaemia and primary hypertriglyceridaemia on platelet in man', *Scandinavian Journal of Haematology, 12,* 329-40

Nordøy, A., Brox, J.H. and Killis, J.E. (1981) 'Influence of cod liver oil and corn oil on platelets and vessel wall in man', *Thrombosis and Haemostasis, 46,* 179

Nossel, H.L., Wilner, G.D. and LeRoy, E.C. (1969) 'Importance of polar groups for initiating blood coagulation and aggregating platelets', *Nature, 221,* 75-6

O'Brien, J.R. (1955) 'Relation of blood-coagulation to lipaemia', *Lancet, ii,* 690-3

O'Brien, J.R. (1956) 'Effect of a meal of eggs and different fats on blood coagulability', *Lancet, ii,* 232-4

O'Brien, J.R. (1957) 'Some postprandial effects of eating various phospholipids and triglycerides', *Lancet, i,* 1213-6

O'Brien, J.R., Etherington, M.D. and Jamieson, S. (1976a) 'Acute platelet changes after large meals of saturated and unsaturated fats', *Lancet, i,* 878-80

O'Brien, J.R., Etherington, M.D., Jamieson, S., Vergroesen, A.J. and Ten Hoor, F. (1976b) 'Effect of a diet of unsaturated fats on some platelet function tests', *Lancet, ii,* 995-7

Ogston, D. (1962) 'Serum mucoproteins and plasma fibrinolytic activity in coronary-artery disease', *British Medical Journal, 1,* 1242-4

Ogston, D. (1982) 'Inhibition of the plasminogen activator urokinase by alpha-tocopherol', *Acta Haematologica, 67,* 114-8

Ogston, D. and Fullerton, H.W. (1962) 'Effect of alimentary lipaemia on plasma fibrinolytic activity', *British Medical Journal, 2,* 1288-90

Ogston, D. and McAndrew, G.M. (1964) 'Fibrinolysis in obesity', *Lancet, ii,* 1205-7

Ogston, D. and West, L.L. (1981) 'The effect of alpha-tocopherol on plasmin and plasminogen activators', *Journal of Physiology, 316,* 11P

Ogston, D., Herbert, R.J., Akinsete, F.I. and Douglas, A.S. (1973) 'The influence of sodium salts of fatty acids on the fibrinolytic enzyme system', *Thrombosis Research, 3,* 489-99

Okuma, M., Takayama, H. and Uchino, H. (1980) 'Generation of prostaglandin-like substance and lipid peroxidation in vitamin E-deficient rats', *Prostaglandins, 19,* 527-36

Olson, R.E. (1973) 'Vitamin E and its relation to heart disease', *Circulation, 48,* 179-84

Pappenhagen, A.R., Koppel, J.L. and Olwin, J.H. (1963) 'Observations on the *in vitro* effects of chylomicra, low-density lipoproteins and phospholipid on human plasma euglobulin lysis', *Thrombosis et Diathesis Haemorrhagica, 9,* 164-74

Pennington, S.N. and Smith, C.P. (1979) 'The effects of ethanol on thromboxane synthesis by blood platelets', *Prostaglandins and Medicine, 2,* 43-50

Peyman, M.A., Nitzberg, S.I., Goldstein, R., Nothman, M. and Proger, S. (1960) 'The effect of long term feeding of essential fatty acids on blood coagulation and fibrinolysis', *American Journal of Medicine, 28,* 884-94

Phillips, C. and Poyser, N.L. (1978) 'Inhibition of platelet aggregation by onion extracts', *Lancet, i,* 1051-2

Philp, R.B. and Wright, H.P. (1965) 'Effect of adenosine on platelet adhesiveness in fasting and lipaemic bloods', *Lancet, ii*, 208-9

Pilkington, T.R.E., (1957) 'The effects of alimentary lipaemia on the calcium clotting time of human plasma', *Clinical Science, 16*, 261-8

Pollard, H.B., Tack-Goldman, K., Pazoles, C.J., Creutz, C.E. and Shulman, N.R. (1977) 'Evidence for control of serotonin secretion from human platelets by hydroxyl ion transport and osmotic lysis', *Proceedings of the National Academy of Sciences, 74*, 5295-9

Poole, J.C.F. and Robinson, D.S. (1956a) 'A comparison of the effects of certain phosphatides and of chylomicra on plasma coagulation in the presence of Russell's viper venom', *Quarterly Journal of Experimental Physiology, 41*, 31-5

Poole, J.C.F. and Robinson, D.S. (1956b) 'Further observations on the effects of ethanolamine phosphatide on plasma coagulation', *Quarterly Journal of Experimental Physiology, 41*, 295-300

Post, R.M. and Desforges, J.F. (1968) 'Thrombocytopenia and alcoholism', *Annals of Internal Medicine, 68*, 1230-6

Prost-Dvojakovic, R.J. and Samama, M. (1973/74) 'Clot-promoting and platelet-aggregating effects of fatty acids', *Haemostasis, 2*, 73-84

Purcell, I.M. and Constantine, J.W. (1971) 'Platelets and experimental scurvy', *Nature, 235*, 389-91

Rao, G.H.R., Cox, C.A., Gerrard, J.M. and White, J.G. (1981) 'Alpha tocopherol quinone (αTQ): a potent inhibitor of platelet function', *Progress in Lipid Research, 20*, 549-52

Ratnoff, O.D. (1977) 'The haemostatic defects of liver disease', in D. Ogston and B. Bennett (eds.) *Haemostasis: Biochemistry, Physiology and Pathology*, Wiley, London, pp. 446-66

Renaud, S., Dumont, E., Godsey, F. and Suplisson, A. (1978) 'Platelet functions in relation to dietary fats in farmers from two regions of France', *Thrombosis and Haemostasis, 40*, 518-31

Renaud, S., Morazain, R., Godsey, F., Dumont, E., Symington, I.S., Gillanders, E.M. and O'Brien, J.R. (1981) 'Platelet functions in relation to diet and serum lipids in British farmers', *British Heart Journal, 46*, 562-70

Riding, I.M. and Ellis, D. (1964) 'Antiplasmin activity of β-lipoprotein', *Journal of Atherosclerosis Research, 4*, 189-93

Robinson, D.S. and Poole, J.C.F. (1956) 'The similar effect of chylomicra and ethanolamine phosphatide on the generation of thrombin during coagulation', *Quarterly Journal of Experimental Physiology, 41*, 36-50

Sagel, J., Colwell, J.A., Crook, L. and Laimins, M. (1975) 'Increased platelet aggregation in early diabetes mellitus', *Annals of Internal Medicine, 82*, 733-8

Sarji, K.E., Kleinfelder, J., Brewington, P., Gonzales, J., Hempling, H. and Colwell, J.A. (1979) 'Decreased platelet vitamin C in diabetes mellitus: possible role in hyperaggregation', *Thrombosis Research, 15*, 639-50

Shattil, S.J., Anaya-Galindo, R., Bennett, J., Colman, J.W. and Cooper, R.A. (1975) 'Platelet hypersensitivity induced by cholesterol incorporation', *Journal of Clinical Investigation, 55*, 636-43

Shattil, S.J., Bennett, J.S., Colman, R.W. and Cooper, R.A. (1977) 'Abnormalities of cholesterol-phospholipid composition in platelets and low-density lipoproteins of human hyperlipoproteinemia', *Journal of Laboratory and Clinical Medicine, 80*, 341-53

Shaw, D.A. and MacNaughton, D. (1963) 'Relationship between blood fibrinolytic activity and body fatness', *Lancet, i*, 352-4

Sheehy, T.W. and Eichelberger, J.W. (1958) 'Alimentary lipemia and the coagulability of blood. Analysis by thromboelastography and silicone clotting time', *Circulation, 42*, 927-35

324

Shimizu, M., Hatta, Y., Hayashi, H., Itokawa, M., Yanagisawa, Y., Otani, T. and Nakachi, N. (1970) 'Effect of ascorbic acid on fribrinolysis', *Acta Haematologica Japonica, 33*, 137-48

Shore, P.A. and Alpers, H.S. (1963) 'Platelet damage induced in plasma by certain fatty acids', *Nature, 200*, 1331-2

Siess, W., Roth, P., Scherer, B., Kurzmann, I., Böhlig, B. and Weber, P.C. (1980) 'Platelet-membrane fatty acids, platelet aggregation, and thromboxane formation during a mackerel diet', *Lancet, i*, 441-4

Silver, M.J., Smith, J.B., Ingerman, C. and Kocsis, J.J. (1973) 'Arachidonic acid-induced human platelet aggregation and prostaglandin formation', *Prostaglandins, 4*, 863-75

Skrzydlewski, Z., Niewiarowski, S. and Skrzydlewska, J. (1966) 'Inhibition of proteolytic enzymes by β-lipoprotein', *Journal of Athersclerosis Research, 6*, 273-6

Spöttl, F., Holzknecht, F. and Braunsteiner, H. (1969) 'Enhancement of the fibrinolytic activity by venous occlusion in patients with primary "carbohydrate-induced" hypertriglyceridemia', *Acta Haematologica, 41*, 154-61

Spöttl, F., Holzknecht, F., Knapp, E., Steinmetz, U. and Braunsteiner, H. (1970) 'A basic study on the global coagulation and fibrinolysis of hyperlipaemic and atherosclerotic patients Part 2. Plasminogen activator, plasminogen, α_2-macroglobulin in atherosclerosis and endogenous hypertriglyceridaemia', *Atherosclerosis, 12*, 427-37

Steele, J.C.H., (1979) 'The effect of plasmin on human plasma low density lipoprotein', *Thrombosis Research, 15*, 573-9

Steiner, M. and Anastasi, J. (1976) 'Vitamin E. An inhibitor of the platelet release reaction', *Journal of Clinical Investigation, 57*, 732-7

Stuart, M.J., Gerrard, J.H. and White, J.G. (1980) 'Effect of cholesterol on production of thromboxane B_2 by platelets *in vitro*', *New England Journal of Medicine, 302*, 6-10

Subbarao, K., Kuchibhotla, J. and Green, D. (1977) 'Pyridoxine-induced inhibition of platelet aggregation and the release reaction', *Circulation, 56*, Part II, 77

Subbarao, K., Kakkar, V.V. and Ganguly, P. (1978) 'Binding of pyridoxal phosphate to human platelets: its effect on platelet function', *Thrombosis Research, 13*, 1017-29

Subbarao, K., Kuchibhotia, J. and Kakkar, V.V. (1979) 'Pyridoxal 5'-phosphate — a new physiological inhibitor of blood coagulation and platelet function', *Biochemical Pharmacology, 28*, 531-4

Sullivan, L.W. and Herbert, V. (1964) 'Suppression of hematopoiesis by ethanol', *Journal of Clinical Investigation, 43*, 2048-62

Surgala, M.J., Beesley, E.D. and Brubaker, R.R. (1967) 'Influence of unsaturation on fibrinolytic activity of salts of fatty acids', *Proceedings of the Society for Experimental Biology and Medicine, 126*, 256-8

Sweet, B., Rifkind, B.M. and McNicol, G.P. (1966) 'The relationship between blood lipids and the fibrinolytic enzyme system', *Journal of Atherosclerosis Research, 6*, 359-67

Szanto, S. and Yudkin, J. (1969) 'The effect of dietary sucrose on blood lipids, serum insulin, platelet adhesiveness and body weight in human volunteers', *Postgraduate Medical Journal, 45*, 602-7

Tanser, A.R. (1966) 'Fibrinolytic response to oral glucose', *Lancet, ii*, 147

Tompkins, M.J. and Dayton, S. (1965) 'Relationship of whole blood clotting time to physiological variations in circulating saturated free fatty acids', *Proceedings of the Society for Experimental Biology and Medicine, 119*, 588-90

Tompkins, M.J., Dayton, S. and Pearce, M.L. (1964) 'Effect of long-term feeding of various fats on whole blood clotting times in men', *Journal of Laboratory*

and Clinical Medicine, 64, 763-72
Tremoli, E., Folco, G., Agradi, E. and Galli, C. (1979) 'Platelet thromboxanes and serum cholesterol', *Lancet, i,* 107-8
Tulloch, J.A., Overman, R.S. and Wright, I.S. (1953) 'Failure of ingestion of cream to affect blood coagulation', *American Journal of Medicine, 14,* 674-80
Tsapogas, M.J., Cotton, L.T., Flute, P.T. and Murray, J.G. (1962) 'The effects of chlorpropamide on intermittent claudication and fibrinolysis', *Lancet, i,* 1213-5
Verstraete, M. (1971) 'The effect of lipids on platelets, blood coagulation and fibrinolysis', *Acta Cardiologia, 15,* Suppl. 15, 37-47
Visudhiphan, S., Poolsuppasii, S., Piboonnukarintr, O. and Tumliang, S. (1982) 'The relationship between high fibrinolytic activity and daily capsicum ingestion in Thais', *American Journal of Clinical Nutrition, 35,* 1452-8
Waldron, J.M., Beidelman, B. and Duncan, G.G. (1951) 'The local and systemic effects of cream on blood coagulation: a physiological basis for early feeding in gastrointestinal bleeding', *Gastroenterology, 17,* 360-6
Wallerstein, R.O. and Wallerstein, R.O. Jr. (1976) 'Scurvy', *Seminars in Haematology, 13,* 211-8
Warlow, C.P., McNeill, A., Ogston, D. and Douglas, A.S. (1972) 'Platelet adhesiveness, coagulation, and fibrinolytic activity in obesity', *Journal of Clinical Pathology, 25,* 484-6
Wasantapruer, S., Poolsuppasit, S. and Pibolnukarintr, O. (1974) 'Enhanced fibrinolytic activity after capsicum ingestion', *New England Journal of Medicine, 290,* 1259-60
Willis, A.L., Comal, K., Kuhn, D.C. and Paulsrud, J. (1974) 'Dihomo-γ-linolenate suppresses platelet aggregation when administered *in vitro* or *in vivo*', *Prostaglandins, 8,* 509-19
Wilson, P.A., McNicol, G.P. and Douglas, A.S. (1967) 'Platelet abnormality in human scurvy', *Lancet, i,* 975-8
Yudkin, J., Szanto, S. and Kakkar, V.V. (1969) 'Sugar intake, serum insulin and platelet adhesiveness in men with and without peripheral vascular disease', *Postgraduate Medical Journal, 45,* 608-11
Zahavi, J., Betteridge, J.D., Jones, N.A.G., Glaton, D.J. and Kakkar, V.V. (1981) 'Enhanced *in vivo* platelet release reaction and malondialdehyde formation in patients with hyperlipidemia', *American Journal of Medicine, 70,* 59-64
Zierler, K.L., Grob, D. and Lilienthal, J.L. (1948) 'On the antithrombic and antiproteolytic activity of alpha-tocopheryl phosphate', *American Journal of Physiology, 153,* 127-32

11 The Influence of Physical Activity, Mental Stress and Injury on the Hemostatic Mechanism

Physical Activity

Platelet Numbers and Function

Platelet Numbers. Short periods of strenuous physical activity have been found by most investigators to increase the numbers of platelets in the peripheral blood (Sarajas *et al.*, 1961; Ikkala *et al.*, 1963; Finkel and Cumming, 1965; Dawson and Ogston, 1969; Bennett, 1972; Warlow and Ogston, 1974; Davis *et al.*, 1976; Freedman *et al.*, 1977). This finding, however, is not universal: no change in the platelet count in response to exercise was reported by Prentice and associates (1972) or by Hyers *et al.* (1980). Whether the platelet count alters during prolonged exercise is also controversial. Using a semi-quantitative technique for platelet enumeration Isaacs and Gordon (1924/25) noted a rise in platelet numbers in men competing in a marathon race. Mandalaki and colleagues (1980) also examined the effect of running a marathon on the platelet count: in a comparison between different years they found a significant increase in platelets when the race was on a day of extreme heat (25°C), but no change was found in the subjects taking part in the two years when the run was undertaken at temperatures of 15 to 18°C. This study demonstrates the possible importance of exogenous factors in the response of the platelet count to exercise and points to the need for considerable additional stress in prolonged exercise to induce a rise in the number of circulating platelets. Neither a 9.6 km-walk in 90 minutes (Bennett, 1972) nor a 10-mile march in 90 minutes (Sarajas *et al.*, 1961) resulted in an increase in the platelet count, and a prolonged walk of 50 miles over 20 hours did not alter platelet numbers (Pegrum *et al.*, 1967).

The mechanism of the rise in the platelet count in response to exercise is uncertain. The rapidity of the rise indicates that the platelets must be released from existing stores. The spleen, containing a rapidly mobilisable pool of platelets (Aster, 1966), seemed a likely source, but some investigators have found that the rise in the platelet count after exercise occurs in asplenic individuals (Dawson and Ogston, 1969; Freedman *et al.*, 1977).

Platelet Function. Experiments to define the influence of physical activity on platelet function have not provided decisive conclusions. Differing techniques to assess platelet function and varying severity of the exercise may have contributed to the discordant findings. In the studies of Ikkala *et al.* (1966), Bennett (1972) and Warlow and Ogston (1974) short periods of strenuous exercise produced no alteration in platelet adhesiveness. In contrast, Prentice and co-workers (1972) and Davis and associates (1976) found increased platelet retention on glass bead columns after exercise. Finkel and Cumming (1965) observed increased platelet adhesiveness by a glass wool filter method when the exercise was performed at 25°C, but not after exercise at −20°C. The influence of prolonged exercise was examined by Bennett (1972): unlike the finding after short periods of exercise he found reduced platelet adhesiveness. Pegrum and colleagues (1967) also observed reduced adhesiveness after a 50-mile walk using two techniques for the measurement of platelet adhesiveness.

The findings from studies on the influence of physical activity on platelet aggregation have also been inconclusive. The rate of ADP-induced platelet aggregation was reported to be variably increased after exercise by Ikkala and associates (1966), while Poller *et al.* (1971), using a Chandler tube technique, found an increased rate of irreversible platelet aggregation after exercise. Norepinephrine-induced aggregation, but not ADP-induced aggregation, was noted to be increased after exercise (Harrison *et al.*, 1967): in the same study 'platelet instability' — clumping activity after incubation — was also found after exercise. Warlow and Ogston (1974) found that there is an increased rate of the second phase of ADP and epine-phrine-induced aggregation, 5-hydroxytryptamine-induced aggre-gation and disaggregation, and collagen-induced aggregation after a 15-minute period of strenuous exertion: they could not confirm the finding of Harrison *et al.* (1967), however, since the changes in platelet aggregation seen with increasing time interval after vene-puncture occurred at the same rate in pre- and post-exercise platelet-rich plasma. Increased aggregation in response to ADP was also noted by Prentice *et al.* (1972). After a marathon run ADP and collagen-induced aggregation was increased when the run was made at a temperature of 25°C, but not at 15 to 18°C (Dimitriadou *et al.*, 1977; Mandalaki *et al.*, 1980).

Based on the lack of increase in the level of β-thromboglobulin it was concluded by Hyers and colleagues (1980) that no platelet activation takes place *in vivo* during exercise, but in their study no increase took place in the platelet count in response to physical activity. In contrast, Duncan and Didisheim (1981) reported substantial increases in both β-thromboglobulin and platelet factor 4 following strenuous exercise.

The cause of the increased platelet aggregability after exercise is not established. It seems likely that it is partly, at least, the result of a stress reaction mediated by catecholamines since it has been shown that epinephrine potentiates platelet aggregation induced by ADP (Ardlie *et al.*, 1966) and by thrombin (Thomas, 1967).

Influence on the Coagulation System

Short periods of strenuous physical exertion have been reported to induce a significant shortening of the whole blood clotting time in both glass and siliconised tubes, in the partial thromboplastin time, in the 'Stypven' time and in the thrombin time (Finkel and Cumming, 1965; Ferguson and Guest, 1974; Vogt *et al.*, 1979). In the study of Ferguson and Guest (1974) prolonged moderate exercise for one hour resulted in similar changes. A non-competitive marathon run shortened the activated partial thromboplastin time, but did not affect the prothrombin time (Mandalaki *et al.*, 1980). Brisk exercise did not change the whole blood clotting time or the thromboplastin generation rate in the studies by Keeney and Laramie (1962), or the partial thromboplastin time and prothrombin time in the studies by Cohen and colleagues (1968).

Support for the concept of coagulation activation and thrombin formation during physical activity was provided by a study showing a marked increase in the level of fibrinopeptide A in plasma following exercise, although the quantity represented only a small proportion of the total fibrinogen catabolism (Hyers *et al.*, 1980). In contrast to these findings, Vogt and colleagues (1979) found no increase in the plasma level of fibrinopeptide A following exercise in spite of significant shortening of the activated partial thromboplastin time, thrombin time and Reptilase time. They conclude that exercise does not produce thrombin formation. This view receives support from the demonstration by Collen and associates (1977) that prothrombin turnover is not changed by intermittent cycle exercise over a two-day period.

Specific assays of individual coagulation factors performed before and after exercise have provided general agreement that the plasma levels of factors V, VII, IX, X and prothrombin are unaffected by exertion (Iatridis and Ferguson, 1963; Cohen *et al.*, 1968; Siervogel *et al.*, 1977). Most investigators have also noted that plasma fibrinogen concentrations are not influenced by strenuous exercise (Iatridis and Ferguson, 1963; Hyers *et al.*, 1980), even after brisk walking for a period of four hours (Bennett *et al.*, 1968) or a marathon run (Mandalaki *et al.*, 1980). Herold and Straub (1972) and Ferguson and Guest (1974) did find a small increase in the fibrinogen level after strenuous exercise, but probably only in proportion to hemoconcentration. Antithrombin III levels do not appear to be affected by exercise (Hyers *et al.*, 1980).

The influence of exercise on factor XII levels has been more controversial. In the study of Iatridis and Ferguson (1963) mean factor XII levels rose by over 300 per cent in response to exercise. Herold and Straub (1972) reported a small rise in factor XII for a short period after exhaustive exercise on a bicycle ergometer. In contrast, Egeberg (1963) and Cohen and associates (1968) were unable to detect any change in factor XII levels after exercise.

A universal finding is a substantial rise in the level of factor VIII after physical exertion (Rizza, 1961; Egeberg, 1963; Iatridis and Ferguson, 1963). This rise in factor VIII occurs within minutes of the start of exercise, suggesting that it is the result of release from stores or the activation of molecules already in the circulation rather than the result of increased synthesis of factor VIII. In accordance with this view is the observation that the rise in factor VIII:C can be shown by both one-stage and two-stage assay methods, and the finding that factor VIII antigen is also increased after exercise (Bennett and Ratnoff, 1972).

The observation that factor VIII activity was unchanged after exercise in asplenic individuals led to the suggestion that the spleen has an important role in the exercise-induced increase in factor VIII (Goudemand *et al.*, 1964; Libre *et al.*, 1968). Later studies, however, have been unable to confirm this finding (Rizza and Eipe, 1971; Prentice *et al.*, 1972).

The proposal that the rise in factor VIII during exercise is mediated by β-adrenergic receptor stimulation is supported by the finding that β-blockade with propranolol prevents the exercise-induced

330

increase in factor VIII (Ingram and Vaughan Jones, 1966; Cohen *et al.*, 1968). Additionally, Britton and colleagues (1974) found that the rises in plasma epinephrine and factor VIII levels after exercise to exhaustion were closely correlated.

The influence of physical conditioning on the response of clotting assays to exercise was examined by Ferguson and Guest (1974): they found that the shortening of the whole blood clotting time, partial thromboplastin time, 'Stypven' time and thrombin time also occurred after physical conditioning, but the magnitude of the responses were diminished. Hyers and his colleagues (1980), however, found no difference in plasma fibrinopeptide A levels during exercise between sedentary and highly trained individuals.

Influence on the Fibrinolytic System

The first clear evidence that physical exertion induces an increase in plasma fibrinolytic activity came from the experiments of Biggs and her colleagues (1947): they reported a rise in fibrinolytic activity immediately after running up and down stairs with a decline after resting. The effect of exercise on fibrinolysis has been confirmed repeatedly (Ogston and Fullerton, 1961; Ikkala *et al.*, 1963; Cash, 1966; Menon *et al.*, 1967), and shown to be due to increased plasma levels of plasminogen activator (Sawyer *et al.*, 1960).

The magnitude of the fibrinolytic response is dependent on the duration of the exercise up to certain limits (Ogston and Fullerton, 1961; Bennett *et al.*, 1968) and on its intensity (Rosing *et al.*, 1970; Hawkey *et al.*, 1975; Davis *et al.*, 1976).

It was postulated originally that the increased fibrinolysis associated with exercise resulted from epinephrine secretion (Biggs *et al.*, 1947). However, in contrast to the factor VIII response to exercise, the increase in fibrinolytic activity induced by exertion is not reduced by β-adrenergic blockade (Cohen *et al.*, 1968; Britton *et al.*, 1976). In addition, changes in the blood catecholamine level during exercise do not parallel the increase in plasminogen activator (Hawkey *et al.*, 1975). The precise mechanism of exercise-induced fibrinolytic activity remains an enigma.

Urinary plasminogen activator activity has been reported to be decreased after exercise on a treadmill lasting five minutes (Hedlin *et al.*, 1978), but the mechanism and significance of this observation is obscure.

The relationship of the individual's physical condition to both resting fibrinolytic activity and the fibrinolytic response to a standardised exercise stimulus has been assessed by several groups. Menon and colleagues (1967) claimed that athletes in training had a higher resting plasma fibrinolytic activity than untrained subjects. In other studies, however, there was no statistical difference in resting fibrinolytic activity between an inactive sedentary group of men and a group who had been regularly performing strenuous exercise (Winther, 1966; Winckelmann *et al.*, 1968; Moxley *et al.*, 1970), while there was no correlation between leisure and occupational physical exertion and resting fibrinolytic activity in a sample of 54-year old men (Korsan-Bengsten *et al.*, 1973). Observations on the same group of subjects before and after a period of physical conditioning suggested that the resting fibrinolytic activity is decreased by training (Ferguson and Guest, 1974; Keber *et al.*, 1979; Williams *et al.*, 1980). The fibrinolytic response to exercise in trained relative to untrained subjects has been claimed to be increased (Menon *et al.*, 1967), unchanged (Moxley *et al.*, 1970) or decreased (Winckelmann *et al.*, 1968). In studies where measurements were made before and after a period of physical conditioning the augmentation of fibrino-lytic activity by exercise was found to be essentially the same before and after training (Ferguson and Guest, 1974; Keber *et al.*, 1979). It is probable that the increased plasma activator level induced by exercise is derived from the vascular endothelium (Marsh and Gaffney, 1980; Rijken *et al.*, 1980): it is pertinent, therefore, that a ten-week programme of physical conditioning was found to increase the fibrinolytic reponse to venous occlusion (Williams *et al.*, 1980).

Whether exercise can deplete endothelial stores of activator is speculative, but is compatible with observations that resting fibrino-lytic activity and the fibrinolytic response to further exercise is reduced on the day following prolonged and strenuous exercise (Ogston and Fullerton, 1961). There was a similar finding after a six-week period of strenuous activity climbing in the Andes (Keber *et al.*, 1979), and after a 60-day expedition at altitude in the Himalayas (Coccheri *et al.*, 1982): in the latter study α_2-antiplasmin activity was also reduced, attributed to consumption during the prolonged and strenuous exercise.

The rise in the plasma level of plasminogen activator induced by exertion is not associated with a fall in the plasminogen concentration (Bennett *et al.*, 1968; Ferguson and Guest, 1974; Mandalaki *et al.*, 1980; Marsh and Gaffney, 1980), although Collen and associates (1977) have reported that the turnover of plasminogen is slightly increased by repeated bouts of strenuous exercise over two days.

Evidence for activation of the fibrinolytic system and plasmin formation might also be provided by the finding of a rise in the level of serum fibrin(ogen) degradation products. Some workers have indeed found increases in such breakdown products following physical exertion (Das *et al.*, 1967; Hyers *et al.*, 1980; Mandalaki *et al.*, 1980), but others have been unable to detect any rise (Poortmans *et al.*, 1971; Karp and Bell, 1974; Collen *et al.*, 1977; Vogt *et al.*, 1979). Both Collen and colleagues (1977) and Marsh and Gaffney (1980) have demonstrated an increase in degradation of the Aα chain of fibrinogen, and a small amount of fragment X was detected after exercise in the study by Marsh and Gaffney. Others, however, have found no difference in Aα chain degradation after exercise (Ferguson *et al.*, 1979). In the experiments of Hyers and co-workers (1980) exercise resulted in an increase in the serum D antigen level in parallel with the release of plasminogen activator: the D antigen was considered to be D-dimer, and therefore originating from fibrin, because of the assay dilution slope identity of the serum D antigen with the D-dimer standard curve slope. The general conclusion must be that the rise in plasminogen activator induced by exercise does not result in the appearance of plasma proteolytic activity, but that small amounts of fibrin, present prior to exercise or formed during the exercise, may be degraded by plasmin formed in association with the deposited fibrin.

Mental Stress

In comparison with exercise, mental stress is much less amenable to quantitation. In consequence, studies on the influence of such stresses as anxiety lack precision. Nevertheless, a variety of observations have indicated that mental stress can produce measurable effects on platelet function, coagulation and fibrinolysis. A practical consequence of these effects is that anxiety in experimental subjects may distort the results of physiological studies on hemostatic mechanisms.

Influence on Platelet Function

Studies in this area are very limited. An increase in platelet aggregability in response to ADP has been associated with apprehension prior to cardiac catheterisation and other diagnostic procedures (Gordon *et al.*, 1973).

Influence on the Coagulation System

Mental stress has been known to be capable of shortening the whole blood clotting time for many decades (Cannon and Mendenhall, 1914; Macht, 1952; Friedman *et al.*, 1958; Dreyfuss, 1956). The recalcified plasma clotting time has also been found to be shorter in patients adjudged to be anxious by crude assessment at the time of venepuncture than in those assessed as non-anxious (Ogston *et al.*, 1962).

Stress of longer duration, in the form of a 77-hour period of such activities as firing an electronic rifle during exposure to noise and answering questionnaires, was reported to induce significant decreases in factors V, VIII, IX and fibrinogen and it was suggested that these changes represent an expression of adaptation to prolonged stress (Palmblad *et al.*, 1977): in this study there was no alteration in the activated partial thromboplastin time or factor XII level.

Influence on the Fibrinolytic System

The initial observation that anxiety might influence fibrinolysis was made by Macfarlane and Biggs (1946) who attributed the increased fibrinolytic activity found prior to operation to fear of the impending surgery. Similar observations were made by Latner (1947) and Truelove (1951) who reported additionally that the production of a state of mental agitation by hypnotic suggestion could increase plasma fibrinolytic activity. In later studies it was found that patients deemed to be anxious prior to venepuncture had higher plasma fibrinolytic activity than those considered not to be anxious (Ogston *et al.*, 1962). In the study by Palmblad and colleagues (1977) stress of long duration did not alter fibrinolytic activity as assessed by the euglobulin clot lysis time; this is in contrast to the changes in clotting factors found in the same subjects (see above).

The changes observed in the hemostatic system in situations of mental stress are, in general, compatible with catecholamine-mediated effects.

Injury

Physical stress, exemplified by surgical or accidental trauma, produces a characteristic pattern of changes in platelets, coagulation factors and the fibrinolytic system.

Influence on Platelet Numbers and Function

In the first two to four days after operation or accidental injury there is a fall in the platelet count (Sharnoff *et al.*, 1960; Egeberg, 1962; Innes and Sevitt, 1964; Ham and Slack, 1967; Ygge, 1970; Letheby *et al.*, 1974; Rennie *et al.*, 1974). The fall appears to take place after the injury, Slichter and colleagues (1974) observing that the platelet count is not altered during operation. The reduction in the platelet numbers is followed by a phase of thrombocytosis.

Platelet function has also been shown to be altered following trauma. Platelet adhesiveness measured by the rotating bulb technique was first shown to be increased after operation by Wright (1942). Others have subsequently confirmed that a phase of increased adhesiveness occurs after injury, using either the same technique (Bennett, 1967; Ham and Slack, 1967) or the glass bead column method (Bygdeman *et al.*, 1966; Bennett, 1967), although there have been differences in the timing and duration of the observed change in adhesiveness. O'Brien and colleagues (1971) reported that platelet aggregation in response to ADP and collagen decreased during and immediately after prolonged major operations, and proposed that the temporary state of refractiveness was due to ADP liberated from damaged tissue or red cells.

Influence on the Coagulation System

Surgical or accidental injury is associated with a rapid and marked increase in the level of factor VIII (Egeberg, 1962; Davidson and Tomlin, 1963; Ygge, 1970). Its concentration has been observed to rise both during operation and in the post-operative period (Britton *et al.*, 1974). Both factor VIII:C and VIIIR:Ag levels are increased after surgery, but the ratio VIII:C to VIIIR:Ag has been found to decrease (Holmberg and Nilsson, 1974). It was proposed by Denson (1977) that this is due to an alteration in factor VIII:C by thrombin. Fibrinogen, an acute phase reactant, also rises substantially after injury (Innes and Sevitt, 1964; Bennett *et al.*, 1967; Ygge, 1970; Rennie *et al.*, 1974).

Changes in other clotting factors following physical stress are less striking, but early reductions in factors V and VII and prothrombin have been reported (Innes and Sevitt, 1964) with a later rise in factor V (Brozović, 1977). Increases in the levels of factors IX, X and XI have also been found following surgery (Egeberg, 1962; Davidson and Tomlin, 1963). Antithrombin III levels, both activity and antigen, may fall during operation (Dombrose *et al.*, 1981). A reduction in antithrombin III level has also been found after surgery, but this is probably due, at least in part, to changes in the plasma volume: Korvald and associates (1974) found a 16 per cent fall in the antithrombin III concentration on the third post-operative day when the hematocrit had fallen by 14 per cent. Patients studied in the post-operative period were found to have a significant reduction in plasma factor XIII (Letheby *et al.*, 1974).

Influence on the Fibrinolytic System

Innes and Sevitt (1964) observed a phase of increased blood fibrinolytic activity immediately following accidental trauma: this was succeeded by a more prolonged stage of reduced fibrinolytic activity which lasted several days. The period of reduced fibrinolysis has been confirmed repeatedly, both after surgery and accidental trauma (Bennett *et al.*, 1967; Chakrabarti *et al.*, 1969; Rennie *et al.*, 1974), but the phase of increased fibrinolytic activity has not been detected in all studies, probably because of its early and transient nature. Observations made during surgical operations have shown a marked increase in fibrinolytic activity at that time with a post-operative reduction in fibrinolysis (Mansfield, 1972; Britton *et al.*, 1974).

The plasma plasminogen concentration has been found to fall following surgery, but when the level is corrected for changes in the plasma volume a progressive rise in the circulating quantity is apparent (Bennett *et al.*, 1967). This increase in plasminogen has also been seen after accidental trauma (Rennie *et al.*, 1974).

Blood antiplasmin activity increases following trauma (Bennett *et al.*, 1967; Ygge, 1970). Although there is a marked rise in α_1-antitrypsin after injury, the increased inhibitory activity against plasmin results from a rise in the level of α_2-antiplasmin (Carlin and Saldeen, 1982). After surgery the plasma level of histidine-rich glycoprotein initially falls followed by an increase (Jacobsson *et al.*, 1982).

Mechanism of Hemostatic Component Changes

The mechanisms contributing to the observed changes in the components of the hemostatic mechanism have been delineated by Brozović (1977): she defined three composite processes — pre- and post-operative stress, consumption of clotting factors in the wound, and metabolic changes affecting plasma proteins in general and acute phase reactants in particular.

The increased factor VIII and plasminogen activator levels which appear very rapidly after the stress of injury suggest that adrenergic influences may be important in their production. Compatible with this interpretation is the finding that the rise in fibrinolytic activity immediately after surgical operations and the succeeding phase of reduced activity is not seen in patients anesthetised with epidural analgesia extending from T4 to S5, a procedure which prevents a large part of the endocrine-metabolic response to surgery (Dalsgaard-Nielsen and Gormsen, 1982): it was concluded that neurogenic stimuli play a significant role in the production of the sequence of changes in fibrinolytic activity.

Consumption of hemostatic components in the injured area is probably responsible for some of the changes observed. Slichter and colleagues (1974) found that the number of platelets and the level of fibrinogen were unaltered during operation in spite of presumed utilisation; this was attributed to mobilisation of platelets from storage pools and release of fibrinogen from hepatic reserves, both possibly catecholamine-mediated. Accumulation of platelets and fibrin at the site of the surgical wound could be detected for several days after operation, consumption being held to be responsible for the fall in the platelet count. Any reduction in fibrinogen due to increased utilisation would be masked by the increased synthesis (see below). The reduced level in factor XIII after surgery has also been attributed to utilisation at the site of tissue damage (Letheby *et al.*, 1974). The fall in antithrombin III may also be partly due to consumption by activated clotting factors, but hemodilution probably plays a significant part in its reduced plasma concentration.

Following injury there is a rise in the plasma concentration of a number of proteins due to increased hepatic synthesis. These acute phase reactants include fibrinogen and α_1-antitrypsin, but other coagulation components probably share in the reaction, notably factor VIII. Changes in the measured concentration of these proteins will reflect a balance between increased synthesis and hemodilution due to the plasma volume changes which follow injury. In addition, acute phase reactants which are unrelated to the hemostatic mechanism may influence its function. For example, the effect of trauma on platelet function may be mediated through C-reactive protein, one of the classical acute phase reactants. Fiedel and Gewurtz (1976) reported that C-reactive protein inhibits thrombin-induced platelet aggregation, and later showed that this effect resulted from inhibition of the release of the platelet granule ADP and serotonin (Fiedel *et al.*, 1977), possibly through interference with prostaglandin metabolism.

References

Ardlie, N.G., Glew, G. and Schwartz, C.J. (1966) 'Influence of catecholamines on nucleotide-induced platelet aggregation', *Nature, 212*, 415-7

Aster, R.H. (1966) 'Pooling of platelets in the spleen: role in the pathogenesis of "hypersplenic" thrombocytopenia', *Journal of Clinical Investigation, 45*, 645-57

Bennett, B. and Ratnoff, O.D. (1972) 'Changes in antihemophilic factor (AHF, factor VIII) procoagulant activity and AHF-like antigen in normal pregnancy, and following exercise and pneumoencephalography', *Journal of Laboratory and Clinical Medicine, 80*, 256-63

Bennett, N.B., Ogston, C.M. and Ogston, D. (1967) 'Studies on the blood fibrinolytic enzyme system following acute myocardial infarction', *Clinical Science, 32*, 27-37

Bennett, N.B., Ogston, C.M. and Ogston, D. (1968) 'The effect of prolonged exercise on the components of the blood fibrinolytic enzyme system', *Journal of Physiology, 198*, 479-85

Bennett, P.N. (1967) 'Post-operative changes in platelet adhesiveness', *Journal of Clinical Pathology, 20*, 708-9

Bennett, P.N. (1972) 'Effect of physical exercise on platelet adhesiveness', *Scandinavian Journal of Haematology, 9*, 138-41

Biggs, R., Macfarlane, R.G. and Pilling, J. (1947) 'Observations on fibrinolysis. Experimental activity produced by exercise or adrenaline', *Lancet, i*, 402-5

Britton, B.J., Hawkey, C., Wood, W.G. and Peele, M. (1974) 'Stress — a significant factor in venous thrombosis', *British Journal of Surgery, 61*, 814-20

Britton, B.J., Wood, W.G., Smith, M., Hawkey, C. and Irving, M.H. (1976) 'The effect of beta adrenergic blockade upon exercise-induced changes in blood coagulation and fibrinolysis', *Thrombosis and Haemostasis, 35*, 396-402

Brozović, M. (1977) 'Physiological mechanisms in coagulation and fibrinolysis', *British Medical Bulletin, 23*, 231-8

Bygdeman, S., Eliasson, R. and Johnson, S.-R. (1966) 'Relationship between postoperative changes in adenosine-diphosphate induced platelet adhesiveness and venous thrombosis', *Lancet, i*, 1301-2

Cannon, W.B. and Mendenhall, W.L. (1914) 'Factors affecting the coagulation time of blood. IV. The hastening of coagulation in pain and emotional excitement', *American Journal of Physiology, 34*, 251-61

Carlin, G. and Saldeen, T. (1982) 'Studies on posttraumatic fibrinolysis inhibition by use of tissue plasminogen activator', *Haemostasis, 11*, Suppl. 1, 85

Cash, J.D. (1966) 'Effect of moderate exercise on the fibrinolytic system in normal young men and women', *British Medical Journal, 2*, 502-6

Chakrabarti, R., Hocking, E.D. and Fearnley, G.R. (1969) 'Reaction pattern to three stresses — electroplexy, surgery and myocardial infarction — of fibrinolysis and plasma fibrinogen', *Journal of Clinical Pathology, 22*, 659-62

Coccheri, S., Palareti, G., De Rosa, V., Poggi, M. and Rodorigo, G. (1982) 'Reduced fibrinolytic activity associated to α_2-antiplasmin depletion after a high altitude expedition', *Haemostasis, 11*, Suppl. 1, 37

Cohen, R.J., Epstein, S.E., Cohen, L.S. and Dennis, L.H. (1968) 'Alterations of fibrinolysis and blood coagulation induced by exercise, and the role of beta-adrenergic-receptor stimulation', *Lancet, ii*, 1264-6

Collen, D., Semeraro, N., Tricot, J.P. and Vermylen, J. (1977) 'Turnover of fibrinogen, plasminogen, and prothrombin during exercise in man', *Journal of Applied Physiology, 42*, 865-73

Dalsgaard-Nielsen, J. and Gormsen, J. (1982) 'Fibrinolytic activity in patients with

halothane and epidural analgesia', *Haemostasis, 11.* Suppl. 1, 13
Das, P.C., Allan, A.G.E., Woodfield, D.G. and Cash, J.D. (1967) 'Fibrin
    degradation products in sera of normal subjects', *British Medical Journal, 4,*
    718-20
Davidson, E. and Tomlin, S. (1963) 'The levels of the plasma coagulation factors
    after trauma and childbirth', *Journal of Clinical Pathology, 16,* 112-4
Davis, G.L., Abildgaard, C.F., Bernauer, E.M. and Britton, M. (1976) 'Fibrinolytic
    and haemostatic changes during and after maximal exercise in males', *Journal
    of Applied Physiology, 40,* 287-92
Dawson, A.A. and Ogston, D. (1969) 'Exercise-induced thrombocytosis', *Acta
    Haematologica, 42,* 241-6
Denson, K.W.E. (1977) 'The role of factor VIII-related antigen and factor VIII
    biological activity as an index of hypercoagulability and intravascular clotting',
    Thrombosis Research, 10, 107-19
Dimitriadou, C., Dessypris, A., Louizou, C. and Mandalaki, T. (1977) 'Marathon
    run. II: Effects on platelet aggregation', *Thrombosis and Haemostasis, 37,*
    451-5
Dombrose, F.A., Seyfer, A.E. and Callahan, J. (1981) 'Antithrombin III and
    plasminogen levels in moderate and severe trauma', *Thrombosis and
    Haemostasis, 46,* 287
Dreyfuss, F. (1956) 'Coagulation time of the blood, level of blood eosinophils and
    thrombocytes under emotional stress', *Journal of Psychosomatic Research, 1,*
    252-7
Duncan, A. and Didisheim, P. (1981) 'The influence of physiological parameters
    on the plasma concentration of β-thromboglobulin (β-TG) and platelet factor 4
    (PF-4) in normal healthy subjects', *Thrombosis and Haemostasis, 46,* 28
Egeberg, O. (1962) 'Changes in the coagulation system following major surgical
    operations', *Acta Medica Scandinavica, 171,* 679-85
Egeberg, O. (1963) 'Changes in the activity of antihemophilic A factor (factor
    VIII) and in the bleeding time associated with muscular exercise and adrenaline
    infusion', *Scandinavian Journal of Laboratory and Clinical Investigation, 15,*
    539-49
Ferguson, E.W. and Guest, M.M. (1974) 'Exercise, physical conditioning, blood
    coagulation and fibrinolysis', *Thrombosis et Diathesis Haemorrhagica, 31,*
    63-71
Ferguson, E.W., Barr, C.F. and Bernier, L.L. (1979) 'Fibrinogenolysis and
    fibrinolysis after strenuous exercise', *Journal of Applied Physiology, 47,*
    1157-61
Fiedel, B.A. and Gewurtz, H. (1976) 'Effects of C-reactive protein on platelet
    function. I. Inhibition of platelet aggregation and release reactions', *Journal of
    Immunology, 116,* 1289-94
Fiedel, B.A., Simpson, R.M. and Gewurtz, H. (1977) 'Effect of C-reactive protein
    on platelet function. III. The role of cAMP, contractile elements, and
    prostaglandin metabolism in CRP-induced inhibition of platelet aggregation
    and secretion', *Journal of Immunology, 119,* 877-82
Finkel, A. and Cumming, G.R. (1965) 'Effects of exercise in the cold on blood
    clotting and platelets', *Journal of Applied Physiology, 20,* 423-4
Freedman, M., Altszuler, N. and Karpatkin, S. (1977) 'Presence of a nonsplenic
    platelet pool', *Blood, 50,* 419-25
Friedman, M., Rosenman, R.H. and Carroll, V. (1958) 'Changes in the serum
    cholesterol and blood clotting time in men subjected to cyclic variation of
    occupational stress', *Circulation, 17,* 852-61
Gordon, J.L., Bowyer, D.E., Evans, D.W. and Mitchinson, M.J. (1973) 'Human
    platelet reactivity during surgical diagnostic procedures ', *Journal of Clinical*

Pathology, 26, 958-62

Goudemand, M., Foucaut, M. Habay, D. and Parquet-Gernez, A. (1964) 'Les variations du taux de facteur VIII du cours de l'exercise musculaire. Essai d'interprétation', *Nouvelle Revue Française d'Hématologie, 4*, 315-9

Ham, J.M. and Slack, W.W. (1967) 'Platelet adhesiveness after operation', *British Journal of Surgery, 54*, 385-9

Harrison, M.J.G., Emmons, P.R. and Mitchell, J.R.A. (1967) 'The variability of human platelet aggregation', *Journal of Atherosclerosis Research, 7*, 197-205

Hawkey, C.M., Britton, B.J., Wood, W.G., Peele, M. and Irving, M. (1975) 'Changes in blood catecholamine levels and blood coagulation and fibrinolytic activity in response to graded exercise in man', *British Journal of Haematology, 29*, 377-84

Hedlin, A.M., Milojevic, S. and Korey, A. (1978) 'Plasminogen activator levels in plasma and urine during exercise and oral contraceptive use', *Thrombosis and Haemostasis, 39*, 743-50

Herold, R. and Straub, P.W. (1972) '*In vivo* role of factor XII (Hageman factor) in hypercoagulability and fibrinolysis', *Journal of Laboratory and Clinical Medicine, 79*, 397-412

Holmberg, L. and Nilsson, I.M. (1974) 'AHF related protein in clinical praxis', *Scandinavian Journal of Haematology, 12*, 221-31

Hyers, T.M., Martin, B.J., Pratt, D.S., Dreisin, R.B. and Franks, J.J. (1980) 'Enhanced thrombin and plasmin activity with exercise in man', *Journal of Applied Physiology, 48*, 821-5

Iatridis, S.G. and Ferguson, J.H. (1963) 'Effect of physical exercise on blood clotting and fibrinolysis', *Journal of Applied Physiology, 18*, 337-44

Ikkala, E., Myllylä, G. and Sarajas, H.S. (1963) 'Haemostatic changes associated with exercise', *Nature, 199*, 459-61

Ikkala, E., Myllylä, G., Sarajas, H.S. (1966) 'Platelet adhesiveness and ADP-induced platelet aggregation in exercise,' *Annales Medicinae Experimentalis et Biologiae Fenniae, 44*, 88-92

Ingram, G.I.C. and Vaughan Jones, R. (1966) 'The rise in clotting factor VIII induced in man by adrenaline: effect of α- and β-blockers', *Journal of Physiology, 187*, 447-54

Innes, D. and Sevitt, S. (1964) 'Coagulation and fibrinolysis in injured patients', *Journal of Clinical Pathology, 17*, 1-13

Isaacs, R. and Gordon, B. (1924/25) 'The effect of exercise on the distribution of corpuscles in the blood stream', *American Journal of Physiology, 71*, 106-11

Jacobsson, H., Carlin, g. and Saldeen, T. (1982) 'Histidine-rich glycoprotein — of importance for posttraumatic fibrinolysis inhibition?' *Haemostasis, 11*, Suppl. 1, 50

Karp, J.E. and Bell, W.E. (1974) 'Fibrinogen-fibrin degradation products and fibrinolysis following exercise in humans', *American Journal of Physiology, 227*, 1212-5

Keber, D., Stegner, M., Keber, I. and Accetto, B. (1979) 'Influence of moderate and strenuous daily physical activity on fibrinolytic activity of blood: possibility of plasminogen activator stores depletion', *Thrombosis and Haemostasis, 41*, 745-55

Keeney, C.E. and Laramie, D.W. (1962) 'Effect of exercise on blood coagulation', *Circulation Research, 10*, 691-5

Korsan-Bengsten, K., Wilhelmsen, L. and Tibblin, G. (1973) 'Blood coagulation and fibrinolysis in relation to degree of physical activity during work and leisure time: a study based on a random sample of 54-year-old men', *Acta Medica Scandinavica, 193*, 73-7

Korvald, E., Abildgaard, U. and Fagerhol, M.K. (1974) 'Major operations, hemostatic parameters and venous thrombosis', *Thrombosis Research, 4,* 147-54

Latner, A.L. (1947) 'Anxiety as a cause of fibrinolysis', *Lancet, i,* 194

Letheby, B.A., Davis, R.B. and Larsen, A.E. (1974) 'The effect of major surgical procedures on plasma and platelet levels of factor XIII', *Thrombosis et Diathesis Haemorrhagica, 31,* 20-9

Libre, E.P., Cowan, D.H., Watkins, S.P. and Shulman, N.R. (1968) 'Relationships between spleen, platelets and factor VIII levels', *Blood, 31,* 358-68

Macfarlane, R.G. and Biggs, R. (1946) 'Observations on fibrinolysis. Spontaneous activity associated with surgical operations, trauma, etc.', *Lancet, ii,* 862-4

Macht, D.I. (1952) 'Influence of some drugs and of emotion on blood coagulation', *Journal of the American Medical Association, 148,* 265-70

Mandalaki, T., Dessypris, A., Louizou, C., Panayotopoulou, C. and Dimitriadou, C. (1980) 'Marathon run. III: Effects on coagulation, fibrinolysis, platelet aggregation and serum cortisol levels', *Thrombosis and Haemostasis, 43,* 49-52

Mansfield, A.O. (1972) 'Alteration in fibrinolysis associated with surgery and venous thrombosis', *British Journal of Surgery, 59,* 754-7

Marsh, N. and Gaffney, P. (1980) 'Some observations on the release of extrinsic and intrinsic plasminogen activators during exercise in man', *Haemostasis, 9,* 238-47

Menon, I.S., Burke, F. and Dewar, H.A. (1967) 'Effect of strenuous and graded exercise on fibrinolysis activity', *Lancet, i,* 700-3

Moxley, R.T., Brakman, P. and Astrup, T. (1970) 'Resting levels of fibrinolysis in blood in inactive and exercising men', *Journal of Applied Physiology, 28,* 549-52

O'Brien, J.R., Etherington, M. and Jamieson, S. (1971) 'Refractory state of platelet aggregation with major operations', *Lancet, ii,* 741-3

Ogston, D. (1962) 'The influence of anxiety in tests of blood coagulability and fibrinolytic activity', *Lancet, ii,* 521-3

Ogston, D. and Fullerton, H.W. (1961) 'Changes in fibrinolytic activity produced by physical activity', *Lancet, ii,* 730-3

Ogston, D., McDonald, G.A. and Fullerton, H.W. (1962) 'The influence of anxiety in tests of blood coagulability and fibrinolytic activity', *Lancet, ii,* 521-3

Palmblad, J., Blombäck, M., Egeberg, N., Fröberg, J., Karlsson, C.-G. and Levi, L. (1977) 'Experimentally induced stress in man: effects on blood coagulation and fibrinolysis', *Journal of Psychosomatic Research, 21,* 87-92

Pegrum, G.D., Harrison, K.M., Shaw, S., Haselton, A. and Wolff, S. (1967) 'Effect of prolonged exercise on platelet adhesiveness', *Nature, 213,* 301-2

Poller, L., Priest, C.M. and Thomson, J.M. (1971) 'Platelet aggregation and strenuous exercise', *Journal of Physiology, 213,* 525-31

Poortmans, J., Luke, K.H., Zipursky, A. and Bienstock, J. (1971) 'Fibrinolytic activity and fibrinogen split products in exercise proteinuria', *Clinica Chimica Acta, 35,* 449-54

Prentice, C.R.M., Hassanein, A.A., McNicol, G.P. and Douglas, A.S. (1972) 'Studies on blood coagulation, fibrinolysis and platelet function following exercise in normal and splenectomized people', *British Journal of Haematology, 23,* 541-2

Rennie, A.M., Ogston, D., Cooke, R.J. and Douglas, A.S. (1974) 'The fibrinolytic enzyme system after trauma and in patients with fat embolism', *Journal of Bone and Joint Surgery, 56B,* 421-6

Rijken, D.C., Wijngaards, G. and Welbergen, J. (1980) 'Relationship between tissue plasminogen activator and the activators in blood and vascular wall', *Thrombosis Research, 19,* 815-30

Rizza, C.R. (1961) 'Effect of exercise on the level of antihaemophilic globulin in human blood', *Journal of Physiology, 156,* 128-35

Rizza, C.R. and Eipe, J. (1971) 'Exercise, factor VIII and the spleen', *British Journal of Haematology, 20,* 629-35

Rosing, D.R., Brakman, P., Redwood, D.R., Goldstein, R.E., Beiser, G.D., Astrup, T. and Epstein, S.E. (1970) 'Blood fibrinolytic activity in man. Diurnal variation and the response to varying intensities of exercise', *Circulation Research, 17,* 171-84

Sarajas, H.S.S., Konttinen, A. and Frick, M.H. (1961) 'Thrombocytosis evoked by exercise', *Nature, 192,* 721-2

Sawyer, W.D., Fletcher, A.P., Alkjaersig, N. and Sherry, S. (1960) 'Studies on the thrombolytic activity of human plasma', *Journal of Clinical Investigation, 39,* 426-34

Sharnoff, J.G., Bagg, J.F., Breen, S.R., Rogliano, A.G., Walsh, A.R. and Scardino, V. (1960) 'The possible indication of postoperative thromboembolism by platelet counts and blood coagulation studies in the patient undergoing extensive surgery', *Surgery, Gynecology and Obstetrics, 111,* 469-74

Siervogel, R.M., Lester, R.H., Elston, R.C. and Graham, J.B. (1977) 'The effect of some common environmental factors on the level of blood clotting factor X', *Thrombosis Research, 12,* 187-90

Slichter, S.J., Funk, D.D., Leandoer, L.E. and Harker, L.A. (1974) 'Kinetic evaluation of haemostasis during surgery and wound healing', *British Journal of Haematology, 27,* 115-25

Thomas, D.P. (1967) 'Effect of catecholamines on platelet aggregation caused by thrombin', *Nature, 215,* 298-9

Truelove, S.C. (1951) 'Fibrinolysis and the platelet count', *Clinical Science, 10,* 229-40

Vogt, A., Hofmann, V. and Straub, P.W. (1979) 'Lack of fibrin formation in exercise-induced activation of coagulation', *American Journal of Physiology, 236,* H577-9

Warlow, C.P. and Ogston, D. (1974) 'Effect of exercise on platelet count, adhesion and aggregation', *Acta Haematologica, 52,* 47-52

Williams, R.S., Logue, E.E., Lewis, J.L., Barton, T., Stead, W.W., Wallace, A.G. and Pizzo, S.V. (1980) 'Physical conditioning augments the fibrinolytic response to venous occlusion in healthy adults', *New England Journal of Medicine, 302,* 987-91

Winckelmann, G., Meyer, G. and Roskamm, H. (1968) 'Der Einfluss körperlicher Belastung auf Blutgerinnung und Fibrinolyse bei untrainierten Personen und Hochleitungssportlern', *Klinische Wochenschrift, 46,* 712-6

Winther, O. (1966) 'Exercise and blood-fibrinolysis', *Lancet, ii,* 1195-6

Wright, H.P. (1942) 'Changes in the adhesiveness of blood platelets following parturition and surgical operations', *Journal of Pathology and Bacteriology, 54,* 461-8

Ygge, J. (1970) 'Changes in blood coagulation and fibrinolysis during the postoperative period', *American Journal of Surgery, 119,* 225-32

12 The Hemostatic Mechanism in Pregnancy, Labor and the Puerperium

Pregnancy is accompanied by widespread maternal physiological changes involving most systems of the body; these represent adaptations appropriate to the provision of a suitable environment for the developing fetus and the facilitation of childbirth. The hemostatic mechanism also responds to the stimulus of pregnancy, although the precise function of the changes produced is more speculative. The mediation of the maternal responses to pregnancy are presumably hormonal: during the last half of pregnancy large quantities of estrogen and progesterone are produced by the placenta, but detailed information on the relationship of these to the hemostatic changes is lacking.

Platelet Numbers and Function in Pregnancy, Labor and the Puerperium

Platelet Numbers

Studies on the platelet count in pregnancy have provided discordant results. While most investigators have observed no significant change in the platelet count over the course of normal pregnancy (Todd *et al.*, 1965; Nilsson and Kullander, 1967; Fenton *et al.*, 1977; Yamazaki *et al.*, 1979; Inglis *et al.*, 1982), others have reported a reduction in platelets during pregnancy (Ward and MacArthur, 1948; Shaper *et al.*, 1968; Sejeney *et al.*, 1975; O'Brien, 1976). The fall in platelet numbers was attributed by Hathaway and Bonnar (1978) to the inclusion of women with abnormal pregnancies and to the physiological hemodilution of pregnancy by Wallenburg and van Kessel (1978). In contrast to all these findings, Mor and colleagues (1960) reported a marked and progressive increase in the platelet count over pregnancy with a sharp fall at the time of placental detachment.

The platelet count is generally agreed to rise in the puerperium

(Wright, 1942; Ward and MacArthur, 1948; Kennan and Bell, 1957; Shaper *et al.*, 1968; Bonnar *et al.*, 1970a; Gjønnaess *et al.*, 1975).

Platelet Life-span

The introduction of a non-radioactive technique to measure platelet life-span (malondialdehyde production after the administration of acetylsalicylic acid) has allowed studies to be made during pregnancy. The reported findings agree that platelet life-span is unchanged in uncomplicated pregnancy (Wallenburg and van Kessel, 1978; Rakoczi *et al.*, 1979).

Platelet Function

Platelet Adhesiveness. Using the rotating bulb technique Wright (1942) found no change in platelet adhesiveness in pregnancy, but a rise was seen in the puerperium. Others have confirmed that platelet adhesiveness does not differ between pregnant and non-pregnant women, using both the glass bead column and the rotating bulb technique (Nilsson and Kullander, 1967; Shaper *et al.*, 1968).

Platelet Aggregation. The reported findings on platelet aggregability during pregnancy have not been wholly conclusive. Whigham *et al.* (1978) found no significant difference between pregnant and non-pregnant women's platelet aggregability in response to collagen, epinephrine, vasopressin, 5-hydroxytryptamine, ADP and arachidonic acid. Most, however, have found that platelets are more susceptible to aggregating agents during pregnancy. Examples are the finding of a marked increase in the rate of appearance of secondary aggregation induced by ADP or epinephrine in platelet-rich plasma from pregnant women (Yamazaki *et al.*, 1979) and of an increase in aggregability in response to ADP over the course of pregnancy (Inglis *et al.*, 1982).

Using the minimum dose of ADP causing irreversible platelet aggregation as their index Lewis *et al.* (1980) found increased aggregability in late pregnancy and the puerperium. In the same study the plasma level of 6-oxo-$PGE_1\alpha$, reflecting the circulating concentration of prostacyclin, was reported to be increased in late pregnancy. A possible explanation for this anomaly of increased aggregability in the presence of a rise in the prostacyclin level is provided by the report of Ylikorkala and Viinikka (1980); they found an increase in the plasma concentration of thromboxane B_2 (the stable metabolite of thromboxane A_2) and in thromboxane B_2 release in response to thrombin-induced platelet aggregation during spontaneous clotting in the blood of women during pregnancy and the puerperium.

A marked decrease in ADP and epinephrine-induced aggregation, assessed by the screen filtration pressure method, was found during labor, followed by a significant increase in blood samples obtained five to 12 weeks postpartum (Sanderson *et al.*, 1973).

Platelet Factor 4 and β-thromboglobulin Release. Using a serial heparin dilution technique for the assay of platelet factor 4 Gjesdal (1974) found no change in platelet factor 4 activity in five women in their third trimester of pregnancy compared with non-pregnant women. The possibility of *in vivo* platelet activation during pregnancy is raised, however, by the finding of an increase in the β-thromboglobulin level towards the end of pregnancy (Inglis *et al.*, 1982).

The Coagulation System in Pregnancy

Many observations have been reported on the changes in components of the coagulation system which take place in pregnancy, and their causation and physiological significance have been the subject of speculation. The conclusions drawn from some of the reported studies merit less weight, either because of the small number of women on whom measurements were made or because the coagulation assays were not performed on the same group of women at the different stages of pregnancy.

Fibrinogen. It is a long-standing and consistent observation that a marked and progressive rise in the plasma fibrinogen concentration takes place in pregnancy (Plass and Matthew, 1926; Ratnoff *et al.*, 1954; Gillman *et al.*, 1959; Nilsson and Kullander, 1967, and many others). When the plasma volume changes are taken into account the levels in late pregnancy represent a doubling of the quantity of circulating fibrinogen. It has been shown from turnover studies that the rise in the fibrinogen concentration is the result of increased synthesis (Regoeczi and Hobbs, 1969).

Small increases in soluble fibrin-fibrinogen complexes have been found in the plasma of women during pregnancy (Graeff *et al.*, 1973; McKillop *et al.*, 1976; Fletcher *et al.*, 1979), indicating thrombin generation and the conversion of fibrinogen into fibrin.

Prothrombin. The majority of investigators who have performed specific prothrombin assays have found either no change or only a minor rise in the level of prothrombin during pregnancy (Ratnoff *et al.*, 1954; Alexander *et al.*, 1956; Todd *et al.*, 1956; Pechet and Alexander, 1961; Kasper *et al.*, 1964).

Factor VII. All reported studies have shown substantial increases in the level of factor VII in pregnancy (Alexander *et al.*, 1956; Fresh *et al.*, 1959; Kasper *et al.*, 1964; Todd *et al.*, 1965).

Factor X. Factor X levels are also progressively increased over the course of normal pregnancy (Kasper *et al.*, 1964; Rutherford *et al.*, 1964; Todd *et al.*, 1965).

Factor V. There is general agreement that the level of factor V is unchanged over the course of pregnancy (Alexander *et al.*, 1956; Kennan and Bell, 1957; Kasper *et al.*, 1964; Talbert and Langdell, 1964; Todd *et al.*, 1965; Nilsson and Kullander, 1967; Hahn, 1974).

Factor IX. Some controversy exists on the changes in factor IX in pregnancy. A marked increase in its level was reported by Ratnoff and Holland (1959), Rutherford *et al.* (1964) and Kasper *et al.* (1964) whereas others have found little change in this factor in pregnant women (Ross, 1963; Todd *et al.*, 1965; Nilsson and Kullander, 1967).

Factor VIII. With the exception of a few early reports (Fresh *et al.*, 1956; Ratnoff and Holland, 1959) there has been abundant evidence produced that factor VIII coagulant activity rises dramatically over the course of pregnancy to reach a mean of about 200 per cent in the third trimester (for example, Strauss and Diamond, 1963; Kasper *et al.*, 1964; Talbert and Langdell, 1964; Nilsson and Kullander, 1967, Bennett and Ratnoff, 1972; Fournie *et al.*, 1981). Factor VIII-related antigen also increases during pregnancy, whether proportionately to the increase in factor VIII:C (Bennett and Ratnoff, 1972) or slightly more (van Royen and Ten Cate, 1973; Inglis *et al.*, 1982) is debated. The interpretation of the latter is that there is a thrombin-induced loss of factor VIII coagulant activity as a consequence of activation of the coagulation system.

Factor XI. Low levels of factor XI were noted in the last trimester of pregnancy by Hilgartner and Smith (1963) while Nossel and co-workers found low levels at the time of parturition. Phillips and associates (1973) determined factor XI levels at different stages of pregnancy and found significantly reduced levels in the second and third trimesters.

Factor XII. No change in the factor XII level during pregnancy was found by Ratnoff and Holland (1959) whereas Biland and Duckert (1973) claimed that it increased in late pregnancy.

Factor XIII. Coopland and associates (1969) reported a progressive decrease in the factor XIII level as pregnancy advanced to reach approximately 50 per cent of non-pregnant values at term; in this study the factor XIII was assayed with [^{125}I]fibrinogen-labelled clots to determine the degree of clot solubility. Ebert *et al.* (1981) also found a reduction in factor XIII during pregnancy while the mean level at term was shown to be considerably reduced in the study by Biland and Duckert (1973). In the sole dissenting study Padar (1972) concluded that a number of pregnant women had an elevated level of factor XIII, and the proportion was greater in the second half of pregnancy.

Coagulation Inhibitors. The changes in the level of antithrombin III in pregnancy are, at most, small. A slight decrease was noted by Biland and Duckert (1973) and by Bonnar (1976). Hahn (1974) also found a small reduction, but this did not reach statistical significance.

$\overline{C1}$ inactivator levels have also been reported to fall during pregnancy (Donaldson, 1966; Amir *et al.*, 1972; Johansen *et al.*, 1981). Our own studies have confirmed that some women have a reduced concentration in late pregnancy, but when changes in plasma volume are taken into account, the total circulating $\overline{C1}$ inactivator actually rises (Ogston *et al.*, 1981). A further inhibitor of factor XIa, presently uncharacterised, has also been reported to be reduced in pregnancy (Amir *et al.*, 1972).

The Coagulation System in Labor and the Puerperium

The physical stresses and physiological hemostatic requirements during labor, delivery and the puerperium would suggest that coagulation system adaptations were likely to take place. The reported

347

changes found in the coagulation system, however, are not wholly consistent, making their interpretation difficult. There can be little doubt that these inconsistencies can be traced, at least partly, to the varying methodology used in the studies and in the timing of the blood sampling.

In a study on peripheral blood changes during and after childbirth Bonnar and colleagues (1970a) found increases in factors VIII and V and a decrease in fibrinogen at the time of placental separation, changes attributed to activation of the coagulation mechanism. In contrast, Bennett and co-workers (1974) reported that some women during normal labor and delivery had reduced factor VIII:C without change in factor VIIIR:Ag, and they suggested that their findings could be due to inactivation of factor VIII coagulant activity by thrombin during a state of mild intravascular coagulation. These findings are not mutually incompatible, however, since the pro-coagulant activity of factor VIII is first increased and then inactivated by thrombin. In a further study by Bonnar and his associates (1970b) samples of blood were obtained simultaneously from a peripheral and a uterine vein before, during and after placental separation in deliveries by Caesarean section. The uterine vein blood draining from the placental site showed a marked shortening of the whole blood clotting time in plastic tubes, a shortening of the recalcified plasma clotting time, and an increase in factor VIII coagulant activity at the time of placental separation. The factor IX level was unaltered indicating that the changes in factor VIII activity were not the result of the presence of thromboplastin in the assay system. It was suggested that thrombin produced by the extrinsic system following local release of tissue factor from the placental site is the cause of the sudden increase in factor VIII activity. The changes were transitory and much more pronounced than in the peripheral blood: they were interpreted as reflecting local activation of the clotting mechanism, performing an essential function in the control of uterine bleeding. Similar findings have been recorded by Hahn (1974).

Further alterations in the level of coagulation factors in the early days of the puerperium are not marked. A secondary rise in the fibrinogen concentration has been reported (Kennan and Bell, 1957; Bonnar *et al.*, 1970a; Gjønnaess *et al.*, 1975), although a fall was seen by Ygge (1969). Factor VIII remains high (Bonnar *et al.*, 1970a) or shows a moderate decline (Nilsson and Kullander, 1967; Ygge, 1969). The levels of factor X, factor V and prothrombin were found to fall (Bonnar *et al.*, 1970a) while factor VII fell sharply towards

348

normal in the puerperium (Alexander *et al.*, 1956). Factor XI levels tended to rise (Phillips *et al.*, 1973). The antithrombin III concentration has been reported to increase in the early puerperium without change in α_2-macroglobulin (Gjønnaess *et al.*, 1975). Increased levels of soluble fibrin monomer complexes have been noted in the early puerperium compared to that seen three months postpartum (Graeff *et al.*, 1976).

The Fibrinolytic Enzyme System in Pregnancy

Blood Fibrinolytic Activity. There is general agreement that fibrinolytic activity is reduced in the later months of pregnancy when the euglobulin clot lysis time or the dilute whole blood clot lysis time is used to assess fibrinolysis (Biezenski and Moore, 1958; Shaper *et al.*, 1965; Brakman, 1966; Nilsson and Kullander, 1967; Woodfield *et al.*, 1968; Bonnar *et al.*, 1969a; Menon *et al.*, 1970; Condie and Ogston, 1976). In contrast, Arias and associates (1979) have observed greater fibrinolytic activity during pregnancy, measured in whole blood by a solid state ^{125}I-labelled fibrin substrate assay, perhaps suggesting a contribution from cell-mediated fibrinolytic activity. In support of this finding of increased fibrinolytic activity in pregnancy is the increase in high molecular weight fibrin(ogen) complexes in the plasma of pregnant women, reflecting the rate of fibrin formation/fibrinolysis *in vivo* (Fletcher *et al.*, 1979).

The reduced fibrinolytic activity in pregnancy, if it exists, has been attributed to a variety of causes. Vein biopsy samples obtained during pregnancy have shown a reduced plasminogen activator content (Astedt *et al.*, 1970; Astedt, 1972), while the release of activator in response to exercise (Woodfield *et al.*, 1968) and venous occlusion (Astedt, 1972) has been reported to be impaired in pregnancy. These findings would be compatible with the hypothesis that reduced plasma fibrinolytic activity in pregnancy is a consequence of impaired synthesis of activator in the vessel wall. Astedt and colleagues (1970) proposed that it is the hormonal changes of pregnancy which are associated with the decreased synthesis or release of activator from the vessel wall, but no evidence was provided. Adsorption of activator to fibrin deposits in the placenta has also been suggested as a cause of reduced blood fibrinolytic activity (Fletcher *et al.*, 1979), while the production and release of a specific activator inhibitor from the placenta has received some experimental support as an explanation for the reduced fibrinolysis (Arias *et al.*, 1979). Inhibition of fibrinolysis in pregnancy is discussed below.

Plasminogen. Plasma plasminogen levels have been found by most investigators to rise during pregnancy (Phillips and Skrodelis, 1958; Hedner and Nilsson, 1965; Mitchell and Cope, 1965; Ygge, 1969); the raised levels can be detected as early as the fourth month (Bonnar *et al.*, 1969a).

Antiplasmins. Some investigators have reported an increase in total antiplasmin activity during pregnancy (Guest, 1954; Phillips and Skrodelis, 1958), while others have found no change (Biezenski, 1960). Such studies, however, did not clearly differentiate between inhibition of plasmin, plasminogen activator or streptokinase. Using specific assays no change in overall plasmin inhibition by pregnancy plasma was detected by Brakman and Astrup (1963) whereas Nilsson and Kullander (1967) and Hedner and Astedt (1971) observed a moderate increase in antiplasmin activity in pregnancy, and Hahn (1974) a significant decrease.

A number of reports have appeared on the level of individual plasma protease inhibitors in pregnancy. As measured by the chromogenic substrate S-2251 Teger-Nilsson (1979) found that the level of α_2-antiplasmin was not significantly changed in pregnancy; our own data suggest that the total circulating quantity of α_2-antiplasmin is increased (Walker *et al.*, 1982). The findings on the serum concentration of α_2-macroglobulin have been variable: Ganrot and Bjerre (1967) and Hahn (1974) were unable to detect any change in level during pregnancy whereas Teger-Nilsson (1979) found a slight increase in late pregnancy and Hedner and Astedt (1971) claimed a substantial rise in level. It is now recognised that α_1-antitrypsin does not function as a physiological plasmin inhibitor; there is general agreement that its level rises progressively throughout pregnancy (Ganrot and Bjerre, 1967; Condie and Ogston, 1976; Gilabert *et al.*, 1978).

Antiactivator Activity. Disagreement on the influence of pregnancy plasma on urokinase-induced lysis can be traced to methodological differences. Those investigators using clot lysis techniques to measure urokinase inhibition have reported no alteration in pregnancy (Bonnar *et al.*, 1969a; Menon *et al.*, 1970; Hedner and Astedt, 1971). In contrast, using fibrin plates to measure residual urokinase

350

activity in mixtures of plasma and urokinase, Brakman and Astrup (1963) found substantial increases in selective inhibition of urokinase-induced fibrinolysis in the plasma of pregnant women in the second and third trimesters. This finding has been confirmed repeatedly (Thorsen, 1973; Clemmensen *et al.*, 1976; Campbell *et al.*, 1980). It has been claimed that the urokinase inhibitor is identical to α_1-antitrypsin (Clemmensen *et al.*, 1976), but our own studies indicate that the urokinase inhibitor can be separated from α_1-antitrypsin.

Urokinase inhibitors have been isolated from the placenta (Kawano *et al.*, 1968: Uszynski and Abildgaard, 1971) and placental fractions have been shown to possess the capacity to inhibit urokinase-induced fibrinolysis (Arias *et al.*, 1979). These findings have led to the hypothesis that the placenta releases a specific inhibitor of urokinase into the circulation. The relevance of high levels of a urokinase inhibitor in pregnancy plasma must be speculative in the present uncertainty on the importance of urokinase-like plasminogen activators in physiological blood fibrinolytic activity. Differing from the situation with urokinase, the inhibition of tissue activator-induced lysis by plasma has been found to be unaltered in pregnancy (Brakman and Astrup, 1963; Thorsen, 1973). Increased inhibitory activity against tissue activator has been demonstrated in plasminogen-depleted pregnancy plasma (Campbell *et al.*, 1980), but the physiological significance of this observation is unknown.

Fibrin/Fibrinogen Degradation Products. Findings on the serum level of degradation products of fibrin and fibrinogen during normal pregnancy have been inconsistent. The differences may partly stem from methodological variations: the Laurell technique is, for example, less sensitive than the tanned red cell hemagglutination inhibition immunoassay. A number of investigators have found a small rise in the level of these degradation products during the later months of pregnancy (Woodfield *et al.*, 1968; Hyde *et al.*, 1973; Condie and Ogston, 1976; Gilabert *et al.*, 1978), but others have been unable to detect a significant change (Bonnar *et al.*, 1969b; Hedner and Astedt, 1970; Naish *et al.*, 1973). Using a radioimmunoassay to detect the specific degradation fragment E Gordon and colleagues (1975) found a gradual increase in the level of this fragment until the 38th to 40th week of pregnancy.

The origin of the fibrin/fibrinogen degradation products in pregnancy is uncertain. The breakdown of deposited fibrin by locally produced plasmin activity is a probable explanation, but the finding of reduced vein wall activator in pregnancy (Astedt, 1972) raises a question over its source. An alternative explanation is that the increase in degradation products is secondary to the increased plasma fibrinogen level, the same proportion being partly degraded and unclottable and therefore measured in serum.

The Fibrinolytic Enzyme System in Labor, Parturition and the Puerperium

Blood Fibrinolytic Activity. During labor, despite its associated mental and physical stress, the plasma fibrinolytic activity, as assessed by the euglobulin clot lysis time or the dilute whole blood clot lysis time, remains low (Biezenski and Moore, 1958; Bonnar *et al.*, 1969a). There is general agreement that fibrinolytic activity rises abruptly on delivery (Biezenski and Moore, 1958; Shaper *et al.*, 1965; Bonnar *et al.*, 1969a). Such a rise has been claimed to take place before the umbilical cord is clamped and while the placenta is *in situ* (Shaper *et al.*, 1966), but it is placental separation that is more generally considered to be associated with the substantial rise in fibrinolytic activity that takes place at parturition. Uszynski and Uszynska-Folejewska (1967) reported a rapid fall in the activator content of the myometrium during labor; to explain the increase in blood fibrinolytic activity at the time of parturition Hahn (1974) proposed that the hypoxia during myometrial contractions results in the release of plasminogen activator into the circulation, but that the plasma fibrinolytic activity is only slightly increased as long as the placenta is present to exert its inhibitory activity.

Plasminogen. The plasma plasminogen concentration appears to decrease in the early postpartum period (Nilsson and Kullander, 1967; Ygge, 1969), and Bonnar and associates (1970a) observed this decrease in the third stage of labor. No consistent trend in the plasminogen level was seen by Kleiner *et al.* (1970).

Fibrinolytic Inhibitors. Reported changes in antiplasmin or anti-activator levels during labor or the early phases of the puerperium are neither consistent nor striking, and conclusions from them are hampered by the wide range of assay techniques used. Anti-fibrinolytic activity, assessed by a technique which does not clearly distinguish between plasmin and urokinase inhibition, was found to increase after delivery (Ygge, 1969), while Kleiner *et al.* (1970) observed no changes in the level of inhibitors of plasmin or urokinase during delivery or on the first postpartum day. An increase in anti-plasmin activity was noted by Gjønnaess and associates (1975), but little change was seen in the level of inhibitors of urokinase-induced lysis using a clot lysis assay (Bonnar *et al.*, 1970a). In our own studies the level of anti-urokinase inhibition in fractions of gel-filtered plasma declined from the high level in pregnancy within 60 minutes of parturition (Campbell *et al.*, 1980). Serum inhibition of tissue activator, found to rise during pregnancy, was observed to decline gradually after delivery (Nilsson and Kullander, 1967).

Assays of specific protease inhibitors have demonstrated a further increase in α_1-antitrypsin in the puerperium (Condie and Ogston, 1976; Gilabert *et al.*, 1978). A small fall in the level of α_2-antiplasmin during delivery with a rise in the puerperium was reported by Teger-Nilsson (1979): no changes were seen in the level of α_2-macro-globulin (Gjønnaess *et al.*, 1975).

Fibrin/Fibrinogen Degradation Products. While there is some differ-ence of opinion on the changes in fibrin(ogen) degradation pro-ducts in pregancy there is agreement that their level rises during labor with a further rise in the puerperium (Bonnar *et al.*, 1969b; Kleiner *et al.*, 1970; Stiehm *et al.*, 1970; Condie and Ogston, 1976; Gilabert *et al.*, 1978).

Physiological Significance of the Hemostatic Changes in Pregnancy and the Puerperium

Neither the mechanism of production nor the function of the hemo-static alterations in pregnancy is clear. The level of coagulation factors in plasma represents a balance between synthesis and con-sumption. In view of the increased concentration of most of the clot-ting factors during pregnancy synthesis presumably exceeds the rate of consumption. The stimuli for the synthesis of these factors are not fully elucidated (for control of fibrinogen synthesis see Chapter 3), but the similarities between the changes in women taking estrogen-containing oral contraceptives and those present in pregnancy (Table 12.1) would suggest an important hormonal role. It is pertinent to bear in mind in making such a comparison that there is a very sub-stantial increase in the plasma volume in pregnancy with consequent dilution effects on the plasma proteins.

Table 12.1: Comparison of Changes in Hemostatic Components in Women Receiving Estrogen-containing Oral Contraceptives and During Pregnancy

Hemostatic component	Estrogen-containing contraceptive	Pregnancy
Fibrinogen	unchanged/increase	marked increase
Prothrombin	increase	unchanged/increase
Factor V	unchanged	unchanged
Factor VII	increase	increase
Factor VIII	increase	marked increase
Factor IX	increase	unchanged/increase
Factor X	increase	increase
Factor XI	unchanged	decrease
Factor XII	increase	unchanged/increase
Factor XIII	unchanged	decrease
Antithrombin III	decrease	slight decrease
Platelet numbers	unchanged	unchanged/decrease
Plasminogen activator	increase	marked decrease
Plasminogen	increase	increase
Alpha$_2$-antiplasmin	unchanged	unchanged

The puerperium represents a phase of return to normality after pregnancy, but superimposed are the changes which follow the stress of injury (see Chapter 11). For this reason it might be expected that a similar pattern of hemostatic component changes to those seen after trauma would be seen following parturition, although partly masked by the trend towards non-pregnant values with the removal of the placental hormone stimulation.

The function of the changes in hemostatic components presumably relate, at least in part, to the need for an efficient mechanism, in conjunction with blood vessel constriction, to ensure minimal loss of blood at the time of placental separation. The placenta and decidua are rich in thromboplastic tissue factor, providing an initiating mechanism to activate the coagulation mechanism. Fibrin formation may also be of importance in the maintenance of adequate placental blood supply during pregnancy. Sheppard and Bonnar (1974) have demostrated fibrin deposition in the walls of the spiral arteries supplying the placenta. These arteries undergo changes in which a fibrin-containing matrix replaces the elastic lamina and smooth muscle of the artery: such changes would allow expansion of the lumen of the artery to accommodate increased blood flow. The vascular changes in the uterus during pregnancy and their relation to the hemostatic mechanism have been reviewed by Bonnar (1978). Accelerated fibrin formation and inhibition of fibrinolytic activity may both contribute to this physiological fibrin deposition within the media of the spiral arteries: Sheppard and Bonnar (1978) have shown that endothelial cells are replaced by an intimal lining of cytotrophoblast and there is cytotrophoblast in the media. The cytotrophoblast has much less fibrinolytic activity than intimal endothelial cells, suggesting that it may be a source of fibrinolytic inhibitors.

References

Alexander, B., Meyers, L., Kenny, J., Goldstein, R., Gurewich, V. and Grinspaan, L. (1956) 'Blood coagulation in pregnancy. Proconvertin and prothrombin, and the hypercoagulable state', *New England Journal of Medicine, 254*, 358-63

Amir, J., Pensky, J. and Ratnoff, O.D. (1972) 'Plasma inhibition of activated plasma thromboplastin antecedent (factor XIa) in pregnancy', *Journal of Laboratory and Clinical Medicine, 79*, 106-12

Arias, F., Andrinopoulos, G. and Zamora, J. (1979) 'Whole-blood fibrinolytic activity in normal and hypertensive pregnancies and its relation to the placental concentration of urokinase inhibitor', *American Journal of Obstetrics and Gynecology, 133*, 624-9

Astedt, B. (1972) 'On fibrinolysis', *Acta Obstetrica et Gynaecologica*, Suppl. 18

Astedt, B., Isacson, S., Nilsson, I.M. and Pandolfi, M. (1970) 'Fibrinolytic activity of veins during pregnancy', *Acta Obstetrica et Gynaecologica, 49*, 171-3

Bennett, B. and Ratnoff, O.D. (1972) 'Antihemophilic factor (AHF, factor VIII) procoagulant activity and AHF-like antigen in normal pregnancy, and following exercise and pneumoencephalography', *Journal of Laboratory and Clinical Medicine, 80*, 256-63

Bennett, B., Oxnard, S.C., Douglas, A.S. and Ratnoff, O.D. (1974) 'Studies on antihemophilic factor (AHF, factor VIII) during labor in normal women, in patients with premature separation of the placenta, and in a patient with von Willebrand's disease', *Journal of Laboratory and Clinical Medicine, 84*, 851-60

Biezenski, J.J. (1960) 'Antifibrinolytic activity in normal pregnancy', *Journal of Clinical Pathology, 13*, 220-3

Biezenski, J.J. and Moore, H.C. (1958) 'Fibrinolysis in normal pregnancy', *Journal of Clinical Pathology, 11*, 306-10

Biland, L. and Duckert, F. (1973) 'Coagulation factors of the newborn and his mother', *Thrombosis et Diathesis Haemorrhagica, 29*, 644-51

Bonnar, J. (1976) 'Coagulation disorders', *Journal of Clinical Pathology, 29*, Suppl. 10, 35-41

Bonnar, J. (1978) 'Fibrinolysis in human pregnancy', in P.J. Gaffney and S. Balkuv-Ulutin (eds.) *Fibrinolysis. Current Fundamental and Clinical Concepts*, Academic Press, London, pp. 173-86

Bonnar, J., McNicol, G.P. and Douglas, A.S. (1969a) 'Fibrinolytic enzyme system and pregnancy', *British Medical Journal, 3*, 387-9

Bonnar, J., Davidson, J.F., Pidgeon, C.F., McNicol, G.P. and Douglas, A.S.

(1969b) 'Fibrin degradation products in normal and abnormal pregnancy and parturition', *British Medical Journal, 3*, 137-40

Bonnar, J., McNicol, G.P. and Douglas, A.S. (1970a) 'Coagulation and fibrinolytic mechanisms during and after normal childbirth', *British Medical Journal, 2*, 200-3

Bonnar, J., Prentice, C.R.M., McNicol, G.P. and Douglas, A.S. (1970b) 'Haemostatic mechanism in the uterine circulation', *British Medical Journal, 2*, 564-7

Brakman, P. (1966) 'The fibrinolytic system in human blood during pregnancy', *American Journal of Obstectrics and Gynecology, 94*, 14-20

Brakman, P. and Astrup, T. (1963) 'Selective inhibition in human pregnancy blood of urokinase induced fibrinolysis', *Scandinavian Journal of Clinical and Laboratory Investigation, 15*, 603-9

Campbell, D.M., Ogston, D. and Walker, J.E. (1980) 'The inhibition of urokinase and tissue plasminogen activator by human pregnancy plasma', *Journal of Physiology, 303*, 44-5P

Clemmensen, I., Thorsen, S. and Müllertz, S. (1976) 'Purification of a plasminogen activator inhibitor indistinguishable from α_1-antitrypsin and an urokinase inhibitor in pregnancy plasma', *Haemostasis, 5*, 218-30

Condie, R.G. and Ogston, D. (1976) 'Sequential studies on components of the haemostatic mechanism in pregnancy with particular reference to the development of pre-eclampsia', *Journal of Obstetrics and Gynaecology of the British Commonwealth, 83*, 938-42

Coopland, A., Alkjaersig, N. and Fletcher, A.P. (1969) 'Reduction in plasma factor XIII (fibrin stabilising factor) concentration during pregnancy', *Journal of Laboratory and Clinical Medicine, 73*, 144-53

Donaldson, V.H. (1966) 'Serum inhibitor of Cl-esterase in health and disease', *Journal of Laboratory and Clinical Medicine, 68*, 369-82

Ebert Ch., Berg, H. and Zimmermann, R.E. (1981) 'The hormone-dependent concentration changes of factor XIII (fibrinoligase) during pregnancy', *Thrombosis and Haemostasis, 46*, 345

Fenton, V., Saunders, K. and Cavill, I. (1977) 'The platelet count in pregnancy', *Journal of Clinical Pathology, 30*, 68-9

Fletcher, A.P., Alkjaersig, N.K. and Burstein, R. (1979) 'The influence of pregnancy upon blood coagulation and plasma fibrinolytic enzyme function', *American Journal of Obstetrics and Gynecology, 134*, 743-51

Fournie, A., Monrozies, M., Pontonnier, G., Boneu, B. and Bierme, R. (1981) 'Factor VIII complex in normal pregnancy, pre-eclampsia and foetal growth retardation', *British Journal of Obstetrics and Gynaecology, 88*, 250-4

Fresh, J.W., Adams, H. and Morgan, F.M. (1959) 'Vitamin K-blood clotting studies during pregnancy and prothrombin and proconvertin levels in the newborn', *Obstetrics and Gynecology, 13*, 37-40

Ganrot, P.O. and Bjerre, B. (1967) 'α_1-antitrypsin and α_2-macroglobulin concentration in serum during pregnancy', *Acta Obstetrica et Gynaecologica Scandinavica, 46*, 126-37

Gilabert, J., Aznar, J., Parrilla, J.J., Regañon, E., Vila, V. and Estelles, A. (1978) 'Alteration in the coagulation and fibrinolysis system in pregnancy, labour and puerperium, with special reference to a possible transitory state of intravascular coagulation during labour', *Thrombosis and Haemostasis, 40*, 387-96

Gillman, T., Naidoo, S.S. and Hathorn, M. (1959) 'Plasma fibrinogen activity in pregnancy', *Lancet, ii*, 70-1

Gjesdal, K. (1974) 'Platelet factor 4 (PF-4). A study in methodology and physiology', *Scandinavian Journal of Haematology, 12*, 51-9

Gjønnaess, H., Fagerhol, M.K. and Stormorken, H. (1975) 'Studies on coagulation and fibrinolysis in blood from puerperal women with and without oestrogen treatment', *British Journal of Obstetrics and Gynaecology, 82*, 151-7

Gordon, Y.B., Ratky, S.M., Sola, C.M., Lewis, J., Baker, L.R.I. and Chard, T. (1975) 'Circulating levels of fibrin/fibrinogen degradation fragment E in normal pregnancy, and in association with intrauterine growth retardation and perinatal asphyxia', *British Journal of Obstetrics and Gynaecology, 82*, 958-63

Graeff, H., von Hugo, R. and Hafter, R. (1973) '*In vivo* formation of soluble fibrin monomer complexes in human plasma', *Thrombosis Research, 3*, 465-76

Graeff, H., Wiedemann, A., von Hugo, R. and Hafter, R. (1976) 'Amount and distribution patterns of soluble fibrin monomer complexes during the early puerperium', *American Journal of Obstetrics and Gynecology, 124*, 21-4

Guest, M.M. (1954) 'Profibrinolysin, antifibrinolysin, fibrinogen and urine fibrinolytic factors in the human subject', *Journal of Clinical Investigation, 33*, 1553-9

Hahn, L. (1974) 'On fibrinolysis and coagulation during parturition and menstruation', *Acta Obstetrica et Gynaecologica Scandinavica*, Suppl. 28

Hathaway, W.E. and Bonnar, J. (1978) *Perinatal Coagulation*, Grune and Stratton, New York, p. 38

Hedner, U. and Astedt, B. (1970) 'Fibrinolytic split products in serum and urine in pregnancy', *Acta Obstetrica et Gynaecologica Scandinavica, 49*, 363-6

Hedner, U. and Astedt, B. (1971) 'Studies on fibrinolytic inhibitors during pregnancy', *Acta Obstetrica et Gynaecologica Scandinavica, 50.* 99-103

Hedner, U. and Nilsson, I.M. (1965) 'Determination of plasminogen in human plasma by a casein method', *Thrombosis et Diathesis Haemorrhagica, 14*, 545-61

Hilgartner, M.W. and Smith, C.H. (1965) 'Plasma thromboplastin antecedent (factor XI) in the neonate', *Journal of Pediatrics, 66*, 747-52

Hyde, E., Joyce, D., Gurewich, V. and Flute, P.T. (1973) 'Intravascular coagulation during pregnancy and the puerperium', *Journal of Obstetrics and Gynaecology of the British Commonwealth, 80*, 1059-66

Inglis, T.C.M., Stuart, J., George, A.J. and Davies, A.J. (1982) 'Haemostatic and rheological changes in normal pregnancy and pre-eclampsia', *British Journal of Haematology, 50*, 461-5

Johansen, K.A., Williams, J.H. and Stark, J.M. (1981) 'Acute-phase C 56- forming ability and concentrations of complement components in normotensive and hypertensive pregnancies', *British Journal of Obstetrics and Gynaecology, 88*, 504-12

Kasper, C.K., Hoag, M.S., Aggeler, P.M. and Stone, S. (1964) 'Blood clotting factors in pregnancy: factor VIII concentrations in normal and AHF-deficient women', *Obstetrics and Gynecology, 24*, 242-7

Kawano, T., Morimoto, K. and Uemura, Y. (1968) 'Urokinase inhibitor in human placenta', *Nature, 217*, 253-4

Kennan, A.L. and Bell, W.B. (1957) 'Blood coagulation during normal pregnancy, labor and the puerperium', *American Journal of Obstetrics and Gynecology, 73*, 57-64

Kleiner, G.J., Merskey, C., Johnson, A.J. and Marcus, W.B. (1970) 'Defibrination in normal and abnormal parturition', *British Journal of Haematology, 19*, 159-78

Lewis, P.J., Boylan, P., Friedman, L.A., Hensby, C.N. and Downing, I. (1980) 'Prostacyclin in pregnancy', *British Medical Journal, 3*, 1581-2

McKillop, C., Howie, P.W., Forbes, C.D. and Prentice, C.R.M. (1976) 'Soluble fibrinogen/fibrin complexes in pre-eclampsia', *Lancet, i*, 56-8

Menon, I.S., Rannie, G.H. and Dewar, H.A. (1970) 'A comparative study of blood fibrinolytic activity in normal women, pregnant women and women on oral contraceptives', *Journal of Obstetrics and Gynaecology of the British Commonwealth, 77*, 752-6

Mitchell, P. and Cope, I. (1965) 'The estimation of plasminogen in human serum or plasma', *Clinica Chimica Acta, 11*, 414-23

Mor, A., Yang, W., Schwarz, A. and Jones, W.C. (1960) 'Platelet counts in pregnancy and labour', *Obstetrics and Gynecology, 16*, 338-43

Naish, P., Clark, A.D., Winston, R.M.L. and Peters, D.K. (1973) 'Serum and urine fibrinogen derivatives in normal pregnancy and pre-eclampsia', *Obstetrics and Gynecology, 42*, 861-7

Nilsson, I.M. and Kullander, S. (1967) 'Coagulation and fibrinolytic studies during pregnancy', *Acta Obstetricia et Gynaecologica Scandinavica, 46*, 273-85

Nossel, H.L., Lanzkowsky, P. and Levy, S. (1966) 'A study of coagulation factor levels in women during labour and in their newborn infants', *Thrombosis et Diathesis Haemorrhagica, 16*, 185-97

O'Brien, J.R. (1976) 'Platelet counts in normal pregnancy', *Journal of Clinical Pathology, 29*, 174

Ogston, D., Walker, J. and Campbell, D.M. (1981) 'Cl inactivator level in pregnancy', *Thrombosis Research, 23*, 453-5

Padar, E. (1972) 'Fibrin cross-linkage, thrombosis and pregnancy', *Annals of the New York Academy of Sciences, 202*, 230-4

Pechet, L. and Alexander, B. (1961) 'Increased clotting factors in pregnancy', *New England Journal of Medicine, 265*, 1093-7

Phillips, L.L. and Skrodelis, V. (1958) 'The fibrinolytic enzyme system in normal, hemorrhagic and disease states', *Journal of Clinical Investigation, 37*, 965-73

Phillips, L.L., Rosano, L. and Skrodelis, V. (1973) 'Changes in factor XI (plasma thromboplastin antecedent) levels during pregnancy', *American Journal of Obstetrics and Gynecology, 116*, 1114-6

Plass, E.D. and Matthew, C.W. (1926) 'Plasma protein fractions in normal pregnancy, labour and puerperium', *American Journal of Obstetrics and Gynecology, 12*, 346-58

Rákóczi, I., Tallián, F., Bagdány, S. and Gáti, I. (1979) 'Platelet life-span in normal pregnancy and pre-eclampsia as determined by a non-radiographic technique', *Thrombosis Research, 15*, 553-6

Ratnoff, O.D. and Holland, T.R. (1959) Coagulation components in normal and abnormal pregnancies, *Annals of the New York Academy of Sciences, 75*, 626-33

Ratnoff, O.D., Colopy, B.S. and Pritchard, J.A. (1954) 'The blood clotting mechanism during normal parturition', *Journal of Laboratory and Clinical Medicine, 44*, 408-15

Redman, C.W.G., Bonnar, J. and Beilin, L. (1978) 'Early platelet consumption in pre-eclampsia', *British Medical Journal, 1*, 467-9

Regoeczi, E. and Hobbs, K.R. (1969) 'Fibrinogen turnover in pregnancy', *Scandinavian Journal of Haematology, 6*, 175-8

Ross, R.A. (1963) 'Blood coagulation defects in obstetrics and gynecology', *American Journal of Obstetrics and Gynecology, 86*, 77-84

Rutherford, R.N., Hougie, C., Banks, A.L. and Coburn, W.A. (1964) 'The effects of sex steroids and pregnancy on blood coagulation factors', *Obstetrics and Gynecology, 24*, 886-92

Sanderson, J.H., Taylor, D.S. and Delamore, I.W. (1973) 'Changes in platelet thrombotic tendency during parturition', *Journal of Obstectrics and Gynaecology of the British Commonwealth, 80*, 633-8

Sejeney, S.A., Eastham, R.D. and Baker, S.R. (1975) 'Platelet counts during normal pregnancy', *Journal of Clinical Pathology, 28*, 812-3

Shaper, A.G., Kear, J., Macintosh, D.M., Kyobe, J. and Njama, D. (1968) 'The

platelet count, platelet adhesiveness and aggregation and the mechanism of
fibrinolytic inhibition in pregnancy and the puerperium', *British Journal of
Obstetrics and Gynaecology of the British Commonwealth*, 75, 433-41
Shaper, A.G., Macintosh, D.M., Evans, C.M. and Kyobe, J. (1965) 'Fibrinolysis
and plasminogen levels in pregnancy and the puerperium', *Lancet, ii*, 706-8
Shaper, A.G., Macintosh, D.M. and Kyobe, J. (1966) 'Fibrinolytic activity in
pregnancy, during parturition, and in the puerperium', *Lancet, ii*, 874-6
Sheppard, B.L. and Bonnar, J. (1974) 'The ultrastructure of the arterial supply of
the human placenta in early and late pregnancy', *Journal of Obstetrics and
Gynaecology of the British Commonwealth*, 81, 497-511
Sheppard, B.L. and Bonnar, J. (1978) 'Fibrinolysis in decidual spiral arteries in late
pregnancy', *Thrombosis et Diathesis Haemorrhagica*, 39, 751-8
Stiehm, E.R., Kennan, A.L. and Schelble, D.T. (1970) 'Split products of fibrin in
maternal serum in the perinatal period', *American Journal of Obstetrics and
Gynecology*, 108, 941-5
Strauss, H.S. and Diamond, L.K. (1963) 'Elevation of factor VIII (antihemophilic
factor) during pregnancy in normal persons and in a patient with von
Willebrand's disease', *New England Journal of Medicine*, 269, 1251-2
Talbert, L.M. and Langdell, R.D. (1964) 'Normal values of certain factors in the
blood clotting mechanism in pregnancy', *American Journal of Obstetrics and
Gynecology*, 90, 44-9
Teger-Nilsson, A.-C. (1979) 'Use of chromogenic substrates for screening of
inhibitors of coagulation and fibrinolysis in patients', in M.F. Scully and V.V.
Kakkar (eds.) *Chromogenic Peptide Substrates: Chemistry and Clinical Usage*,
Churchill Livingstone, Edinburgh, pp. 269-76
Thorsen, S. (1973) 'The inhibition of tissue plasminogen activator and
urokinase-induced fibrinolysis by some natural proteinase inhibitors and by
plasma and serum from normal and pregnant subjects', *Scandinavian Journal of
Clinical and Laboratory Investigation*, 31, 51-9
Todd, M.E., Thompson, J.H., Bowie, E.J.W. and Owen, C.A. (1965) 'Changes in
blood coagulation during pregnancy', *Mayo Clinics Proceedings*, 40, 370-83
Uszynski, M. and Abildgaard, U. (1971) 'Separation and characterization of two
fibrinolytic inhibitors from human placenta', *Thrombosis et Diathesis
Haemorrhagica*, 25, 580-9
Uszynski, M. and Uszynska-Folejewska, R. (1967) 'The plasminogen activator
content of the myometrium in nonpregnant, pregnant, and parturient women',
American Journal of Obstetrics and Gynecology, 98, 825-31
Van Royen, E.A. and Ten Cate, J.W. (1973) 'Antigen-biological activity ratio for
factor VIII in late pregnancy', *Lancet, ii*, 449-50
Walker, J.E., Campbell, D.M. and Ogston, D. (1982) 'Blood levels of proteinase
inhibitors in pregnancy', *British Journal of Obstetrics and Gynaecology*, 89,
208-10
Wallenburg, H.C.S. and van Kessel, P.H. (1978) 'Platelet lifespan in normal
pregnancy as determined by a nonradioisotopic technique', *British Journal of
Obstetrics and Gynaecology*, 85, 33-6
Ward, C.V. and MacArthur, J.L. (1948) 'Blood platelet studies during pregnancy
and the puerperium', *American Journal of Obstetrics and Gynecology*, 55,
600-7
Whigham, K.A.E., Howie, P.W., Drummond, A.H. and Prentice, C.R.M. (1978)
'Abnormal platelet function in pre-eclampsia', *British Journal of Obstetrics and
Gynaecology*, 85, 28-32
Woodfield, D.G., Cole, S.K. and Cash, J.D. (1968) 'Impaired fibrinolytic response
to exercise stress in normal pregnancy', *American Journal of Obstetrics and
Gynecology*, 102, 440-6

Wright, H.P. (1942) 'Changes in the adhesiveness of blood platelets following parturition and surgical operation', *Journal of Pathology and Bacteriology, 54,* 461-8

Yamazaki, H., Motomiya, T., Kitutani, N., Sakakihara, C., Watanabe, S., Numata, M. and Noguchi, K. (1979) 'Platelet aggregation during menstrual cycle and pregnancy', *Thrombosis Research, 14,* 333-40

Ygge, J. (1969) 'Changes in blood coagulation and fibrinolysis during the puerperium', *American Journal of Obstetrics and Gynecology, 104,* 2-12

Ylikorkala, O. and Viinikka, L. (1980) 'Thromboxane A_2 in pregnancy and puerperium', *British Medical Journal 4,* 1601-2

13 Hemostasis and the Menstrual Cycle

The separate consideration of hemostasis in relation to the menstrual cycle is pertinent for three reasons. Firstly, any alterations in hemostatic components through the cycle must be considered in studies on women in their reproductive years. Secondly, the changes may provide information on the influence of hormonal changes on the hemostatic mechanism. Thirdly, the periodic shedding of the endometrium offers a unique physiological challenge to the hemostatic system.

Hemostatic Components in Different Stages of the Menstrual Cycle

During the first half of the menstrual cycle estrogens are secreted in substantial quantities from the ovaries, culminating in a peak at the time of ovulation. The second half of the cycle is associated with the secretion of both estrogens and progesterone.

Platelet Count and Function

A number of early studies detected a mid-cycle peak in the platelet count, presumably corresponding to ovulation, and a tendency for the count to be lower at the onset of menstruation (Genell, 1936; Pohle, 1939; Pepper and Lindsay, 1956, 1959, 1960). No cyclic pattern in the platelet count was seen in oophorectomised women (Pepper and Lindsay, 1956), and it has been claimed that the mid-cycle peak does not occur in anovulatory cycles (Mobius and Johannes, 1963). McBride and Snodgrass (1968) also found that the highest mean platelet count occurred at the probable time of ovulation, but the difference from other parts of the cycle did not reach statistical significance.

Platelet adhesiveness has also been examined at different stages of the menstrual cycle. Caspary and Peberdy (1965) stated that the stage of the menstrual cycle had no effect on platelet adhesiveness as assessed by the rotating bulb technique, but no experimental data were presented. A similar conclusion was reached by McBride and Snodgrass (1968) from findings obtained using the glass bead column method.

Thrombin-induced platelet aggregation was observed to alter little over the menstrual cycle (Tindall *et al.*, 1981): thrombin-stimulated platelet malondialdehyde production, an index of cyclo-oxygenase activity, was found to fall by some 33 per cent during normal menstruation, but remained constant throughout the remainder of the cycle.

Coagulation System

There is little experimental evidence to suggest that the ovarian cycle is responsible for any significant alteration in the coagulation mechanism. A cyclic variation in the fibrinogen concentration was observed by Turksoy and co-workers (1961) with increased levels ten to eleven and three to four days before menstruation, both peaks being related in time to reported increases in total urinary estrogen excretion. In contrast, Egeberg and Owren (1963), studying five normal women over a single menstrual cycle, found no change in plasma fibrinogen level and, additionally, no alterations in the partial thrombplastin time or in prothrombin and factors V, VII, VIII, IX and XI. Brakman and associates (1966) were also unable to find any alteration in coagulation components between various phases of the menstrual cycle; their measurements included the prothrombin time, prothrombin-proconvertin assay, factor V and fibrinogen. Antithrombin III levels have been found to be unchanged over the menstrual cycle (Teger-Nilsson, 1979; Tindall *et al.*, 1981).

Fibrinolytic Enzyme System

A number of early studies on blood fibrinolytic activity suggested that there was an increase during menstruation (Smith and Smith, 1945; Willson and Munnell, 1946; Marx and Rovatti, 1952). More recent investigations have not demonstrated any significant changes in fibrinolytic activity during the menstrual cycle (Beller *et al.*, 1964; Rao, 1964; Brakman *et al.*,1966), although Menon (1968) found that euglobulin clot lysis times were shorter in women tested during menstruation. A cyclic variation in the plasma plasminogen level was claimed by Turksoy and associates (1961), but this has not been confirmed (Beller *et al.*, 1964; Brakman *et al.*, 1966). Total plasma antiplasmin and anti-urokinase activities have been reported to be unchanged over the cycle (Beller *et al.*, 1964; Brakman *et al.*, 1966). Information on the individual fibrinolytic system inhibitors is scanty, but Jespersen and Kluft (1982) found no significant fluctuation in the level of histidine-rich glycoprotein while Teger-Nilsson (1979) reported slight decreases in the plasma α_2-antiplasmin and α_2-macroglobulin levels in a small group of women during menstruation.

It is clear that measurable plasma components of the hemostatic system are not significantly different in the various phases of the menstrual cycle with the possible exception of an increase in the platelet count at the time of ovulation. This lack of demonstrable changes is not wholly unexpected since the hormonal variations over the menstrual cycle are relatively minor compared to those in the later months of pregnancy when estrogen production reaches a peak of some fifty times that found in the normal menstrual cycle.

The Fluidity of Menstrual Discharge

Explanations for the fluid nature of menstrual blood and the absence of fibrinogen include that of Bell (1913/14) who suggested that endometrial thromboplastin induced clotting so that menstrual blood lacked 'fibrin ferment' and was essentially serum containing red cells. Under this general proposal it was suggested that the fibrin of the clots was degraded by proteolytic enzymes (Whitehouse, 1914). Since these early proposals there have been a number of studies demonstrating that endometrium contains a plasminogen activator (Page *et al.*, 1957; Kullander and Källén, 1961; Rybo, 1966), and there is evidence that cervical mucus provides additional activator (Beller and Weiss, 1966). Beller and his co-workers (1971) reported that the clots found in menstrual blood consisted of red cell aggregates rather than fibrin and proposed that the incoagulability of the blood is due to the presence of breakdown products of fibrinogen, these resulting from the proteolytic activity arising from the cervical mucus. High levels of fibrin/fibrinogen degradation products in menstrual blood have been noted by Hahn (1974). The recent observation by Christiaens and colleagues (1981) that extravascular fibrin and platelet aggregates are present in shed tissue in the uterine cavity during the first 32 hours of menstruation, but not in the menstrual discharge from the vagina, is consistent with the concept of fibrin formation and subsequent dissolution. Sheppard and co-workers (1982), however, have reported finding large quantities of polymerised fibrin in both uterine and vaginal menstrual blood.

Uterine Hemostasis at Menstruation

Observations on the hemostatic mechanism during menstruation using light microscopy, electron microscopy and immunofluorescence have been reported by Christiaens and colleagues (1980, 1981) and a number of special characteristics noted. In the premenstrual phase, up to five days before the start of menstruation, vessel lesions characterised by holes in the endothelium and interruptions in the basement membrane were seen and there was focal escape of red cells into the stroma. An unusual feature was the absence of platelet adherence in spite of exposure of subendothelial collagen; this suggests inhibition of platelet reactivity, perhaps the result of a high local concentration of prostaglandins inhibitory to platelet function.

In the phase of early menstruation the endothelium was progressively engorged with blood from the compact layer towards the deeper layers. Part of the endothelium was shed into the uterine cavity and blood vessels ending in the shedding edge were seen to be filled with platelet-fibrin thrombi, often with alternating layers of fibrin and platelets. No platelet-aggregates or fibrin deposits were seen outside the vessels, possibly due to removal by the fibrinolytic system. Extravascular fibrin was seen in shed tissue.

In the later phases of menstruation, 20 to 24 hours after the start, all the functional endometrium was fully shed. Most vessels at this stage did not contain thrombi, but the vessels ending in the shedding margin were constricted.

It was concluded that there are two mechanisms contributing to uterine hemostasis. The first is hemostatic plug formation and this represents the most important mechanism in the functional endometrium during tissue shedding, the damaged vessels in this area having impaired contractile properties. The second mechanism, vasoconstriction, is more important in the basal layers and ensures adequate hemostasis after most of the functional tissue is shed.

References

Bell, W.D. (1913-14) 'The causes of the non-coagulability of normal menstrual blood and of pathological clotting', *Journal of Pathology and Bacteriology, 18,* 462-7

Beller, F.K. (1971) 'Observations on the clotting of menstrual blood clot formation', *American Journal of Obstetrics and Gynecology, 111,* 535-43

Beller, F.K. and Weiss, G. (1966) 'The fibrinolytic enzyme system in cervical mucus', *Fertility and Sterility, 17,* 654-62

Beller, F.K., Goebelsmann, U., Douglas, G.W. and Johnson, A. (1964) 'The fibrinolytic system during the menstrual cycle', *Obstetrics and Gynecology, 23,* 12-6

Brakman, P., Albrechtsen, O.K. and Astrup, T. (1966) 'A comparative study of coagulation and fibrinolysis in blood from normal men and women', *British Journal of Haematology, 12,* 74-85

Caspary, E.A. and Peberdy, M. (1965) 'Oral contraception and blood-platelet adhesiveness', *Lancet, i,* 1142-3

Christiaens, G.C.M.L., Sixma, J.J. and Haspels, A.A. (1980) 'Morphology of haemostasis in menstrual endometrium', *British Journal of Obstetrics and Gynaecology, 87,* 425-39

Christiaens, G.C.M.L., Sixma, J.J. and Haspels, A.A. (1981) 'Fibrin and platelets in menstrual discharge before and after the insertion of an intrauterine contraceptive device', *American Journal of Obstetrics and Gynecology, 140,* 793-8

Egeberg, O. and Owren, P.A. (1963) 'Oral contraception and blood coagulability', *British Medical Journal, 1,* 220-1

Genell, S. (1936) 'A study of variations in the number of blood-platelets during the menstrual cycle', *Journal of Obstetrics and Gynaecology of the British Empire, 43,* 1124-34

Hahn, L. (1974) 'On fibrinolysis and coagulation during parturition and menstruation', *Acta Obstetricia et Gynaecologica Scandinavica,* Suppl. 28

Jespersen, J. and Kluft, C. (1982) 'Histidine-rich glycoprotein and plasminogen levels in plasma during the normal menstrual cycle and in women on oral contraceptives low in estrogen', *Haemostasis, 11,* Suppl. 1, 49

Kullander, S. and Källén, B. (1961) 'The fibrinolytic activity of human endometrium, studied in tissue culture', *Acta Obstetricia et Gynaecologica Scandinavica, 40,* 1-15

Marx, R. and Rovatti, B. (1952) 'Untersuchung über das Vethalten der Fibrinolyse in Serum während des Menstrualcyklus', *Haematologica, 36,* 685-92

McBride, J.A. and Snodgrass, C.A. (1968) 'The effect of the normal menstrual cycle on the total platelet count, adhesive platelet count and platelet adhesiveness', *Journal of Obstetrics and Gynaecology of the British Commonwealth, 75,* 357-9

Menon, I.S. (1968) 'Fibrinolytic activity in the blood of 99 healthy young British, African and Asian subjects', *Laboratory Practice, 17,* 194-5

Mobius, G. and Johannes, S. (1963) 'Variations in the thrombocyte count during the female genital cycle', *Zentralblatt fur Gynaekologie, 85,* 489-500

Page, E.W., Glendenning, M.B. and Parkinson, D. (1957) 'Cyclic biochemical changes in the human endometrium. With special reference to the fibrinolytic enzyme', *American Journal of Obstetrics and Gynecology, 62,* 1100-5

Pepper, H. and Lindsay, S. (1956) 'Elevation of platelets in mid-cycle: an indication of ovulation', *Science, 124,* 180-1

Pepper, H. and Lindsay, S. (1959) 'Levels of platelets, eosinophils, total leucocytes

and polymorphonuclear leucocytes during the normal menstrual cycle',
Obstetrics and Gynecology, 14, 657-64

Pepper, H. and Lindsay, S. (1960) 'Levels of eosinophils, platelets, leukocytes and 17-hydroxycorticosteroids during normal menstrual cycle', *Proceedings of the Society for Experimental Biology and Medicine, 104*, 145-7

Pohle, F.J. (1939) 'The blood platelet count in relation to the menstrual cycle in normal women', *American Journal of Medical Sciences, 197*, 40-7

Rao, A.R. (1964) 'Plasma fibrinolytic activity and menstrual cycle in women', *Lancet, ii*, 593-4

Rybo, G. (1966) 'Plasminogen activators in the endometrium. II. Clinical aspects', *Acta Obstetricia et Gynaecologica Scandinavica, 45*, 429-50

Sheppard, B.L., Dockeray, C.J. and Bonnar, J. (1982) 'The fibrin component of menstrual blood', *Haemostasis, 11*, Suppl. 1, 98

Smith, O.W. and Smith, G.V.S. (1945) 'Fibrinolytic enzyme in menstruation and late pregnancy', *Science, 102*, 135-7

Teger-Nilsson, A.-C. (1979) 'Use of chromogenic substrates for screening of inhibitors of coagulation and fibrinolysis in patients', in M.F. Scully and V.V. Kakkar (eds.), *Chromogenic Peptide Substrates: Chemistry and Clinical Usage*, Churchill Livingstone, Edinburgh, pp. 269-76

Tindall, H., Zuzal, M., Paton, R.C. and McNicol, G.P. (1981) 'Changes in thrombin-stimulated platelet malondialdehyde production during the menstrual cycle', *Journal of Clinical Pathology, 34*, 595-8

Turksoy, R.N., Phillips, L.L. and Southam, A.L. (1961) 'Influence of ovarian function on the fibrinolytic enzyme system. I. Ovulatory and anovulatory cycles', *American Journal of Obstetrics and Gynecology, 82*, 1311-5

Whitehouse, H.B. (1914) 'The physiology and pathology of uterine haemorrhage', *Lancet, i*, 877-85

Willson, J.R. and Munnell, E.R. (1946) 'Fibrinolytic activity in blood serum during pregnancy complicated by hypertensive toxaemias', *Proceedings of the Society for Experimental Biology and Medicine, 62*, 277-9

platelet content 170
reaction with plasmin 167-8
structure 104
alpha$_2$-plasmin inhibitor *see* alpha$_2$-antiplasmin
alpha-tocopherol *see* vitamin E
altitude
influence on: coagulation 280-1;
fibrinolytic system 281;
platelet numbers 279-80
anabolic steroids, influence on
hemostatic components 234-6
anaphylatoxins 84, 196
induction of platelet aggregation
203
androgens 234
influence on hemostatic
components 234-6
antihemophilic globulin *see* factor
VIII
antiplasmins 165-70
in hyperthyroidism 227
platelet content 169-70
tissue 172-3
antithrombin III 2, 7, 10, 100-4, 168
chemistry 100
effect on plasmin 168
inhibition of: coagulation factors
100-1; factor Xa 111-12;
factor XIIa 112; kallikrein
113; thrombin 100, 110-11
mechanism of action 101-2
plasma concentration 104
plasma level; ethnic variations
269; in hypothyroidism 227; in
infancy 253; in the puerperium
349; influence of: age 254,
anabolic steroids 236,
circadian rhythm 258,
compression-decompression
282, dietary linoleic acid 299,
estrogen/progestogen
preparations 232, injury 335-6,
337, menstrual cycle 362,
pregnancy 347, sex 256
structure 100
synthesis 103-4
turnover 104
antithrombin III-like protein 104
apoprotein III 74-5
arachidonic acid 3, 20-1
dietary content and platelet
function 292
see also platelets

ascorbic acid
influence on blood fibrinolytic
activity 306
platelet aggregation 306
autoprothrombin II-A 109

Bernard-Soulier syndrome 18
beta-lipoprotein *see* low-density
lipoprotein
beta-thromboglobulin 28-9
plasma level 28; in
hyperlipoproteinemias 296;
influence of: age 253, physical
activity 329, pregnancy 345
relationship to low-affinity platelet
factor 4 28
structure 28
black tree fungus, influence on
platelet aggregation 312
bleeding time, in Eskimos 292
blood fibrinolytic activity
circadian rhythm 257-8; influence
of: altitude 281, ascorbic acid
306, beer 310, caffeine 313,
cider 310, epinephrine 222-3,
garlic 311, injury 336, insulin
237, norepinephrine 223,
onions 311, testosterone 234,
wine 310
blood group differences
influence on hemostatic
components 272-3

C1 esterase inhibitor *see* C1̄ inactivator
C1̄ inactivator 7, 10, 106-7, 168-9
effect on plasmin 168-9; inhibition
of: factor XIa 106, 112, factor
XIIa 106-7, 112, kallikrein
107, 112-13
plasma level; in fetal life 250; in
hyperthyroidism 226; influence
of: anabolic steroids 236,
estrogen/progestogen
preparations 232, pregnancy
347
caffeine
influence on blood fibrinolytic
activity 313
calcium 5, 6
see also platelets
calorie malnutrition
influence on coagulation factors
316

factor VIII level 330; factor XII level 330; fibrinolytic system 331-3; platelet adhesiveness 328; platelet aggregation 328; platelet factor 4 level 329; platelet numbers 327

physical conditioning
  influence on: plasma fibrinolytic activity 322; response of coagulation system to exercise 332; response of plasma fibrinolytic activity to exercise 331

PIVKA 67

plasma fibrinolytic activity
  ethnic variations 269-71
  in fetal life 250
  in hyperlipoproteinemias 300-1
  in infancy 253
  in labor and parturition 352
  influence of: age 255; alimentary lipemia 299; anabolic steroids 236; capsicum 313; compression-decompression 283; corticosteroids 225; dietary glucose 303; environmental cold 277; environmental heat 278; epinephrine 222-3; estrogen/progestogen preparations 232-3; hypoxia 281; menstrual cycle 362-3; mental stress 334; obesity 314; physical activity 331; plasma triglyceride level 299; pregnancy 349; progestogen 233; vasopressin 237-8; weight reduction 315

plasma thromboplastin antecedent *see* factor XI

plasmin 9-10, 146-7
  action on: C3 212; fibrin 150-1; fibrinogen 147-50
  activation of C1s 211
  activation of pro-C4 212
  generation of C3 convertase 211
  influence of low-density lipoprotein 301-2
  inhibition by cider 310
  inhibition by vitamin E 308-9
  properties 146-7
  replacement of complement factor D 211
  structure 146-7

plasminogen 9-10, 141-6
  activation 144-6
  activation by C8i 209
  activation by complement factor B 210
  adsorption to fibrin 143, 175
  half-life 143
  generic polymorphism 143
  plasma concentration 143
  plasma level; ethnic variations 271; in fetal life 250; in infancy 253; in labor 352; in type II hyperlipoproteinemia 301; influence of: age 255; estrogen/progestogen preparations 233; injury 336; menstrual cycle 362; physical activity 332; pregnancy 350
  structure 141-2; kringles 142; lysine-binding sites 142; microheterogeneity 141
  synthesis 143-4

plasminogen activator
  islets of Langerhans 237
  ovarian, role in ovulation 234
  pituitary gland 240
  release by: electroconvulsive therapy 223, epinephrine 223, hypothalamic stimulation 238, salbutamol 223

plasminogen activator inhibitors 170-3
  corticosteroid-induced production by cultured cells 226
  influence of pregnancy 350-1
  placental 173, 351
  plasma 170-2
  platelet 172
  tissue 172-3

plasminogen activators 2, 9, 155-65
  aqueous humor 163
  bile 164
  cerebrospinal fluid 163
  circulating 155-7; assay 155-6; properties 156-7
  endothelial, inactivation by thrombin 87
  erythrocytes 164
  factor XII-dependent 160-3; high molecular weight kininogen 161; prekallikrein 161
  intrinsic and extrinsic 162-3
  leukocytes 164-5
  lymph 163

macrophages 165
milk 163
platelets 165
saliva 163, 164
seminal fluid 163, 164
synovial fluid 163
tears 163
tissue 151-3;adsorption to fibrin
  153; endometrium 363;
  inhibition by α_2-antiplasmin
  166; inhibition by vitamin E
  309; properties 152-3;
  relationship to vascular
  activator 155
urokinase 157-60; high molecular
  weight form 158; influence of
  exercise on excretion 331;
  inhibition by: α_2-antiplasmin
  166, 170, α_2-macroglobulin
  170, antithrombin III 170,
  vitamin E 309; low molecular
  weight form 158; origin 157;
  pre-urokinase 160, 163;
  properties 158-9; relationship
  with tissue activator 159-60
vascular wall 9, 153-5; properties
  153-4; release 154-5; release
  by venous occlusion 155
plasminogen-independent fibrinolysis
  173-5
platelet-activating factor 40-1
platelet adhesion 2-3, 18, 34, 36-9
  influence of platelet age
  mechanism 37-8
  role of von Willebrand factor 71
platelet adhesiveness
  assessment 38-9
  in ascorbic acid deficiency 305
  influence of: alimentary lipemia
  288; altitude 280; corticosteroids
  224; dietary fat 290-1; dietary
  sucrose 302; estrogen 229;
  estrogen/progestogen
  preparations 229; linoleic acid
  291; injury 335; malnutrition 316;
  somatostatin 239; menstrual cycle
  361-2; physical activity 328;
  pregnancy 344
  seasonal variation 258
platelet aggregates, circulating
  in type II hyperlipoproteinemia
  296
  influence of alimentary lipemia
  288

influence of dietary fat 291
platelet aggregability, assessment 42-3
platelet aggregation 2, 3, 6, 34, 40-3
  anaphylatoxin-induced 203
  complement role in collagen-
  induced 201
  effect of: ADP 25, 40; collagen
  40; 5-hydroxytryptamine 26, 40
  ethnic influences 267
  in ascorbic acid deficiency 305-6
  in Eskimos 267, 294
  in fetal life 249
  in hyperlipoproteinemias 295-6
  in infancy 250-1
  in labor 345
  influence of: age 253; alimentary
  lipemia 288-9; ascorbic acid 306;
  black tree fungus 312; corticosteroids 224; dietary fat 291;
  eicosapentaenoic acid 292-4;
  epinephrine 220-1; estrogen 229-
  30; estrogen/progestogen preparations 229-30; ethanol ingestion
  309-10; fatty acids 289; garlic
  311-12; ginger 312; glucagon
  236; heat 278; injury 335; linoleic
  acid 289; malnutrition 316;
  mental stress 333; onions 311-12;
  physical activity 328; platelet age
  16; pregnancy 344-5; prostacyclin
  24; pyridoxal 5′-phosphate 303-4;
  sex 255-6; somatostatin 239;
  stearic acid 292; thrombin 40;
  thromboxane A_2 21, 22, 40;
  vitamin B_{12} deficiency 305;
  vitamin E 307-8; vitamin E
  quinone 308; zinc deficiency 316
  inhibition by C1s 202
  inhibition by C1q 201
  inhibition of thrombin-induced by
  complement factor D 202
  mechanism 41-2
  thrombin-induced enhancement
  by C5-9 complex 202
platelet-derived growth factors 29-30
  function 30
  properties 29-30
platelet factor 3 35
  influence of dietary fat 290
  influence of ethanol ingestion 310
platelet factor 4 27-8
  influence on heparin 27
  physiological function 27-8
  plasma level 27; in hyperlipo-

starvation
    influence on factor VIII 315
stearic acid
    influence on platelet aggregation 292
Stuart factor *see* factor X
subendothelium, structure 37
sucrose (dietary)
    influence on platelet adhesiveness 302

testosterone, influence on blood fibrinolytic activity 234
thrombin 2,4,6,7,86-7
    action on: factor V 81; factor VIII:C 71; factor XIII 97; fibrinogen 89; platelets 35, 40; protein C α-, β- and γ-forms 87
    binding to endothelial cells 87
    binding to fibrin 87
    effect on vascular plasminogen activator 154-5
    effect on fibrinogen synthesis 93
    inhibition 110-11
    inhibition by: α_1-antitrypsin 106, 110; α_2-macroglobulin 104-5, 110; antithrombin III 100, 110-11; fibrinogen degradation products 113; pyridoxal 5'-phosphate 304
    structure 86
thrombolysis, mechanism 175-6
thrombopoiesis, regulation 14-15
thrombopoietin 14-15
    isolation 14
    site of production 15
thrombospondin 30, 41
thromboxane A_3, generation in platelets 292

thromboxane production
    influence on dietary fat 292
tissue factor 5,6,69,73,74-5
    interaction with factor VII 77-8
    leukocytes, induction of activity by complement 206-7
tissue fibrinolytic activity
    influence of obesity 315
tissue thromboplastin *see* tissue factor
trauma *see* injury

urokinase 9, 157-60
    excretion, circadian rhythm 258
    see also plasminogen activators

vascular responses to injury 1
vassopressin
    influence on factor VIII 237
    influence on plasma fibrinolytic activity 237-8
vitamin B_6 *see* pyridoxal 5'-phosphate
vitamin B_{12}, role in platelet function 305
vitamin C *see* ascorbic acid
vitamin E
    influence on: coagulation 308; plasmin 308-9; platelet aggregation 307-8
vitamin E quinone, influence on platelet aggregation 308
vitamin K 66-7
von Willebrand factor 2, 18, 38
    function 71
    influence on growth hormone 239
    synthesis 70

zinc deficiency, influence on platelet aggregation 316